DNA: Protein Interactions and Gene Regulation

The University of Texas Graduate School of Biomedical Sciences at Galveston

DNA: Protein Interactions and Gene Regulation

Edited by E. Brad Thompson and John Papaconstantinou

University of Texas Press, Austin

Acknowledgments

We wish to thank the authors who have contributed their work to this book with no compensation save the knowledge that they be assisting others. We also thank Mrs. A. Gibson for collating and coordinating the project and Ms. Donna H. Gerhardt, Mrs. Elsa Garcia, and Mrs. Sharon Butler for retyping many pages of the text.

First Edition, 1987

Library of Congress Cataloging-in-Publication Data

DNA, protein interactions and gene regulation.
(University of Texas Medical Branch series in biomedical science)
Includes bibliographies and index.
1. Genetic regulation. 2. Deoxyribonucleic acid. 3. Ribonucleic acid. 4. Nucleoproteins. I. Thompson, E. Brad (Edward Bradbridge), 1933– . II. Papaconstantinou, John. III. Series. [DNLM: 1. DNA—genetics. 2. Gene Expression Regulation. 3. Proteins—physiology. QU 58 D6293]
QH450.D58 1987 574.87′3282 86-30837
ISBN 0-292-71552-8

Contents

Acknowledgments *iv*

Foreword *ix*

Preface *xi*

PART I. Prokaryotic Systems

1. The Structure and Function of Bacteriophage Lambda Repressor *3*
Jeremy M. Berg, Steven R. Jordan, and Carl O. Pabo

2. Domain Structure and Intersubunit Contacts in the *Lac* Repressor *13*
Kathleen Shive Matthews

3. Cro Repressor Structure and Its Interaction with DNA *21*
B. W. Matthews

4. Activation of Transcription by the cII Protein of Bacteriophage Lambda *29*
D. L. Wulff and M. Rosenberg

5. How the Cyclic AMP Receptor Protein of *Escherichia coli* Works *45*
Susan Garges and Sanker Adhya

6. Prokaryotic Chromatin: Site-Selective and Genome-Specific DNA Binding by a Virus-Coded Type II DNA-Binding Protein *57*
Jonathan R. Greene, Krzysztof Appelt, and E. Peter Geiduschek

7. N4 Virion RNA Polymerase-Promoter Interaction *67*
Alexandra Glucksmann and Lucia B. Rothman-Denes

8. Bacteriophage Lambda N System: Interaction with the Host NusA Protein *77*
Alan T. Schauer, Eric R. Olson, and David I. Friedman

9. Nuclear Magnetic Resonance Techniques for Studies of Protein-DNA Interactions *87*
Kurt Wüthrich

PART II. Eukaryotic Systems

10. The Major Protein Components of hnRNP Complexes *97*
G. Leser and T. E. Martin

11. Interactions between the Mammalian Cell La Protein and Small Ribonucleic Acids *107*
Jack D. Keene, Jasemine C. Chambers, and Barbara Martin

12. RNA-Protein Interactions in the Nuclear Ribonucleoprotein Particles *117*
V. Holoubek

13. cDNA Cloning and Structure-Function Relationships of a Mammalian Helix Destabilizing Protein: hnRNP Particle Core Protein A1 *129*
S. H. Wilson, F. Cobianchi, and H. R. Guy

14. The Nuclear Matrix and Its Association with the RNA-Processing Machinery *147*
R. Verheijen, F. Ramaekers, E. Mariman, H. Kuijpers, P. Vooijs, and W. J. van Venrooij

15. A Sequence-Specific, Methylated DNA-Binding Protein *155*
Melanie Ehrlich and Richard Y.-H. Wang

16. A Review of Protein Factors and Nucleoprotein Complexes Involved in Specific Transcription by Eukaryotic RNA Polymerase II *163*
Randolph J. Hellwig and Salil K. Niyogi

17. Heavy Metals as Probes for Nonhistone Protein-DNA Interactions *175*
Ryszard Olinski, Zainy M. Banjar, Warren N. Schmidt, Robert C. Briggs, and Lubomir S. Hnilica

18. Topoisomerase II-DNA Complexes: Novel Targets of Antineoplastic Drug Action *183*
Leonard A. Zwelling

19. Regulation of Gene Expression by Thyroid Hormone Nuclear Receptors *201*
Herbert H. Samuels, Ana Aranda, Juan Casanova, Richard P. Copp, Zebulun D. Horowitz, Laura Janocko, Angel Pascual, Hadjira Sahnoun, Frederick Stanley, Bruce M. Raaka, and Barry M. Yaffe

20. DNA Sequences Required for Steroid Hormone Regulation of Transcription *219*
M. Pfahl, H. Ponta, P. Skroch, A. Cato, and B. Groner

21. The Hormone Response Element of the MMTV LTR: A Complex Regulatory Region *233*
M. Cordingley, H. Richard-Foy, A. Lichtler, and G. L. Hager

22. Applications for Glucocorticoid-Dependent Mouse Mammary Tumor Virus LTR Chimeras in Conditional Gene Regulation and Mutant Selection Studies *245*
B. Groner, H. Ponta, R. Ball, and M. Pfahl

23. Specific DNA-Binding Proteins and DNA Sequences Involved in Steroid Hormone Regulation of Gene Expression *259*
T. Spelsberg, J. Hora, M. Horton, A. Goldberger, B. Littlefield, R. Seelke, and H. Toyoda

24. Interaction of Steroid Hormone Receptors with DNA *269*
M. Beato

Index *281*

Foreword

The University of Texas Graduate School of Biomedical Sciences at Galveston presents the second volume of its biomedical science series whose aim is to bring to the attention of the working scientist the most up-to-date information possible about specific fields of research. The aim of the series is to reflect the dynamic thrust of discovery in a form that will be useful to all who work in the field. The key to the utility of each volume will be the prompt editing and publishing of the latest findings and their critical review and evaluation by leading scientists. For success in this endeavor thanks are due to the section and volume editors who have agreed to carry out their very difficult assignments expeditiously. Thanks are also due to the University of Texas Press for its understanding and cooperation.

J. Palmer Saunders, Ph.D.
Dean of the Graduate School
Executive Editor

Preface

The regulation of genes, their physical organization, and the replication of DNA all require the interaction of proteins with DNA. Though this rather obvious fact has been known for some time, and despite the principles of protein-DNA interactions longstanding in the literature, the precise ways in which these interactions take place remain the focus of attention. In the regulation of genes, a variety of techniques derived from molecular cloning technology have helped to identify certain sequences as important. These sequences, often referred to as "boxes," encompass the promoter sequences and are usually found proximal to the transcription start site of the regulated gene. Particular proteins are found to interact with the sequences and thereby ensure the accurate transcription of the gene. Distal to these promoter sequences are another class of regulatory elements, the enhancers, whose function is to drive transcription and in many cases to confer tissue specificity upon gene expression. In some of the more extensively studied systems, the exact protein-nucleotide interactions have been determined. In most organisms, and especially in eukaryotes, these interactions remain to be elucidated fully. Moreover, there is a growing body of data suggesting that arrays of these regulatory sequences, each with its distinctive regulatory protein, may interact to determine the basal set point as well as the induced or deinduced level of transcription. The multiplier effect of a relatively small number of regulatory "boxes" and regulatory proteins would allow great economy of numbers in achieving the observed cell-to-cell and tissue-to-tissue variations in gene expression. In eukaryotes, hormones and growth factors as well as certain drugs may influence these regulatory molecules. For some hormones, namely, thyroid hormone and the steroids, the action seems simple. The protein receptors for these hormones are intracellular, DNA-binding proteins. It seems that, as a consequence of binding these hormones, their receptor-DNA interactions are altered so as to mediate gene regulation.

The regulation of the levels of RNA, the actual product of gene expression, is clearly not only the result of DNA-protein interactions. Specific RNA-protein interactions produce the ribonucleoprotein complexes that consist of the primary transcript as well as the small nuclear RNAs. The formation of these complexes is an important component of the nuclear RNA processing interactions and possibly their regulation.

The organization of DNA is basically the problem of fitting a long narrow object into a very small volume, while retaining function. Function in this case requires accessibility to regulatory molecules and the enzymes of transcription and replication. Perhaps related to this are modifications of DNA and the long-term shutting off of many genes, such as those of the X chromosome or of autosomal regions not expressed in differentiated cells. The proteins participating in this "packaging" of DNA to a considerable extent have been identified and the topography of DNA in chromatin has been clarified considerably in just the last decade. But there is still a considerable way to go before the function of regions of chromatin can be explained by protein-DNA interactions. In this respect, it is interesting that DNA modifications by methylation alter the affinity of certain proteins to bind to DNA. These alterations in DNA-protein binding appear to be a part of the overall changes in chromatin structure and possibly associated with the stabilization of chromatin structural changes after the regulatory events have occurred. The nuclear matrix, many now believe, is a real entity, involved in keeping DNA organized and used as a framework for replication and transcription. This protein lattice and its interactions with DNA are just beginning to be understood. Though the general activities of the enzymes involved in opening and closing, twisting and untwisting, the topoisomerases and gyrases, have been discovered, their complete function in specific terms is still being unraveled.

When it was discovered that small DNA viruses, for example, SV40, can cause oncogenic transformation, we thought, "Let us learn all there is to know about the genes and DNA sequence of these viruses, and then we will have understood 'the cancer problem.'" This thought proved to be wrong. The actions of the proteins coded by the viruses must be understood to completely realize the issues involved in oncogenesis. When the molecular cloning and sequencing of regulated genes became widely practicable, we thought, "Let us determine the specific regulatory sequences involved; then we will master the secrets of gene control." Again, as more and more of these sequences are elucidated, it is becoming clear that the proteins involved, and their interactions with nucleic acids, must be thoroughly understood before the problem is mastered.

This book is designed to bring together in one place concepts and examples of protein-DNA and protein-RNA interactions as they pertain to the regulation of structure and function of informational macromolecules in both eukaryotic and prokaryotic systems. The two are not frequently found in the same book, and we hope that this juxtaposition will be thought provoking for workers in both areas.

E. Brad Thompson, M.D.
Chairman and Director,
Division of Human Genetics

John Papaconstantinou, Ph.D.
Professor and Director,
Division of Cell Biology

DEPARTMENT OF HUMAN BIOLOGICAL CHEMISTRY AND GENETICS

Part I. *Prokaryotic Systems*

1. The Structure and Function of Bacteriophage Lambda Repressor

Jeremy M. Berg, Steven R. Jordan, and Carl O. Pabo

Department of Biophysics, Johns Hopkins University School of Medicine, 725 North Wolfe Street, Baltimore, MD 21205

INTRODUCTION

The molecular machinery that controls the mode of growth of bacteriophage lambda may be the most thoroughly studied gene regulatory system of any organism. The DNA sequences of the control regions are known (1). The two key regulatory proteins involved have been isolated and thoroughly characterized, including studies of their DNA-binding properties (2) and three-dimensional structures (3,4). Furthermore, the relationships between the binding of these regulatory proteins to DNA and the binding and activity of RNA polymerase have been investigated (5). Finally, a statistical thermodynamic model has been developed that describes biologically important properties of this system in terms of well-defined interaction energies and rate constants (6). This chapter will focus on one of the principal components of this control system, the lambda repressor, and the relationships between repressor structure and function.

Bacteriophage lambda has two alternative modes of growth. In the lysogenic mode, the viral DNA is integrated into the host chromosome and replicates passively. In the lytic mode, the viral DNA replicates independently, directs the synthesis of viral proteins to produce new viral particles, and eventually lyses the host cell. Two regulatory proteins play central roles in determining the mode of growth: the repressor, encoded by the cI gene, and the cro protein, encoded by the cro gene . The relevant control regions of the viral genome contain two operators, and each operator region has three sites where repressor or cro can bind. Each binding site is seventeen base pairs long and has approximate two-fold symmetry. The consensus sequence of the operator sites is shown in Figure 1. The affinities of repressor and cro for the different sites vary since the sequences of the sites are slightly different, and this variation is very important for biological function. Detailed decriptions of this "molecular switch" can be found in references 6-8.

Repressor is needed to maintain the lysogenic state. By specifically binding to sites in the two operator regions, repressor does the following:

(1) It prevents RNA polymerase from binding at adjoining promoter sites and thereby turns off the transcription of cro and other "early" viral genes.

(2) It stimulates the transcription of its own gene, apparently by contacting RNA polymerase bound at another adjacent promoter.

(3) At high concentrations, it blocks transcription of its own gene.

Destroying repressor switches the phage from lysogenic to lytic growth. In response to an intracellular signal caused by ultraviolet light or other activated mutagens, repressor is cleaved into two fragments in a reaction mediated by the recA protein (9). This decreases its DNA binding affinity and allows transcription of the genes required for lytic growth.

STRUCTURE OF LAMBDA REPRESSOR

The repressor monomer contains 236 amino acids. It is slightly acidic overall, with 31 acidic residues and 25 basic residues (10). It forms a dimer with a dissociation constant of approximately 2×10^{-8} M (11). At substantially higher concentrations, repressor forms larger oligomers. Under physiological conditions ($\sim 10^{-7}$ M in a lysogen) repressor exists primarily as a dimer.

The dimerization of repressor is important for its biological function. Each monomer of the repressor dimer interacts with one half of the two-fold symmetric operator site. Dimerization increases the total binding energy for such symmetric sites and thus contributes to recognition of the specific operator sites.

Each monomer of repressor is folded into two domains of approximately equal size (12). This has been demonstrated by several different experiments. First, calorimetric measurements of repressor revealed two unfolding transitions, one at approximately 49 °C and one at 73 °C. Second, limited proteolysis of repressor with papain produced two fragments which are relatively stable to further digestion. The first contains the amino-terminal portion (residues 1-92) while the second contains the carboxyl-terminal portion (resides 132-236). Calorimetric studies of the purified fragments revealed that the N-terminal domain denatured at the lower temperature and that the C-terminal domain denatured at the higher temperature.

The two domains of repressor have distinct functions. The N-terminal domain binds specifically to operator DNA and mediates the positive and negative control functions of repressor (13). However, the binding constant for the N-terminal fragment is several orders of magnitude smaller than that for intact repressor. The binding constant is lower because the fragment does not form a very stable dimer. The C-terminal domain is responsible for the oligimerization of repressor. The C-terminal fragment (residues 132-236) dimerizes almost as well as intact repressor, but it does not bind DNA (12).

The two domains are connected by a 40 amino acid "linker" which is relatively susceptible to proteolytic digestion. This region contains the site of recA mediated cleavage. The peptide bond between Ala-111 and Gly-112 is broken, separating the N- and C-terminal domains (12). Because it does not dimerize as well as intact repressor, the N-terminal fragment produced does not bind DNA as tightly. Thus, the lysogenic phage is "induced" to begin lytic growth. Interestingly, an autodigestion reaction of lambda repressor has been observed which also involves specific cleavage of this bond (14). This suggests that the origin of the selectivity of the cleavage reaction may lie in the repressor itself. RecA may merely facilitate the reaction.

The crystal structure of the N-terminal fragment of repressor has been determined by x-ray diffraction methods using data extending to 3.2 Å resolution (3). The structure consists of an extended amino terminal arm (residues 1-8) and five α-helices (residues 9-23, 33-39, 44-52, 61-69, and 79-92). Helices 1-4 form a globular domain with a classic hydrophobic core. The N-terminal arm and helix 5 project out from this globular structure. Residues 1-3 are disordered in the crystal and are not observed in the electron density map.

In the crystal, the N-terminal fragment forms a dimer with two-fold rotational symmetry. The dimer is stabilized by interactions between helix 5 from one monomer and helix 5 from another monomer. The contact involves hydrophobic side chains from each helix including residues Ile-84, Met-87, and Tyr-88. The formation of a dimer of the N-terminal fragment at high concentrations (~ 1 mM) was predicted from binding studies (15) and is supported by nuclear magnetic resonance studies (16).

Based on the structure of the N-terminal fragment, a model of the repressor-operator complex was proposed (3,17). Because experiments show that repressor binding does not drastically affect DNA structure, B-form DNA was used for model building. The model was built by assuming that the two-fold axis of the protein dimer coincides with the approximate symmetry axis of the operator site. The rotation of the dimer around the two-fold axis and the distance between the protein and DNA were varied. These were adjusted to maximize the complementarity of the protein and DNA surfaces and to cover all contact sites implicated by chemical protection experiments. Only one reasonable arrangement was found. Figure 1 shows this structure.

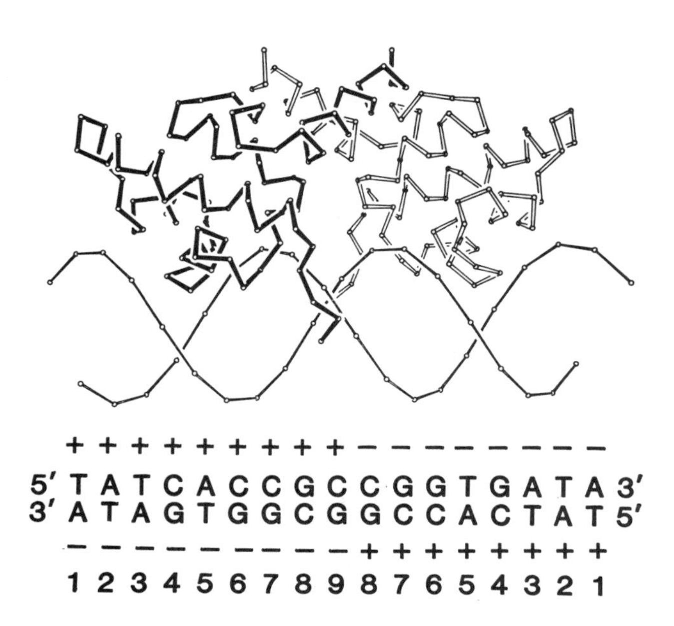

Figure 1. Sketch of the proposed complex between a dimer of the N-terminal fragment and an operator site. The α-carbons of residues 4-92 are shown with solid lines for one monomer and hollow lines for the other. The consensus sequence of the six lambda operator sites is shown below the complex. The two-fold symmetry axis passes through base-pair 9.

Since the initial model seemed quite reasonable, detailed model building was attempted. To maximize hydrogen bonding, side chain angles of surface residues were adjusted and minor (~1 Å) shifts of the protein dimer were allowed. A number of chemically reasonable contacts between the repressor and operator were proposed (17). These include sequence-specific contacts between amino acid side chains and the edges of bases in the major groove of the DNA duplex. The following specific contacts were proposed:

Gln-44 is hydrogen-bonded to adenine+2 in a bidentate manner
Ser-45 is hydrogen-bonded to guanine-4
Ala-49 makes van der Waals contacts with thymine +3 and thymine-5
Asn-55 is hydrogen-bonded to guanine-6 in a bidentate manner

In addition, sequence-independent contacts were proposed involving Gln-33, Asn-52, Asn-58, Tyr-60, and Asn-61. Furthermore, it was proposed that the N-terminal arm wraps around the double helix and contacts the major groove on the "back" of the operator site (18). This proposal seemed reasonable since a proteolytic fragment containing residues 4-92 is unable to make the contacts on the back of the operator site (18). However, the flexibility of the N-terminal arm prevented more detailed predictions of the contacts made by these residues.

Most of the contacts with the operator site are made by residues from helix 2, helix 3, or from the region immediately following helix 3. In the model, helix 3 lies in the major groove of the operator while the N-terminus of helix 2 contacts the sugar-phosphate backbone. This helix-turn-helix motif has been observed in other sequence-specific DNA-binding proteins that have been studied by diffraction methods. These include the lambda cro protein (4) and the catabolite activator protein from E. coli (19), referred to as CAP. Detailed comparisons of these proteins have been made (20,21). The overall structures are quite different. However, the helix-turn-helix units are strikingly similar. Quantitative comparison revealed that the α-carbons of residues 31-53 of lambda repressor can superimposed on those of residues 14-36 of cro. The best orientation gives a root-mean-square error of 0.7 Å (21). A comparable fit was seen between analogous regions of cro and CAP (20). This structure has only been observed in DNA-binding proteins.

The structural homologies aided the discovery of sequence homologies among a large number of DNA-binding and potential DNA-binding proteins (22-25). These sequence homologies suggest that a conserved bihelical unit is used for DNA binding. Although the structure of this unit seems to be quite similar in the three proteins of known structure, model building studies suggest that it is used in slightly different ways by the three proteins. These differences seem to result from differences in the arrangement of the protein dimers that affect the orientation and spacing of the bihelical units. Differences in the amino acid sequences also affect binding. It is important to note that the conserved residues in these bihelical units tend to be buried. The exposed residues which are responsible for sequence recognition are quite different.

MUTANTS OF LAMBDA REPRESSOR

Mutants of the lambda repressor have been extremely helpful in evaluating the model building studies and in locating other functional sites in repressor. Hecht, Nelson, and Sauer have performed a detailed search for repressor mutants which are deficient in operator binding (26-28). A total of 52 different single amino acid substitutions in the N-terminal domain have been characterized. The mutants were classified according to activity. The genes for the mutant repressors were on a plasmid and were expressed in E. coli so that ~1% of the total protein was repressor. Those mutants which were still sensitive to lambda infection despite this high concentration of repressor were classified as having strong mutations. Those which were resistant to infection by wild-type lambda but were sensitive to infection by virulent lamdba phages were classified as having weak mutations. The structure of the N-terminal fragment was used to divide the repressor mutations into two classes. Mutations in the first class change side chains that are fully or partially buried in the wild-type structure. Mutations in the second class change solvent exposed side chains. It was assumed that mutations which change buried residues might act by disrupting the global structure of repressor. This hypothesis was later supported by spectroscopic, calorimetric, and proteolytic studies (28). The second class of mutations are more relevant to the mechanism of site-specific DNA binding. Of the 15 (non-proline) strong mutations in this class, 12 involve residues identified as contact points in the model building studies (26,27). These mutations occur in the helix-turn-helix unit and in the N-terminal arm. The remaining 3 mutations involve glycines in the wild-type structure, and unfavorable interactions may be present with a larger side chain. In summary, isolation of mutants defective in operator binding has led to identification of the same DNA-binding surface proposed from the structural studies. Analogous genetic studies have been performed for several DNA-binding proteins with unknown three-dimensional structures (29,30). These studies have implicated the regions with sequences homologous to the bihelical unit, and support the hypothesis that such regions are used by other DNA-binding proteins.

The importance of repressor's N-terminal arm has been supported by studies of deletion mutants (31). Genes encoding residues 2-236, 4-236, 7-236, and 8-236 were constructed and expressed in E. coli. The deletion of the first amino acid appeared to have no effect of the properties of repressor _in vivo_ and _in vitro_. In contrast, deletion of the entire N-terminal arm (mutant 7-236 or 8-236) had dramatic effects. Bacteria containing these proteins were sensitive to lambda infection, even when the protein was present at a level 150 times over that in a normal lysogen. _In vitro_, no protection of the operator sites was observed even at 8000 times the concentration normally needed to observe protection. These results indicate that interactions between the arm and the operator make a large contribution to the observed binding energy. The 4-236 mutant was partially defective in binding. At high concentrations it rendered cells insensitive to infection by wild-type lambda, but they were still sensitive to virulent strains. _In vitro_, approximately 30 times the concentration of 4-236 was required to provide the same protection as for wild-type. Further studies suggested that the arm may have some role in determining operator site preference.

The N-terminal domain of repressor mediates positive regulation of cI gene transcription. Evidently, a repressor molecule that is bound at one operator site contacts RNA polymerase bound to an adjacent promoter. This picture is supported by chemical modification studies which indicated that bound repressor and bound polymerase contact adjoining sites on the DNA. Ptashne and coworkers recently isolated mutants of lambda repressor which retain the ability to bind operator DNA but no longer stimulate transcription from the cI promoter (32). Mutants with changes at residues 34, 38, and 43 were isolated. The three residues form a patch on the surface of repressor. In the proposed repressor-operator complex, this patch is close to the DNA backbone and is on the surface of repressor closest to the RNA polymerase binding site. Thus, the positive control mutations are consistent with the proposed complex structure and support the proposed mechanism of positive control.

SUMMARY

Structural, biochemical, and genetic studies have led to a clear basic picture of the mechanisms by which lambda repressor regulates gene function. The structural information about repressor is summarized in Figure 2. However, a number of important questions still need to be answered. These include:

(1) What are the detailed interactions between repressor and the lambda operator sites? Are the predicted contacts correct? What are the geometries of the hydrogen bonds? Are any of the interactions mediated by water molecules?
(2) What is the structure of the N-terminal arm in the complex? What contacts does that arm make with the operator site? How much do these contacts contribute to sequence-specific recognition?
(3) How do sequence-dependent structural variations of B-DNA contribute to protein-DNA recognition?
(4) What structural changes in the protein and in the DNA occur upon repressor-operator complex formation?
(5) What is the structure of the carboxyl-terminal domain? How are the two domains positioned with respect to each other?
(6) What are the mechanisms of the recA-mediated cleavage and of repressor's autodigestion reaction?

The first four questions can be addressed most directly through the study of co-crystals containing repressor and an operator site. Anderson, Harrison, and Ptashne reported a low resolution (7 Å) structure of the DNA-binding domain of phage 434 repressor and a 14-base pair operator site (33). However, information about detailed interactions requires a higher resolution structure determination. Recently, we have grown crystals containing the N-terminal fragment of lambda repressor (residues 1-92) and a 20 base DNA duplex with a sequence corresponding to the operator site O_L1 which diffract to 2.5 Å resolution (34). Studies of these crystals should further improve our understanding of the lambda repressor and of the mechanisms of action of gene regulatory proteins.

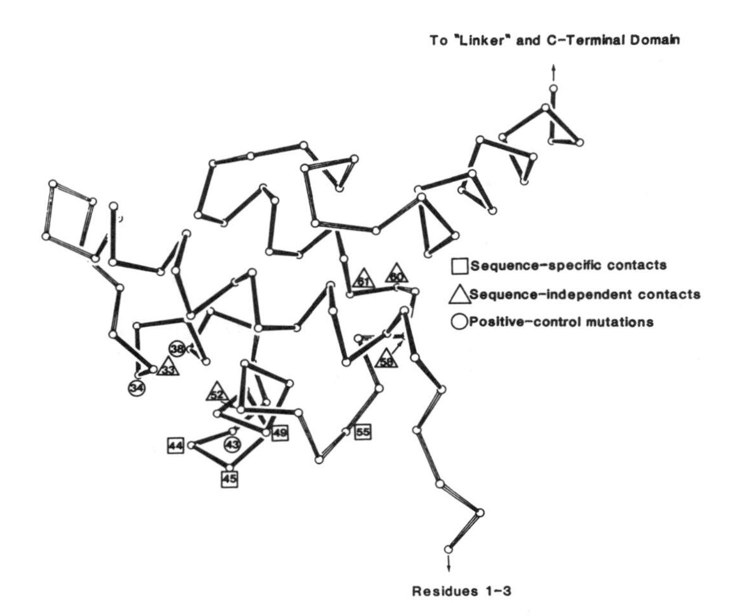

Figure 2. Sketch of a monomer of lambda repressor. The α-carbons of the N-terminal domain are shown with dark lines indicating the five α-helices. The contact residues implicated by model building and genetic studies are noted. All of the sequence-specific contacts as well as the contacts at residues 33 and 52 have been found by both model building and genetic studies. The contacts at residues 58, 60, and 61 are based on model building.

ACKNOWLEDGEMENTS

This research was supported by N.I.H. Grant GM-31471 and by a Junior Faculty Research Award from the American Cancer Society. J.M.B. was supported by a fellowship from the Jane Coffin Childs Memorial Fund and S.R.J. was supported by a training grant from the N.I.H. to the Molecular Biology and Genetics Department.

REFERENCES

1. Humayun, Z., A. Jeffrey, and M. Ptashne: Completed DNA Sequences and Organization of the Repressor-binding Sites in

the Operators of Phage Lambda. J Mol Biol 112:265, 1977.
2. Johnson, A.D., C.O. Pabo, and R.T. Sauer: Bacteriophage λ Repressor and cro Protein: Interactions with Operator DNA. Meth Enzymol 65:839, 1980.
3. Pabo, C.O. and M. Lewis: The Operator-binding Domain of λ Repressor: Structure and DNA Recognition. Nature 298:443, 1982.
4. Anderson, W.F., D.H. Ohlendorf, Y. Takeda, and B.W. Matthews: Structure of the cro Repressor from Bacteriophage λ and its Interaction with DNA. Nature 290:754, 1981.
5. Hawley, D.K., A.D. Johnson, and W.R. McClure: Functional and Physical Characterization of Transcription Initiation Complexes in the Bacteriophage λ O_R Region. J Biol Chem 260:8616, 1985.
6. Shea, M.A. and G.K. Ackers: The O_R Control System of Bacteriophage Lambda-A Physical-Chemical Model for Gene Regulation. J Mol Biol 181:211, 1985.
7. Ptashne, M., A. Jeffrey, A.D. Johnson, M. Maurer, B.J. Meyer, C.O. Pabo, T.M. Roberts, and R.T. Sauer: How the λ Repressor and Cro Work. Cell 19:1, 1980.
8. Ptashne, M., A.D. Johnson, and C.O. Pabo: A Genetic Switch in a Bacterial Virus. Sci Amer 247:128, 1982.
9. Roberts, J.W., C.W. Roberts, and N.L. Craig: Escherichia Coli RecA Gene Product Inactivates Phage λ Repressor. Proc Natl Acad Sci USA 75:4714, 1978.
10. Sauer, R.T. and R. Anderegg: Primary Structure of the λ Repressor. Biochemistry 17:1092, 1978.
11. Sauer, R.T.: Ph.D. Thesis, Harvard University, 1979.
12. Pabo, C.O., R.T. Sauer, J.M. Sturtevant, and M. Ptashne: The λ Repressor Contains Two Domains. Proc Natl Acad Sci USA 76:1608, 1979.
13. Sauer, R.T., C.O. Pabo, B.J. Meyer, M. Ptashne, and K.C. Backman: Regulatory Functions of the λ Repressor Reside in the Amino-terminal Domain. Nature 279:396, 1979.
14. Little, J.W.: Autodigestion of lexA abd Phage λ Repressor. Proc Natl Acad Sci USA 81:1375, 1984.
15. Pabo, C.O.: Ph.D. Thesis, Harvard University, 1980.
16. Weiss, M.A., M, Karplus, D.J. Patel, and R.T. Sauer: Solution NMR Studies of Intact Lambda Repressor. J Biomol Struc Dynam 1:151, 1983.
17. Lewis, M., A. Jeffrey, J. Wang, R. Ladner, M. Ptashne, and C.O. Pabo: Structure of the Operator-binding Domain of Bacteriophage λ Repressor: Implications for DNA Recognition and Gene Regulation. Cold Spring Harbor Symp Quant Biol 47:435, 1983.
18. Pabo, C.O., W. Krovatin, A. Jeffrey, and R.T. Sauer: The N-Terminal Arms of λ Repressor Wrap around the Operator DNA. Nature 298:441, 1982.
19. McKay, D.B., I.T. Weber, and T.A. Steitz: Structure of the Catabolite Gene Activator Protein at 2.9 Å Resolution-Incorporation of Amino Acid Sequence and Interactions with Cyclic AMP. J Biol Chem 257:9518, 1982.
20. Steitz, T.A., D.H. Ohlendorf, D.B. McKay, W.F.

Anderson, and B.W. Matthews: Structural Similarity in the DNA-binding Domains of Catabolite Gene Activator and cro Repressor Proteins. Proc Natl Acad Sci USA 79:3097, 1982.
21. Ohlendorf, D.H., W.F. Anderson, M.Lewis, C.O. Pabo, and B.W. Matthews: Comparison of the Structures of Cro and λ Repressor Proteins from Bacteriophage λ. J Mol Biol 169:757, 1983.
22. Sauer, R.T., R.R. Yocum, R.F. Doolittle, M.Lewis, and C.O. Pabo: Homology among DNA-binding Proteins Suggests Use of a Conserved Super-secondary Structure. Nature 298:447, 1982.
23. Matthews, B.W., D.H. Ohlendorf, W.F. Anderson, and Y. Takeda: Structure of the DNA-binding Region of lac Repressor Inferred from its Homology with cro Repressor. Proc Natl Acad Sci USA 79:1428, 1982.
24. Weber, I.T., D.B. McKay, and T.A. Steitz: Two Helix DNA Binding Motif of CAP Found in lac Repressor and gal Repressor, Nucl Acids Res 10:5085, 1982.
25. Laughon, A. and M.P. Scott: Sequence of a Drosophila Segmentation Gene: Proteins Structure Homology with DNA-binding Proteins. Nature 310:25, 1984.
26. Nelson, H.M.C., M.H. Hecht, and R.T. Sauer: Mutations Defining the Operator-binding Sites of Bacteriphage λ Repressor. Cold Spring Harbor Symp Quant Biol 47:441, 1982.
27. Hecht, M.C., H.M.C. Nelson, and R.T. Sauer: Mutations in λ Repressor's Amino-terminal Domain: Implications for Protein Stability and DNA-binding. Proc Natl Acad Sci USA 80:2676, 1983.
28. Hecht, M.H., J.M. Sturtevant, and R.T. Sauer: Effect of Single Amino Acid Replacement in the Thermal Stability of the NH_2-Terminal Domain of λ Repressor. Proc Natl Acad Sci USA 81:5685, 1984.
29. Kelley, R.L. and C. Yanofsky: Mutational Studies of the trp Repressor of Escherichia coli Support the Helix-Turn-Helix Model of Repressor Recognition of Operator DNA. Proc Natl Acad Sci USA 82:483, 1985.
30. Isackson, P.J. and K.P. Bertrand: Dominant Negative Mutations in the Tn10 tet Repressor: Evidence for the Use of the Conserved Helix-Turn-Helix Motif in DNA Binding. Proc Nat Acad Sci USA 82:6226, 1985.
31. Eliason, J.L., M.A. Weiss, and M. Ptashne: NH_2-terminal Arm of Phage λ Repressor Contributes Energy and Specificity to Repressor Binding and Determines the Effects of Operator Mutations. Proc Natl Acad Sci USA 82:2339, 1985.
32. Hochschild, A., N. Irwin, and M. Ptashne: Repressor Structure and the Mechanism of Positive Control. Cell 32:319, 1983.
33. Anderson, J.E., M. Ptashne, and S.C. Harrison: A Phage Repressor-Operator Complex at 7 Å Resolution. Nature 316:596, 1985.
34. Jordan, S.R., T.V. Whitcombe, J.M. Berg, and C.O. Pabo: Systematic Variation in DNA Length Yields Highly Ordered Repressor-Operator Co-crystals. Science in press.

2. Domain Structure and Intersubunit Contacts in the *Lac* Repressor

Kathleen Shive Matthews

Department of Biochemistry, Rice University, P.O. Box 1892, Houston, TX 77251

INTRODUCTION

The lactose repressor protein controls the expression of the lac metabolic enzymes in E. coli by interacting with its target operator sequence to preclude transcription of the coding regions (1). Modulation of operator affinity by the binding of small sugar ligands to the protein provides responsivity to the environment. Inducer sugars decrease the affinity of the protein for the operator relative to nonspecific DNA sequences and thereby elicit transcription of the lac mRNA (1). Evidence for domain structure within the protein derives from genetic studies (2-5), proteolytic digestion to produce monomeric NH_2-terminal (M_r ~6000) and tetrameric core (M_r ~120,000) domains (6), and calorimetric analysis (7). The placement of the NH_2-terminal domains at the ends of an elongated core tetramer has been suggested by low angle x-ray and neutron scattering studies (8-11). Nuclease protection and methylation perturbation studies have indicated a region of ~35 bp in the operator region to be the contact site for the repressor protein (12-15). Detailed investigations of the interaction of the protein with operator and nonspecific DNA have provided broad understanding of the mechanism and parameters for this interaction (1). However, minimal information is available regarding the intersubunit contacts within the protein.

RESULTS AND DISCUSSION

Sites of Subunit Interaction

Mutations have been demonstrated in the carboxyl-terminal region of the core domain (amino acid residues 250-360) which result in predominantly protein monomers (2, Fig. 1). We have recently isolated and characterized one of these monomers (T-41, tyr 282 → ser); the T-41 protein binds to IPTG with wild-type affinity and kinetic properties and undergoes a conformational alteration on binding to inducer, but does not bind to phosphocellulose or appreciably to DNA (16). Monoclonal antibodies which bind only to denatured or dissociated wild-

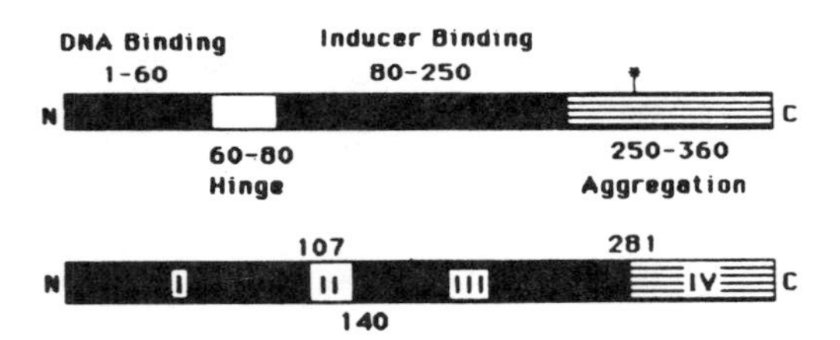

Figure 1. Upper, Regions within the protein in which indicated mutations predominate; star indicates site of T-41 mutation. Lower, Peptides produced by cyanylation. Fragment IV reacts with monoclonal antibody to lac repressor.

type repressor react readily with the mutant monomer (17); these antibodies react with the carboxyl-terminal peptide produced by cyanylation reaction with the repressor cysteines (Fragment IV, amino acid residues 281-360; Fig. 1). Based on the previous genetic data and these results, we conclude that the carboxyl-terminal region of the protein forms a subunit interface which is unavailable to antibody in the wild-type tetramer but reacts readily in the mutant monomer.

Reaction at Subunit Interface Affects Inducer Binding Rates

Modification of the repressor with methyl methanethiosulfonate (MMTS) results in the reaction of all three cysteine residues at high molar ratios of reagent to monomer. At low molar ratios, reaction primarily affects cys 107 and 140 (Fig. 2). Although reaction with MMTS

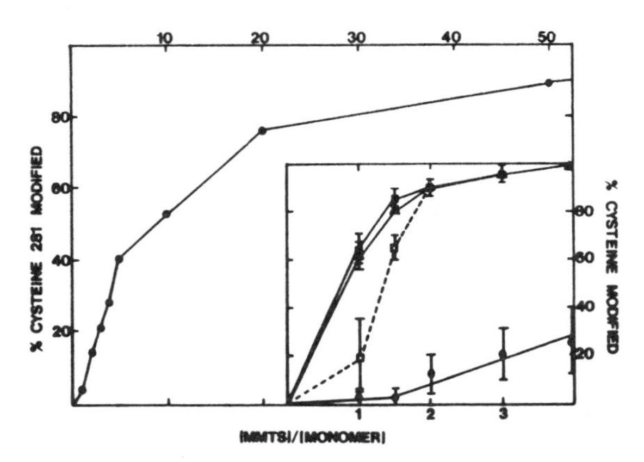

Figure 2. Reaction of cys residues in lac repressor with MMTS. Reaction was measured by subsequent BNP modification of denatured MMTS-protein to provide a visible probe for unreacted residues; this reaction was followed by protease digestion and separation of peptides by HPLC. -●-, cys 281; —▪—, cys 140; —▫—, cys 140 + IPTG; -▲-, cys 107. Reaction at cys 107 and 281 is unaffected by IPTG.

decreases operator affinity by approximately 25-fold at a 150-fold ratio of MMTS/monomer, the repressor-inducer equilibrium dissociation constant is unaffected. However, significant decreases are observed in the kinetic rate constants for inducer binding; both the association and dissociation rate constants are decreased approximately 50-fold. The alteration in kinetic parameters correlates to the reaction of cys 281 (Fig. 3). 2-Bromoacetamido-4-nitrophenol (BNP) reacts with

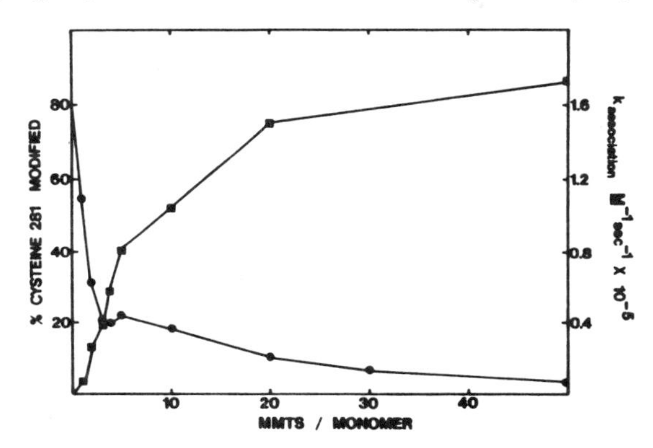

Figure 3. Correlation of MMTS reaction at cys 281 (—▪—) with decreased inducer association rate constant (—●—).

the lactose repressor at cys 107 (~35%) and 140 (~90%) without effect on inducer binding properties; the extent of reaction at cys 107 can be increased to ~85% by reaction in the presence of inducer (18). Using protein modified with BNP in the presence and absence of IPTG followed by secondary reaction with MMTS, it was possible to confirm that reaction at cys 281 was responsible for the decreased kinetic rate constants (Table 1). The availability of cys 281 for reaction at low molar ratios of MMTS in the T-41 monomer (Fig. 4) indicates that this residue is protected by the quaternary structure of the wild-type tetramer. The perturbation of the dynamic properties of inducer binding associated with reaction of cys 281 in *lac* repressor requires the tetrameric structure of the protein, since complete MMTS modification of the mutant monomer minimally alters inducer equilibrium or kinetic binding parameters (Fig. 4, Table 2). By implication, the effects of modification on kinetic rate constants for inducer binding are mediated by alterations at the subunit interface.

Effect of pH on Inducer Binding

Further evidence for alteration of subunit interactions by MMTS modification is provided by pH studies of inducer binding. The affinity of the *lac* repressor is decreased 10-fold at pH 9.2 relative to pH 7.5 (19), and the binding is cooperative (16). The alteration in the inducer equilibrium binding constant can be largely attributed to a decrease in the association rate constant at high pH (Table 2). In contrast, inducer binding parameters for monomer T-41 and MMTS-

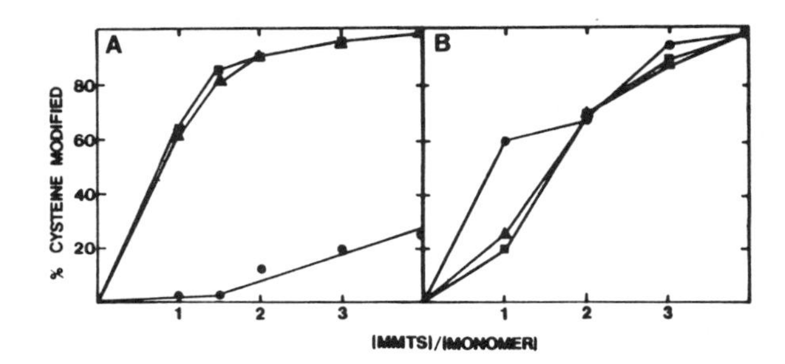

Figure 4. Reaction of cys residues with MMTS. A, Wild-type repressor. B, T-41 mutant. For methods, see Fig. 2. —●—, cys 281; —■—, cys 140; —▲—, cys 107.

modified repressor are minimally affected by increased pH (Table 2), and no cooperativity in inducer binding is observed. Thus, the effects of pH and MMTS reaction on inducer binding require the tetrameric structure and suggest that subunit interactions are influenced by an ionization occurring between pH 7 and 9. The uptake of 2 protons and an associated conformational change has been indicated by temperature jump experiments and DNA binding properties (20,21). The reaction of cys 281 with MMTS abrogates pH influence on inducer binding and results in association rate constants closer to those observed for unmodified repressor at high pH. Apparently both pH increase and MMTS reaction at cys 281 influence intersubunit contacts in the tetrameric repressor protein. Consistent with these results, recent studies using increased hydrostatic pressure to dissociate oligomeric structure have shown that the modification of the lac repressor with MMTS results in increased stability of the tetramer compared to unmodified protein (22).

Model for the Repressor Structure

Homology between the lactose repressor and sugar binding protein sequences has been demonstrated (23), and the known x-ray crystallographic structure of arabinose binding protein (ABP) has been utilized to predict a sugar binding site for the lac repressor (24). The arabinose binding protein consists of a two-domain structure, with the sugar binding site in a cleft between the two domains (25). Assembly of four ABP-like subunits into a square planar tetramer, a structure indicated by electron microscope studies on repressor microcrystals (26), and placement of the NH_2-termini at the ends of such a tetramer as demonstrated by low angle x-ray and neutron scattering studies (8-11) yields a structure schematically presented in Figure 5. This model suggests that the tetramer consists of twelve subdomains (2 in each core monomer plus 4 NH_2-termini) which have a range of motion somewhat independent of the remainder of the protein. This arrangement provides an explanation for the significant structural mobility observed by NMR and fluorescence studies (27,28). The consequent

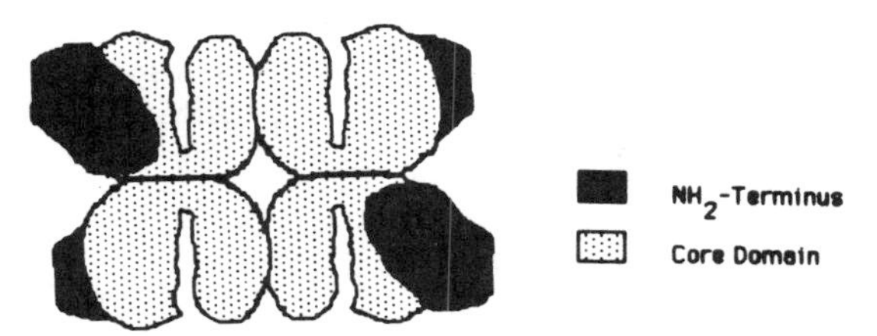

Figure 5. Schematic representation of repressor tetramer with bilobate core monomer structure based on homology to ABP.

structural microheterogeneity may account for the inability to crystallize this protein. It is noteworthy that by homology with ABP, cys 281 is found in the connecting (hinge) region between the two core subdomains of each individual subunit and therefore is positioned in a predicted subunit interface in this model. Obviously, other models with different arrangements of the subunits can also be generated, but maintaining the binding cleft free for sugar access requires the connecting hinge to participate in subunit interactions; the influence of modification in this region on inducer kinetic binding constants can be readily attributed to an impediment to the opening and closing of the cleft and consequent influence on rates of access to and egress from the binding site with no effect on equilibrium binding parameters.

SUMMARY

Subunit interfaces in the lactose repressor protein are formed by residues in the carboxy-terminal region of the sequence. Reaction of cys 281 in this region significantly influences the rate constants for inducer binding and the effect of pH on inducer binding parameters; these alterations are observed only in the tetrameric repressor. From pressure studies, it is apparent that MMTS modification results in an increase in the stability of the tetramer and presumably a greater number and/or more stable contacts between protomers. Cys 281 and/or the region surrounding this residue in the tertiary structure of the protein exert a significant influence on the dynamics and stability of the protein quaternary structure via perturbations at the subunit interface.

ACKNOWLEDGEMENTS

The contributions of the following people to this work are gratefully acknowledged: Thomas Daly, John S. Olson, Catherine Royer, and Clarence Sams. This work was supported by grants from the National Institutes of Health (GM 22441) and the Robert A. Welch Foundation (C-576).

REFERENCES

1. Miller, J.H. and W.S. Reznikoff (Eds.): The Operon. 2nd Edition, Cold Spring Harbor Laboratory, 1980.
2. Schmitz, A., U. Schmeissner, J.H. Miller, and P. Lu: Mutations

affecting the quaternary structure of the lac repressor. J Biol Chem 251:3359, 1976.
3. Pfahl, M., C. Stockter, and B. Gronenborn: Genetic analysis of the active sites of lac repressor. Genetics 76:669, 1974.
4. Miller, J.H., C. Coulondre, M. Hofer, U. Schmeissner, H. Sommer, A. Schmitz, and P. Lu: Genetic studies of the lac repressor IX. Generation of altered proteins by the suppression of nonsense mutations. J Mol Biol 131:191, 1979.
5. Miller, J.H.: Genetic studies of the lac repressor XII. Amino acid replacements in the DNA binding domain of the Escherichia coli lac repressor. J Mol Biol 180:205, 1984.
6. Platt, T., J.G. Files, and K. Weber: Lac repressor: Specific proteolytic destruction of the NH_2-terminal region and loss of the deoxyribonucleic acid binding activity. J Biol Chem 248:110, 1973.
7. Sturtevant, J., S. Manly, and K.S. Matthews: The thermal denaturation of the core protein of lac repressor. Biochemistry, in press and unpublished data.
8. McKay, D.B., C.A. Pickover, and T.A. Steitz: Escherichia coli lac repressor is elongated with its operator DNA binding domains located at both ends. J Mol Biol 156:175, 1982.
9. Charlier, M., J.C. Maurizot, and G. Zaccai: Neutron scattering studies of lac repressor. Nature 286:423, 1980.
10. Charlier, M., J.C. Maurizot, and G. Zaccai: Neutron-scattering studies of lac repressor: A low-resolution model. J Mol Biol 153:177, 1981.
11. Pilz, I., K. Goral, O. Kratky, R.P. Bray, N.G. Wade-Jardetzky, and O. Jardetzky: Small-angle x-ray studies of the quaternary structure of the lac repressor from Escherichia coli. Biochemistry 19:4087, 1980.
12. Ogata, R.T., and W. Gilbert: DNA-binding site of the lac repressor probed by dimethylsulfate methylation of lac operator. J Mol Biol 132:709, 1979.
13. Schmitz, A., and D.J. Galas: The interaction of RNA polymerase and lac repressor with the lac control region. Nucl Acids Res 6:111, 1979.
14. Manly, S.P. and K.S. Matthews: Lac operator DNA modification in the presence of proteolytic fragments of the repressor protein. J Mol Biol 179:315, 1984.
15. Manly, S.P., G.N. Bennett, and K.S. Matthews: Enzymatic digestion of operator DNA in the presence of the lac repressor tryptic core. J Mol Biol 179:335, 1984.
16. Daly, T.J., and K.S. Matthews. Manuscripts in preparation.
17. Sams, C.F., V.B. Hemelt, F.D. Pinkerton, G.J. Schroepfer, Jr., and K.S. Matthews: Exposure of antigenic sites during immunization: Monoclonal antibodies to monomer of lactose repressor protein. J Biol Chem 260:1185, 1985.
18. Yang, D.S., A.A. Burgum, and K.S. Matthews: Modification of the cysteine residues of the lactose repressor protein using chromophoric probes. Biochim Biophys Acta 493:24, 1977.
19. Friedman, B.E., J.S. Olson, and K.S. Matthews: Interaction of lac repressor with inducer. Kinetic and equilibrium measurements. J Mol Biol 111:27, 1977.

20. Wu, F., P. Bandyopadhyay, and C.-W. Wu: Conformational transitions of the lac repressor from E. coli. J Mol Biol 100:459, 1976.
21. DeHaseth, P., T.M. Lohman, and M.T. Record: Nonspecific interaction of lac repressor with DNA: An association reaction driven by counterion release. Biochemistry 16:4783, 1977.
22. Royer, C., G. Weber, T. Daly, and K.S. Matthews: The dissociation of the lactose repressor protein tetramer using high hydrostatic pressure. Manuscript submitted.
23. Müller-Hill, B.: Sequence homology between lac and gal repressors and three sugar-binding periplasmic proteins. Nature 302:163, 1983.
24. Sams, C.F., N.K. Vyas, F.A. Quiocho, and K.S. Matthews: Predicted structure of the sugar-binding site of the lac repressor. Nature 310:429, 1984.
25. Newcomer, M.E., G.L. Gilliland, and F.A. Quiocho: L-arabinose-binding protein complex at 24 Å resolution. Stereochemistry and evidence for a structural change. J Biol Chem 256:13213, 1981.
26. Steitz, T.A., T.J. Richmond, D. Wise, and D. Engelman: Lac repressor protein: Molecular shape, subunit structure and proposed model for operator interaction based on structural studies of micro-crystals. Proc Natl Acad Sci USA 71:593, 1974.
27. Bandyopadhyay, P.K., F.Y.-H. Wu, and C.-W. Wu: Local mobility of the lac repressor molecule. J Mol Biol 145:363, 1981.
28. Jarema, M.A.C., P. Lu, and J.H. Miller: Genetic assignment of resonances in the NMR spectrum of a protein: lac repressor. Proc Natl Acad Sci USA 78:2707, 1981.

TABLES

TABLE 1. Association Rate Constants for MMTS-Modified Repressors

Protein	k_{assoc} ($\underline{M}^{-1}s^{-1}$)
Repressor	1.5×10^5
MMTS-Repressor	4.6×10^3
BNP-Repressor	1.1×10^5
BNP-IPTG-Repressor	1.4×10^5
MMTS-BNP-Repressor	8.8×10^3
MMTS-BNP-IPTG-Repressor	7.8×10^3

TABLE 2. Inducer Binding Parameters for Modified and Mutant Repressors at Varying pH

Protein	pH	K_d ($\underline{M}$)	k_{assoc} ($\underline{M}^{-1}s^{-1}$)
Repressor	7.5	1.6×10^{-6}	1.5×10^5
Repressor	9.2	1.5×10^{-5}	2.7×10^4
T-41 Monomer	7.5	1.2×10^{-6}	2.5×10^5
T-41 Monomer	9.2	1.7×10^{-6}	3.6×10^5
MMTS-Repressor	7.5	1.6×10^{-6}	4.6×10^3
MMTS-Repressor	9.2	1.6×10^{-6}	1.4×10^4
MMTS-T-41 Monomer	7.5	1.2×10^{-6}	1.7×10^5

3. Cro Repressor Structure and Its Interaction with DNA

B. W. Matthews

Institute of Molecular Biology and Department of Physics, University of Oregon, Eugene, OR 97403

CRO REPRESSOR

Cro is a small dimeric protein of 66 amino acids that binds to specific sites (operators) on the genome of bacteriophage λ, and prevents (represses) transcription (Ptashne et al., 1980; Takeda et al., 1977). There are six Cro-specific sites, each of which consists of a slightly different 17-base-pair DNA segment with approximate two-fold sequence symmetry.

The three-dimensional structure of Cro has been determined by X-ray crystallography and shown to consist of three α-helices ($\alpha 1$, $\alpha 2$, $\alpha 3$) and a three-stranded antiparallel β-sheet (Figure 1). The third α-helix, ($\alpha 3$) protrudes from the surface of the protein and, as described in the next section, is an obvious candidate for an interaction with DNA. Residues 55-61 of each monomer extend and lie against the surface of the other monomer. Phe 58, in particular, makes intimate hydrophobic contact with its partner subunit. The carboxyl-terminal residues 62-66 are disordered in the crystals and, presumably, in solution as well (Anderson et al., 1981; Matthews et al., 1983).

INTERACTION OF CRO WITH DNA

The backbone of the Cro molecule, drawn to the same scale as a segment of right-handed Watson-Crick B-form DNA, is shown in Figure 1. The view shows a pair of Cro molecules related by an axis of two-fold symmetry (additional modes of two-fold association occur in the crystals, but are not thought to persist in solution).

The two-fold related α_3-helices are 34 Å apart and are tilted in such a way that they can be readily placed within successive major grooves of the DNA. This apparent mode of interaction between Cro and DNA was immediately suggested by the three-dimensional structure of the protein, and is consistent with chemical protection and modification studies on specific Cro-DNA complexes (Ptashne et al., 1980). It is presumed that the flexible carboxyl-terminal residues of Cro participate in DNA binding

Figure 1.

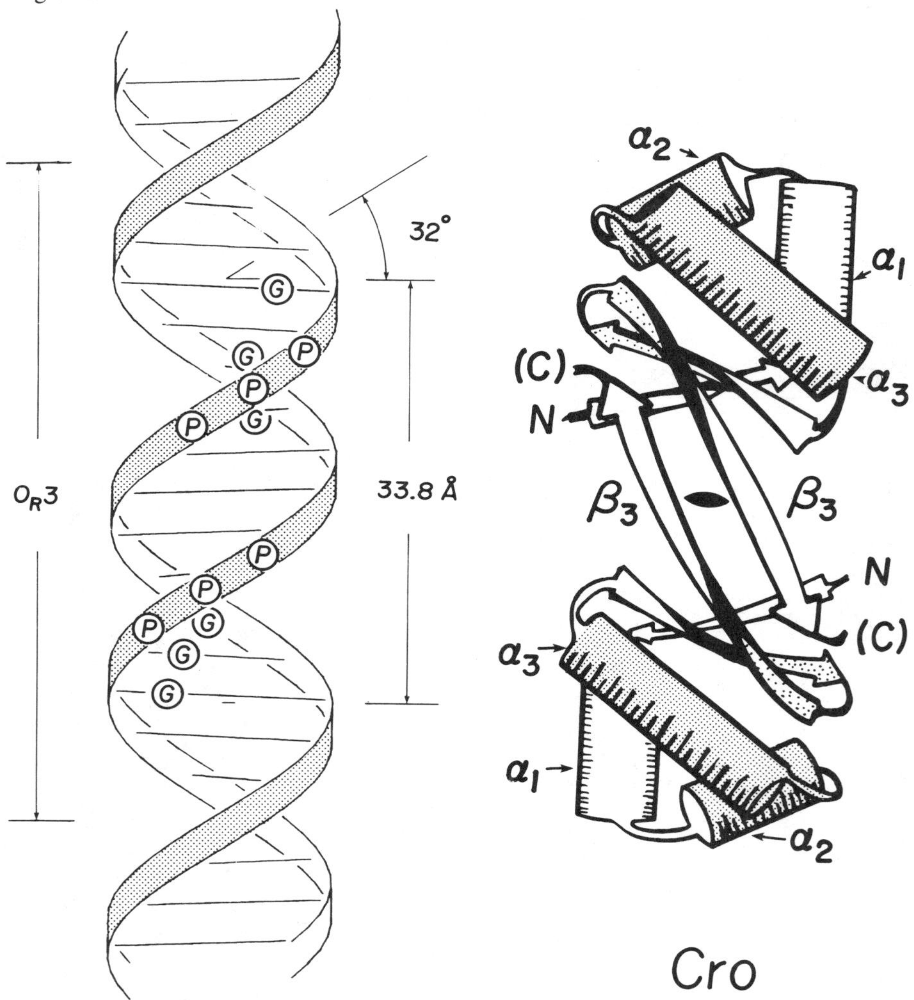

32°
G
G
P
P
P
G
O_R3
33.8 Å
P
P
P
G
G
G
α_2
α_1
(C)
α_3
N
β_3
β_3
N
(C)
α_3
α_1
α_2
Cro

by lying along the minor groove. A characteristic feature of the model is the match between the two-fold symmetry of the protein and the (approximate) two-fold sequence and spatial symmetry of the DNA binding site (Anderson et al., 1981; Matthews et al., 1983; Ohlendorf et al., 1982).

Recognition of specific base sequences on the DNA is thought to be due in large part to a multiple network of specific hydrogen bonds and other interactions between the side chains of the protein and the parts of the DNA base-pairs exposed within the grooves of the DNA.

The presumed interactions with O_R3, the tightest Cro binding site, are summarized in Figure 2 using a schematic representation based on that of Woodbury and von Hippel (1981). In most cases, a given amino acid side chain appears to make multiple interactions with the accessible parts of the base pairs. There is apparent bidentate hydrogen bonding between an arginine and a guanine, and between a glutamine and an adenine. In addition, a serine and a lysine also participate in multiple hydrogen bonding which, no doubt, enhances the specificity of protein-DNA recognition.

A COMMON HELIX-TURN-HELIX DNA-BINDING UNIT

Following the structure determination of Cro, as well as the structures of the catabolite gene activator protein from E. coli (CAP) (McKay and Steitz, 1981) and the λ repressor protein from phage λ (Pabo and Lewis, 1982), it has become apparent that these three proteins have features in common which extend to a number of other DNA binding proteins.

The suggestion that several DNA-binding proteins might have structural similarities came first from comparisons of their amino acid sequences. In some cases, such as Cro and λ repressor, the sequence homology is poor, and was not apparent on first inspection. However, with additional sequences available, the overall homology becomes obvious (Figure 3). The sequence homology includes not only repressor and activator proteins from different phages, but also other DNA-binding proteins such as the lac and trp repressors from E. coli (Anderson et al., 1982; Matthews et al., 1982; Sauer et al., 1982; Ohlendorf et al., 1983; Weber et al., 1982; Pabo and Sauer, 1984).

The region of best sequence homology occurs within the parts of the sequences that align with the α_2 and α_3 helices of Cro and of λ repressor, i.e., within the part of the respective proteins that are assumed to interact with the DNA. Thus, it is reasonable to infer that the homologous proteins contain an α-helical DNA-binding supersecondary structure similar to the α_2-α_3 fold seen in Cro and λ repressor. The locations of known mutants of lac repressor are consistent with such a hypothesis and additional support in this case has subsequently come from NMR studies (Arndt et al., 1982; Zuiderweg et al., 1983; 1984).

As well as the above sequence relationships, it was found that Cro and CAP have a striking structural correspondence in their presumed DNA binding regions (Steitz et al., 1982). The three α-helices (α_D, α_E, α_F,) in the carboxyl-terminal domain of CAP can be approximately super-

Figure 2.

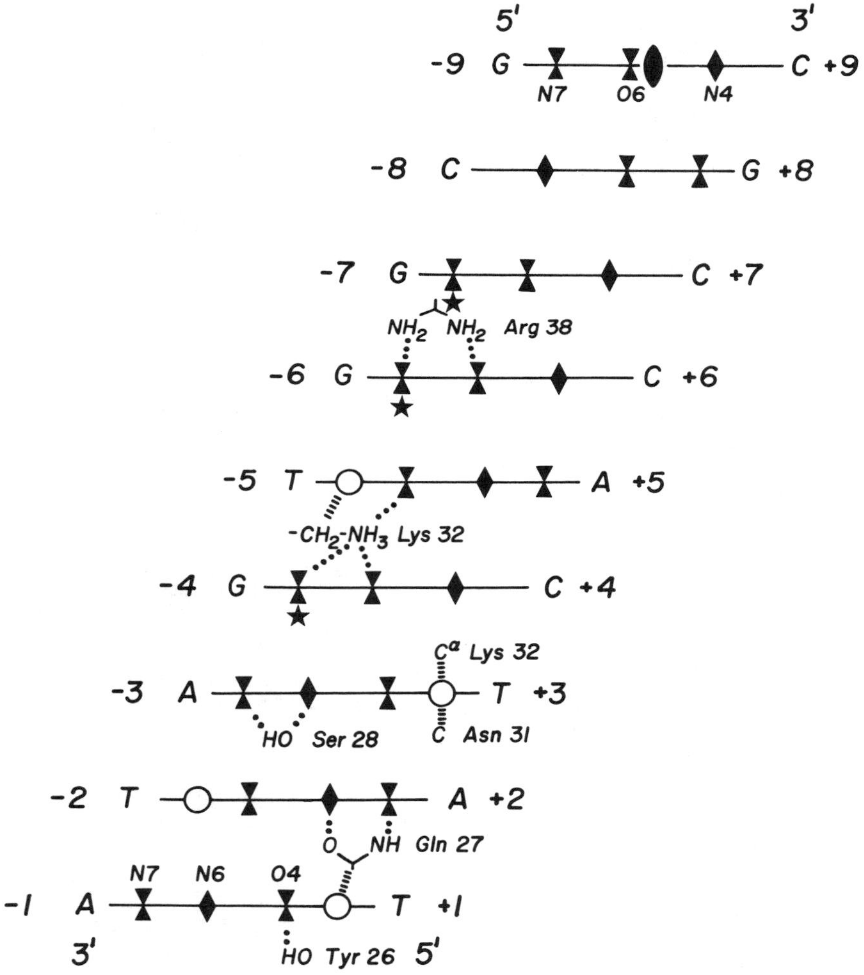

5′
3′
-9 G
C +9
N7
O6
N4
-8 C
G +8
-7 G
C +7
NH2 NH2 Arg 38
-6 G
C +6
-5 T
A +5
-CH2-NH3 Lys 32
-4 G
C +4
Cα Lys 32
-3 A
T +3
HO Ser 28
C Asn 31
-2 T
A +2
O NH Gln 27
N7 N6 O4
-1 A
T +1
3′
HO Tyr 26
5′

Figure 3.

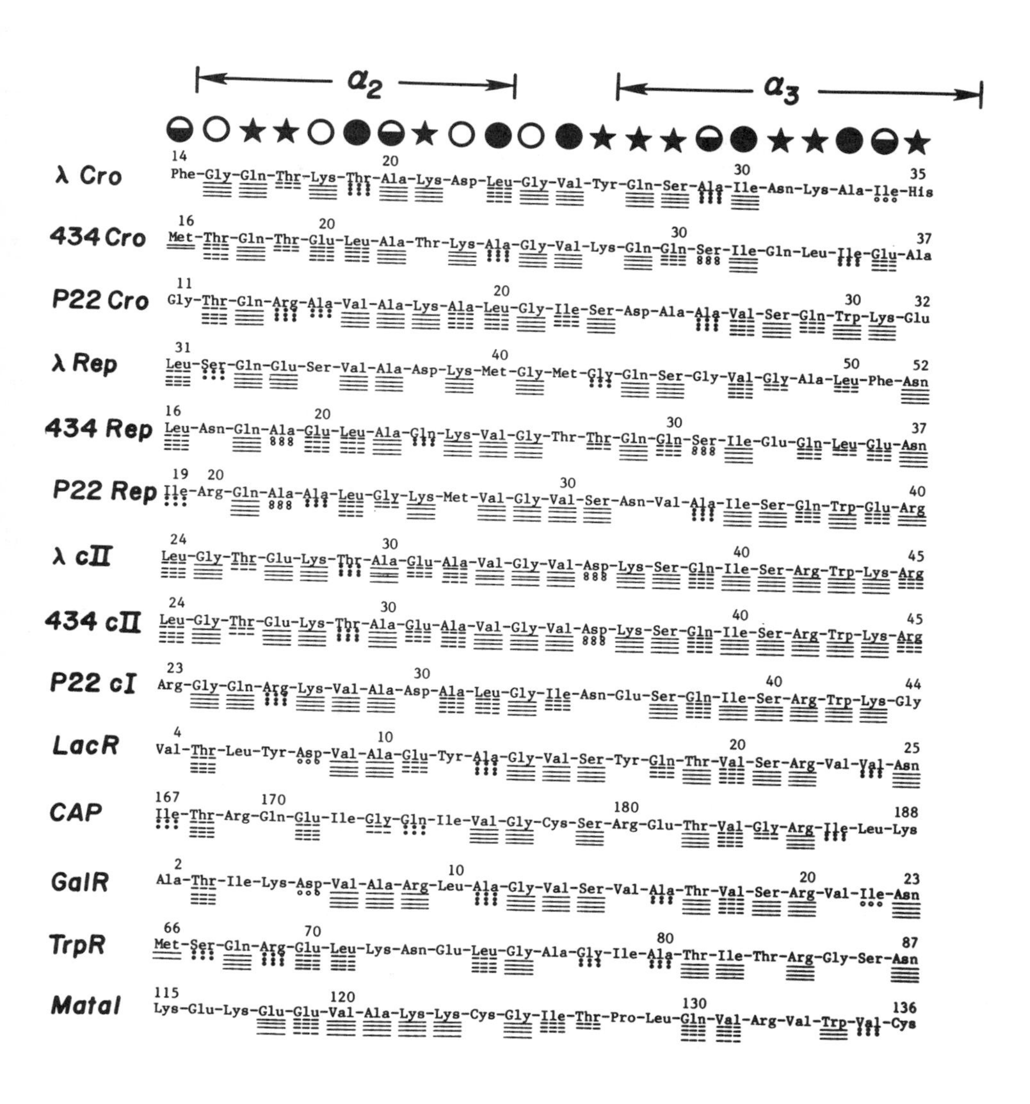
α2
α3
λ Cro
14 20 30 35
Phe-Gly-Gln-Thr-Lys-Thr-Ala-Lys-Asp-Leu-Gly-Val-Tyr-Gln-Ser-Ala-Ile-Asn-Lys-Ala-Ile-His
434 Cro
16 20 30 37
Met-Thr-Gln-Thr-Glu-Leu-Ala-Thr-Lys-Ala-Gly-Val-Lys-Gln-Gln-Ser-Ile-Gln-Leu-Ile-Glu-Ala
P22 Cro
11 20 30 32
Gly-Thr-Gln-Arg-Ala-Val-Ala-Lys-Ala-Leu-Gly-Ile-Ser-Asp-Ala-Ala-Val-Ser-Gln-Trp-Lys-Glu
λ Rep
31 40 50 52
Leu-Ser-Gln-Glu-Ser-Val-Ala-Asp-Lys-Met-Gly-Met-Gly-Gln-Ser-Gly-Val-Gly-Ala-Leu-Phe-Asn
434 Rep
16 20 30 37
Leu-Asn-Gln-Ala-Glu-Leu-Ala-Gln-Lys-Val-Gly-Thr-Thr-Gln-Gln-Ser-Ile-Glu-Gln-Leu-Glu-Asn
P22 Rep
19 20 30 40
Ile-Arg-Gln-Ala-Ala-Leu-Gly-Lys-Met-Val-Gly-Val-Ser-Asn-Val-Ala-Ile-Ser-Gln-Trp-Glu-Arg
λ cII
24 30 40 45
Leu-Gly-Thr-Glu-Lys-Thr-Ala-Glu-Ala-Val-Gly-Val-Asp-Lys-Ser-Gln-Ile-Ser-Arg-Trp-Lys-Arg
434 cII
24 30 40 45
Leu-Gly-Thr-Glu-Lys-Thr-Ala-Glu-Ala-Val-Gly-Val-Asp-Lys-Ser-Gln-Ile-Ser-Arg-Trp-Lys-Arg
P22 cI
23 30 40 44
Arg-Gly-Gln-Arg-Lys-Val-Ala-Asp-Ala-Leu-Gly-Ile-Asn-Glu-Ser-Gln-Ile-Ser-Arg-Trp-Lys-Gly
LacR
4 10 20 25
Val-Thr-Leu-Tyr-Asp-Val-Ala-Glu-Tyr-Ala-Gly-Val-Ser-Tyr-Gln-Thr-Val-Ser-Arg-Val-Val-Asn
CAP
167 170 180 188
Ile-Thr-Arg-Gln-Glu-Ile-Gly-Gln-Ile-Val-Gly-Cys-Ser-Arg-Glu-Thr-Val-Gly-Arg-Ile-Leu-Lys
GalR
2 10 20 23
Ala-Thr-Ile-Lys-Asp-Val-Ala-Arg-Leu-Ala-Gly-Val-Ser-Val-Ala-Thr-Val-Ser-Arg-Val-Ile-Asn
TrpR
66 70 80 87
Met-Ser-Gln-Arg-Glu-Leu-Lys-Asn-Glu-Leu-Gly-Ala-Gly-Ile-Ala-Thr-Ile-Thr-Arg-Gly-Ser-Asn
Matal
115 120 130 136
Lys-Glu-Lys-Glu-Glu-Val-Ala-Lys-Lys-Cys-Gly-Ile-Thr-Pro-Leu-Gln-Val-Arg-Val-Trp-Val-Cys

Figure 4.

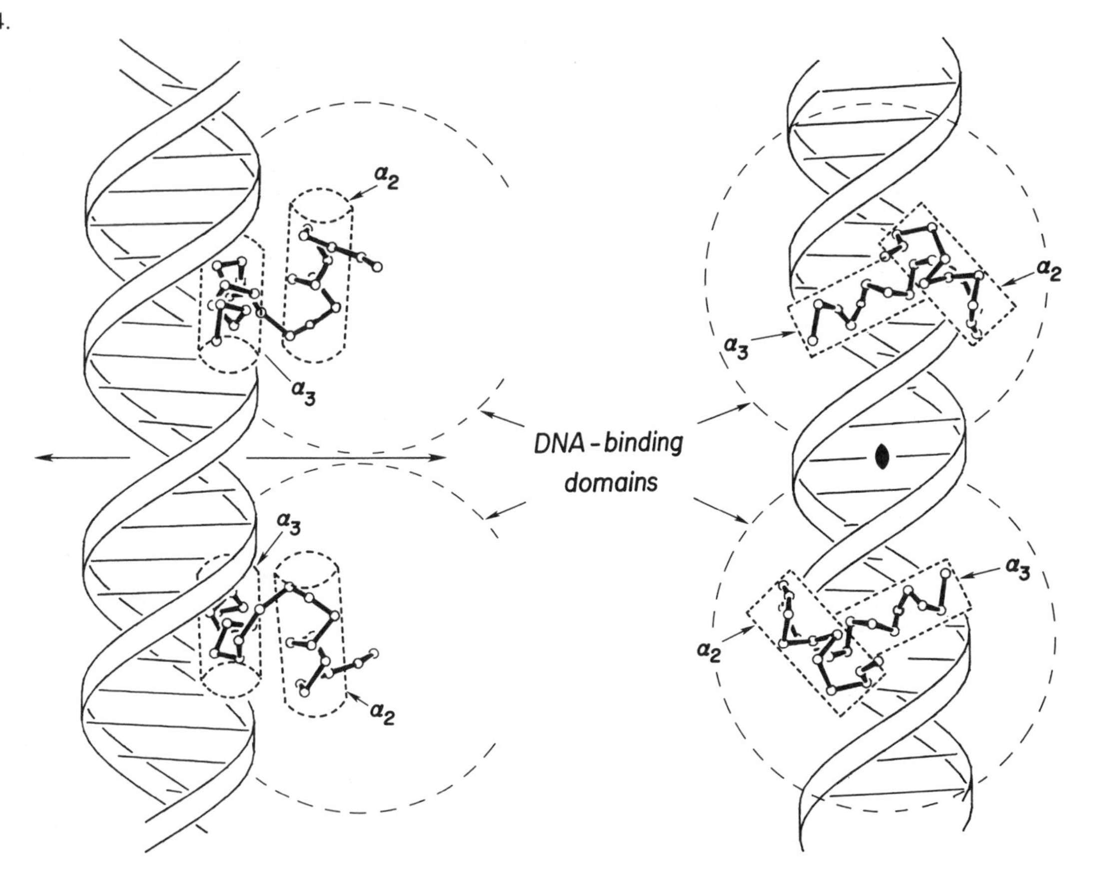

imposed on the α_1, α_2, and α_3 helices of Cro. For the α_E-α_F and α_2-α_3 helical units the superposition is striking. There are 24 α-carbons in the respective units that superimpose with an average discrepancy of 1.1Å.

It has also been shown for Cro and λ repressor that their α_2 and α_3 helices, and parts of their α_1 helices as well, spatially superimpose (Ohlendorf et al., 1983a). Again, as with Cro and CAP, it is the α_2-α_3 helical units of the two proteins that have virtually identical conformations.

The amino acid sequence comparisons and the structural comparisons both point to a special role for the helix-turn-helix "α_2-α_3" unit in DNA recognition and binding. The mode of interaction of this unit with DNA, as inferred from the structure of Cro, is sketched in Figure 4. The α_3 helix occupies the major groove of the DNA with its amino acid side chains positioned so as to make sequence-specific interactions with the exposed parts of the DNA base pairs. Side chains of the α_2 helix are also presumed to contact the DNA, these interactions being primarily to the phosphate backbone.

It is reasonable to anticipate that similar although not necessarily identical modes of DNA binding will be found for a number of other gene regulatory proteins whose sequences have been shown to be homologous with Cro, λ repressor and CAP.

There is now considerable evidence in support of the general features of the model proposed above for the interaction between Cro and DNA (e.g., see Ohlendorf and Matthews, 1983; Takeda et al., 1983; Pabo and Sauer, 1984). However the model still needs to be tested directly, preferably by determining the structure of complexes of Cro with its specific operator sites. Such studies are in progress (Anderson et al., 1983; Brennan et al., 1985).

ACKNOWLEDGEMENTS

This work was supported in part by grants from the NIH, the M.J. Murdock Charitable Trust and the NSF. The structural studies of Cro protein included here were in collaboration with Drs. W.F. Anderson, D.H. Ohlendorf and Y. Takeda.

1. Anderson, W.F., Cygler, M., Vandonselaar, M., Ohlendorf, D.H., Matthews, B.W., Kim, J. and Takeda, Y., 1983, J. Mol. Biol. 168, 903.
2. Anderson, W.F., Ohlendorf, D.H., Takeda, Y. and Matthews, B.W., 1981a, Nature 290, 754.
3. Anderson, W.F., Takeda, Y., Ohlendorf, D.H. and Matthews, B.W., 1982, J. Mol. Biol. 159, 745.
4. Arndt, K., Nick, H., Boschelli, F., Lu, P. and Sadler, J., 1982, J. Mol. Biol. 161, 439.
5. Brennan, R.G., Takeda, Y., Kim, J., Anderson, W.F. and Matthews, W.J., 1985, Manuscript in preparation.
6. Matthews, B.W., Ohlendorf, D.H., Anderson, W.F., Fisher, R.G. and Takeda, Y., 1983, Cold Spring Harbor Symp. Quant. Biol. 47, 427.
7. Matthews, B.W., Ohlendorf, D.H., Anderson, W.F., and Takeda, Y., 1982, Proc. Natl. Acad. Sci. USA 79, 1428.
8. McKay, D.B. and Steitz, T.A., 1981, Nature 290, 744.
9. Ohlendorf, D.H., Anderson, W.F., Fisher, R.G., Takeda, Y. and Matthews, B.W., 1982, Nature 298, 718.
10. Ohlendorf, D.H., Anderson, W.F., Lewis, M., Pabo, C.O. and Matthews, B.W., 1983a, J. Mol. Biol. 169, 757.
11. Ohlendorf, D.H., Anderson, W.F. and Matthews, B.W., 1983b, J. Molec. Evol. 19, 109.
12. Ohlendorf, D.H. and Matthews, B.W., 1983, Ann. Rev. Biophy. Bioen. 12, 259.
13. Pabo, C.O. and Lewis, M., 1982, Nature 298, 443.
14. Pabo, C.O. and Sauer, R.T., 1984, Ann. Rev. Biochemistry, 53, 293.
15. Ptashne, M., Jeffrey, A., Johnson, A.D., Maurer, R., Meyer, B.J., Pabo, C.O., Roberts, T.M. and Sauer, R.T., 1980, Cell 19, 1.
16. Sauer, R.T., Yocum, R.R., Doolittle, R.F., Lewis, M., and Pabo, C.O. 1982, Nature 298, 447.
17. Steitz, T.A., Ohlendorf, D.H., McKay, D.B., Anderson, W.F. and Matthews, B.W., 1982, Proc. Natl. Acad. Sci. USA 79, 3097.
18. Takeda, Y., Folkamanis, A. and Echols, H., 1977, J. Biol. Chem. 252, 6177.
19. Weber, I.T., McKay, D.B. and Steitz, T.A., 1982, Nucl. Acids. Res. 10, 5085.
20. Woodbury, C.P. and von Hippel, P.H., 1981, in "The Restriction Endonucleases" (Chirikjian, J.C., ed.), Vol. 1, pp. 181-207, Elsevier.
21. Zuiderweg, E.R.P., Billeter, M., Boelens, R., Scheek, R.M., Wüthrich, K. and Kaptein, R., 1984, FEBS Lett. 174, 243-247.
22. Zuiderweg, E.R.P., Kaptein, R. and Wüthrich, K., 1983, Proc. Natl. Acad. Sci. USA 80, 5837.

4. Activation of Transcription by the cII Protein of Bacteriophage Lambda

D. L. Wulff[a] *and M. Rosenberg*[b]

[a]Department of Biological Sciences, State University of New York at Albany, Albany, NY 12222; [b]Department of Molecular Genetics, Smith, Kline and French Laboratories, 709 Swedeland Road, Swedeland, PA 19479

INTRODUCTION

The cII protein of bacteriophage λ is required in phage infection for activation of the lysogenic pathway, whereby viral DNA is inserted into the bacterial host chromosome and viral functions are repressed through the action of the viral repressor (cI) protein (10,37). The level of cII protein in any particular infected center is a key determinant of whether that infected center enters the lysogenic or the lytic pathways (10). The cII protein coordinately activates transcription from three promoters, pRE, pI and paQ, located at widely spaced sites on the λ genome. Transcription from pRE results in expression of the repressor (cI) gene (6,22,27,32); transcription from pI results in synthesis of the Int protein, which is required for insertion of the viral DNA into the host chromosome (2,3,17); and transcription from paQ results in antagonism of Q gene expression, the Q gene being required for efficient expression of lytic genes (13,16).

THE pRE PROMOTER

pRE Structure

RNA polymerase does not recognize the pRE promoter in the absence of cII protein, and it is perhaps not surprising that the sequence of the promoter bears poor homology to the prokaryotic consensus sequence (Fig. 1) (27). The -10 region sequence for pRE agrees with the -10 consensus sequence in only three of six positions, including the highly conserved A in the second position and the "invariant" T at the sixth position. The -35 region sequence of pRE shows essentially no homology with consensus. Most notably, this six base sequence is flanked by two TTGC sequences, which constitute a direct repeat sequence of the form 5'-TTGCN_6TTGC-3'. As discussed below, cII protein recognizes and binds primarily to these two TTGC sequences on one face of the DNA double helix, and this facilitates binding of RNA polymerase to the intervening six nucleotides on the opposite face.

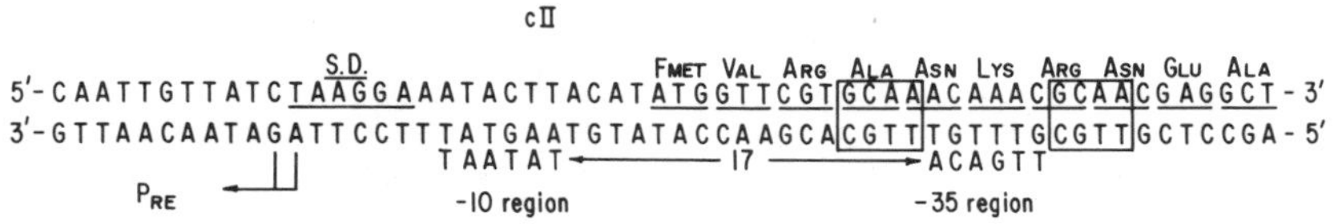

cII

S.D. FMET VAL ARG ALA ASN LYS ARG ASN GLU ALA

5'-CAATTGTTATCTAAGGAAATACTTACATATGGTTCGTGCAAACAAACGCAACGAGGCT-3'

3'-GTTAACAATAGATTCCTTTATGAATGTATACCAAGCACGTTTGTTTGCGTTGCTCCGA-5'

TAATAT ← 17 → ACAGTT

PRE -10 region -35 region

Figure 1. DNA sequence of the pRE promoter and NH_2-terminal region of the cII gene (24,27,29). The 6-base consensus sequences for the -10 and -35 regions of prokaryotic promoters are also indicated (9,23). A 17 nucleotide separation between these sequences is optimal, although small variations among prokaryotic promoters are observed (23). The TTGC sequences recognized by cII protein are outlined. Transcription from pRE initiates at either of 2 sites, as indicated by the arrow (27). The cII gene is transcribed from the pR promoter, which lies several hundrecd nucleotides to its left. The line labeled "S.D." indicates the Shine and Delgarno homology for the cII gene (33). Hyphens have been omitted from the sequence for clarity.

pRE^- Mutations

The pRE promoter overlaps with the NH_2-terminal region and ribosome recognition region of the cII gene (Fig. 1). (The cII gene is transcribed in the opposite direction from pRE, from a promoter which lies well outside the region of overlap.) Mutations in the overlap region may affect pRE function, cII protein activity, or both. Figure 2 shows a current catalog of known sequence changes in this region, comprising altogether 38 different sequence changes in a 50 bp region. Mutations reducing activity of the pRE promoter cluster in the -10 and -35 regions of the promoter (36), as is the case in other prokaryotic promoters (9,23,38). Mutations in the -10 region tend to eliminate conformity to the -10 region consensus sequence in highly conserved bases, similar to mutations in the -10 regions of other promoters. One would infer that RNA polymerase recognizes the pRE -10 region by the same means as it recognizes the -10 regions of other prokaryotic promoters. In contrast to the -10 region mutations, pRE^- mutations in the -35 region do not alter conserved bases, and extend over a thirteen base region, which is far greater than the range over which -35 region mutations occur in other promoters. The -35 region of pRE must be vastly different from the -35 regions of other promoters.

The interaction of cII protein with pRE

The cII gene was cloned into a high copy number plasmid under control of the strong repressible λ pL promoter, allowing substantial quantities of the cII protein to be produced and purified (32). The pure protein, which exists as a tetramer in solution (11), activates transcription from pRE *in vitro* (32). *E. coli* RNA polymerase and cII protein are both necessary and sufficient for transcription to occur.

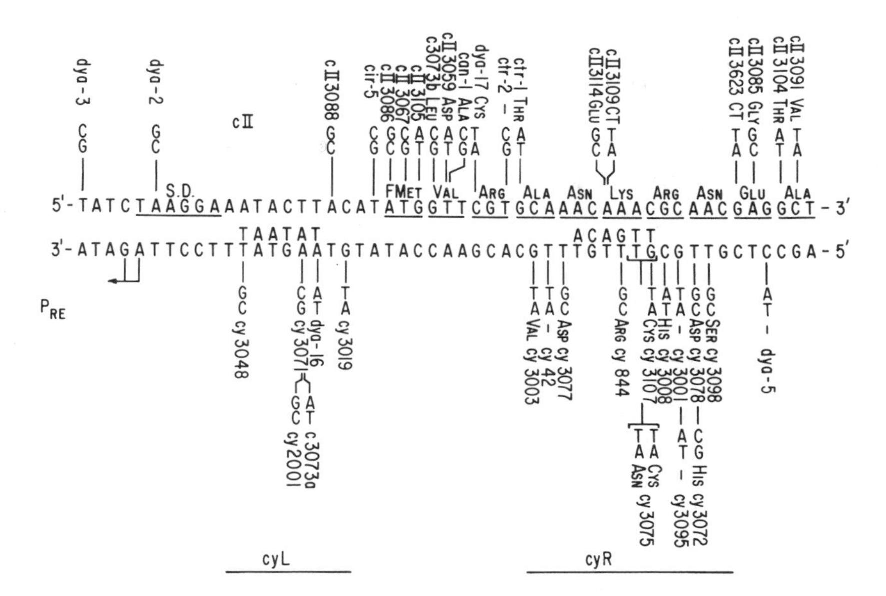

Figure 2. Sequence changes in the pRE region and NH_2-terminal region of the cII gene of bacteriophage λ. Sites with a cII prefix and c3073b are cII⁻ mutations (36,37). Sites with a cy prefix and c3073a are pRE⁻ mutations (36,37). The cir-5 mutation partially reverses the cII⁻ phenotype of λcII3086 (36,37). The can-1 mutation results in a more stable cII protein (18,34,36). The dya-17 mutation partially reverses the cII⁻ phenotype of λcII3073 (e.g., λc3073b) (M. Casio, M. Mahoney and D.W., unpublished observations). The ctr-1 and ctr-2 mutations reverse the cII⁻ phenotype of λcII3059 (21). The dya-2, dya-3, dya-5 and dya-16 mutations all have increased pRE activity (35; M. Mahoney, H. Gross, and D.W., unpublished observations).

DNAse protection experiments showed that cII protein protects a 20-25 bp region of the promoter centered at the 5'-$TTGCN_6TTGC$-3' sequence in the -35 region (12,14). RNA polymerase alone does not protect pRE DNA from nuclease digestion, except at very high concentrations, where some protection at the -10 region is observed (13). These complexes are, however, not active in transcription. In the presence of both RNA polymerase and cII protein, the entire pRE promoter region of about 45 base pairs is protected from nuclease digestion. Half maximal protection occurs at 3×10^{-8} M cII protein (13,14), which is ten-fold less than the required cII protein concentration in the absence of RNA polymerase. The binding of cII protein and RNA polymerase to pRE DNA is therefore a cooperative process.

DNAse protection experiments were used to measure the binding affinity of cII protein to pRE DNA from some 17 different pRE$^-$ mutants (12,14). The pRE$^-$ mutations in the -10 region had no effect on cII binding, as might have been expected. The pRE$^-$ mutations in the -35 region could be subdivided into two groups. Those which altered a base in one of the two TTGC sequences in the 5'-TTGCN$_6$TTGC-3' repeat sequence eliminated binding by cII protein to pRE DNA, whereas mutations in the intervening six nucleotides had little, if any, effect on cII binding. These latter mutations presumably altered RNA polymerase contacts. This same division of pRE$^-$ mutations is indicated by kinetic analyses of transcription (30,31). The primary determinant for cII recognition is therefore the pair of TTGC sequences in the -35 region, while the -35 recognition region for RNA polymerase encompasses the six base sequence between the TTGC sequences, a sequence which RNA polymerase does not recognize in the absence of cII protein.

The interaction of cII protein and RNA polymerase with pRE DNA has also been investigated by chemical protection methods. Binding of cII protein to pRE DNA results in protection from methylation of the four Gs (two Gs on each strand) in the two TTGC sequences (14). No other purine residues are strongly protected by binding of cII protein. Methylation of G residues occurs at the N7 position of the guanine, which faces the major groove of the double helix. Since each base in one TTGC sequence is separated from the corresponding base in the second TTGC sequence by ten nucleotides, or one twist of double helix, cII protein binds to one face of the DNA double helix, making base contacts primarily in the major groove.

While the methylation of A residues is in no case inhibited by binding of cII protein, methylation is greatly enhanced for three of the four A residues in the six nucleotide sequence between the TTGC sequences (14). Methylation of A residues occurs in the minor groove of the double helix at the N3 position of adenine. The dramatic enhancement of A methylation indicates that DNA structure in the region between the TTGC sequences is distorted so that methylation in the minor groove occurs more readily. This distortion may be the means by which cII protein allows RNA polymerase to recognize the pRE -35 region, although protein-protein contacts may also be important.

In the presence of both cII protein and RNA polymerase, the four Gs in the TTGC residues remain protected from methylation, and other purine residues become protected, both outside of and between the repeat sequences (14). The protected sites face both the major and minor grooves of the double helix, and lie primarily on the opposite side from the cII protein. The pRE DNA is therefore sandwiched between the two proteins, which make their primary contacts on opposite faces of the double helix.

An alteration in the repeat sequence which retains pRE activity

The ctr-1 mutation, which was originally isolated on the basis of its compensation for mutations conferring a deficiency in the rate of translation of cII mRNA (21), alters a base in the repeat sequence (the only base in which a pRE⁻ mutation has not been obtained), to yield the sequence $TTGCN_6TTGT$ (Fig. 2). The mutant ctr-1 pRE promoter exhibits full activity in in vitro transcription assays using excess components (21), and shows 50% of pRE^+ activity in vivo (35). Thus, activity of the promoter is relatively independent of whether a C•G or T•A pair is present at this site, even though the methylation experiments show that cII protein makes close contact with the C•G pair in pRE^+ DNA.

Most importantly, the cy3008 mutation, which alters the corresponding position in the upstream sequence to yield the sequence $TTGTN_6TTGC$, is fully pRE^- by plaque morphology and complementation tests (34), nuclease protection assays of cII binding (14), and transcription assays (31). It is apparent that the two TTGC sequences are not recognized in identical fashion by cII protein. If the cII tetramer (11) is arranged with dimeric subunits exhibiting a two-fold axis of symmetry, then the protein-DNA contacts would have to be different for the two TTGC sequences.

Alterations of the spacing between the -10 and -35 regions

A number of substitution mutations have been identified in the region separating the -10 and -35 regions of pRE (these were originally detected as cII^- mutations) (Fig. 2). Phenotypically they are all pRE^+, indicating that the only role of the sequence between the -10 and -35 regions is to establish the proper separation between them. Direct evidence for the importance of spacing has been obtained through the construction of mutant promoters with either a two base pair addition or a two base pair deletion in the sequence between the -10 and -35 regions. Neither of these mutant promoters shows cII dependent pRE activity, either in vivo or in vitro (S. Keilty and M.R., unpublished observation). However, nuclease protection studies show that cII protein binds to the altered promoters in a normal fashion. These observations show that the position of the $TTGCN_6TTGC$ -35 region structure relative to the -10 region and start site sequences is crucial to pRE activity.

Mutations which increase the activity of the pRE promoter

The dya-5 mutation, a C→T change at position -43 of the pRE promoter (Fig. 2), yields a promoter which is 2-to 3-fold more active than pRE^+ in vivo and retains full dependence on cII protein for activity (35; M. Mahoney and D.W., unpublished observations). The dya-5 mutation lies outside of the region which cII protects from methylation, but is on the periphery of the region protected by RNA polymerase. In fact, this mutation affects a position where the major groove of the DNA double helix faces RNA polymerase. Thus, we suspect that dya-5 affects an RNA polymerase contact with DNA, and not a cII protein contact. Mutations of a GC pair to an AT pair at the surrounding positions -41,-44, and -45 are without effect on pRE activity.

The dya-16 mutation alters the pRE -10 region sequence from AAGTAT to TAGTAT (Fig. 2), a change which increases the homology to the consensus sequence. The dya-16 pRE promoter agrees with the consensus sequence in the three most highly conserved positions (TAxxxT) (9,23), whereas pRE^+ agrees with the consensus sequence in only two of these positions (xAxxxT). We do not know as yet whether the dya-16 pRE promoter has become partially independent of cII protein for activity.

A mutant pRE promoter with a perfect -10 consensus sequence was created through directed mutagenesis, and found to function efficiently without cII protein, both *in vivo* and *in vitro* (S. Keilty and M.R., unpublished observation). In the presence of cII protein, promoter activity is enhanced only about two-fold. This constitutive pRE promoter also functions efficiently when additions or deletions are introduced between the -10 and -35 regions, and even when the -35 region is deleted and replaced with foreign DNA. Thus, in the presence of the perfect -10 region consensus sequence, a minimum RNA polymerase recognition and start signal is defined, which requires only the 18 bp between -16 and +2 of the pRE region (S. Keilty and M.R., unpublished observations).

THE cII PROTEIN

cII mutations

A large number of cII^- mutations have been isolated and DNA sequence changes for representative mutations determined (36,37; M. Mahoney and D.W., unpublished observations). A total of 65 different amino acid substitutions have been found, of which 11 retain partial cII activity according to plaque morphology. An additional 16 amino acid substitutions have been found as neutral substitutions associated with other mutations (primarily pRE^- mutations in the region of overlap betweek the pRE promoter and the cII gene), and in selections for cII variants with increased activity (36; M. Mahoney and D.W., unpublished observations).

A representative set of more than twenty cII mutations, spanning the entire cII gene, was selected for biochemical study. The mutant cII genes were cloned into the expression vector system (32) and the mutant cII proteins were isolated, purified, and characterized. Their ability to undergo tetramer formation, interact with DNA, and activate transcription was examined (Y. Ho and M.R., unpublished data). It was possible to divide the protein sequence into a number of distinct domains (Fig. 3) based upon the altered properties of the mutant proteins.

Codon no.	2	9	26	33	37	46	72 97
NH_2 — COOH			helix 2		helix 3		
No. of mutants							
cII^-	2	14	3	2	15	14	2
partial cII^-			3		1	2	5
neutral	11					1	
cII "up"	2		2			1	
Probable function	None	Tetramer formation	Contact with DNA			Tetramer formation	Overall stability

Figure 3. Domain analysis of the cII protein. See text for details.

<u>Domain analysis of the cII protein: the NH_2-terminal end</u>

A first domain encompasses codons 2 through 8, which is the region of overlap with pRE. Only 2 of 15 amino acid substitutions in this region yield inactive proteins. Several variant proteins with cII^+ phenotypes were purified and found to be normal in tetramer formation, DNA binding activity and activation of transcription (Y. Ho, unpublished data). The NH_2-terminal region of the cII protein can apparently withstand a large degree of alteration without affecting protein activity. In this regard the cII protein is very different from the lambda repressor (cI) protein, whose NH_2-terminal end wraps around the DNA and contributes significantly to the strength of interaction between protein and DNA (7,19). It is interesting to note that the repressor (cI) protein has a stronger binding constant to DNA than cII protein by several orders of magnitude, and that when the repressor protein is shorn of its NH_2-terminal arm, it binds DNA with a much lower affinity which is comparable to the affinity of cII protein for DNA.

<u>Domain analysis of cII protein: domains involved in tetramer formation</u>

Two domains appear to be involved primarily in tetramer formation. One of these extends from about codon 9 through codon 25, and the second extends from about codon 46 to codon 71. Tetrameric cII proteins are not formed in cII^- strains with amino acid substitutions in either of these two domains (Y. Ho, unpublished data). Between these two domains is the region of the protein which interacts with DNA (see below).

A domain at the carboxy-terminal end of the protein, extending from about codon 72 to codon 97, is characterized by 7 cII^- mutations, of which 5 retain partial cII activity. Proteins with amino acid substitutions in this domain form tetramers, although this process is somewhat impaired (Y. Ho, unpublished data). The carboxy-terminal domain of the protein is apparently not a vital region, but serves a structural role in stabilizing the protein in the correct conformation.

Domain analysis of cII protein: interaction with the DNA binding site

The lambda cII protein is one of a family of DNA binding proteins which is predicted to use a "helix-turn-helix" motif for specific interaction with the DNA double helix (20,25). This designation is based upon comparison of the primary sequence of the cII protein with that of other DNA binding proteins (λ cI, λ Cro and E. coli CAP) for which three-dimensional structures have been determined by X-ray crystallography. In these proteins, two short regions of α-helix are separated by a non-helical region with a sharp bend. The second of these helices (designated "helix-3" because it is the third α-helical region from the NH_2-terminal end of the λ repressor (cI) protein) extends into the major groove of the DNA and makes specific contacts with bases in the recognition site. The helix-3 region is thus responsible for the specificity of interaction between the protein and the recognition site on the DNA. The first of these helices (designated "helix-2") makes primary contact with the phosphodiester backbone of the DNA, and increases the strength of interaction between protein and DNA, chiefly through contacts between amino acid side chains and phosphate residues.

The helix-3 region of the cII protein (codons 37 through 45) is altered in 16 cII⁻ mutants, and proteins from several of these have been characterized. All fail to bind DNA, yet all form tetramers. Some of these mutant proteins retain their ability to bind non-specifically to DNA, indicating that their sole defects are in recognition of the specific binding site (Y. Ho, unpublished data). This observation strongly supports the designation of this region of the cII protein as the site of specific contact with the bases of the recognition site on the DNA. Other mutant proteins with helix-3 alterations fail to bind non-specifically to DNA in addition to their defect in specific binding. These proteins presumably have alterations which perturb the orientation of the two helices. Some of the helix-3 alterations are rather conservative, such as Lys37→Arg, Ser41→Gly, and Arg42→Lys. The side chains of these amino acids are clearly critical to cII protein function, which again is consistent with their proposed role in recognition of the DNA binding site.

Two cII⁻ alterations have been found in the "turn" between the two helices (codons 33 through 36), and 8 alterations (at least one for each amino acid) have been found in the helix-2 region (codons 26 through 32), the region presumed to make contact with the phosphodiester backbone of the DNA. Three of the 8 involve alterations in charge (Glu27→Lys; Lys28→Glu; and Glu31→Lys). The Lys28→Glu mutation is cII⁻, whereas the Glu27→Lys (dya-8) and Glu31→Lys (dya-1) mutations both show increased cII activity. At each of these three codons, the positively charged lysine yields a more active cII protein than the negatively charged glutamate, which is consistent with the hypothesis that this region makes ionic contact with negatively charged phosphate residues in the DNA phosphodiester backbone. The Lys28→Glu (cII$^-$) protein has been shown to form tetramers normally, and to interact specifically with DNA and activate transcription, but at a ten-fold reduced efficiency (Y. Ho, unpublished data). Thus a non-conservative amino acid alteration has resulted in a protein in which little is changed except for its strength of interaction with DNA. This result is consistent with the proposed role of the cII "helix-2" region in interacting with the DNA phosphodiester backbone.

SPECIFICITY OF INTERACTION

The pI and paQ promoters

The sequence 5'-TTGCN_6TTGC-3' occurs in five places in lambda DNA (13). Three of these are at the sites of the pRE, pI and paQ promoters. In all three promoters, one can detect a semblance of a -10 region promoter sequence, including an "invariant T" at the sixth position of the sequence, at precisely the same interval from the -35 region sequence (Fig. 4). The other two TTGCN_6TTGC sequences on λ are not appropriately spaced from a recognizable -10 region sequence, and cII protein does not activate transcription in these regions of the phage DNA (13). The pRE and paQ promoters have an additional sequence homology of 12 of 13 consecutive residues in their -10 regions, and the pI and paQ promoters have an exact six base pair homology in their -15 to -20 regions (Fig. 4) (13). All three promoters share the same bases at only 14 out of 50 positions (including the TTGC repeat sequences) between -45 and +5 of the promoters. Yet, the interactions of these three promoters with cII protein and RNA polymerase are remarkably similar. The binding constants for cII protein to pRE, pI and paQ are essentially identical, and in all three cases cII protein binds at ten-fold lower concentrations in the presence of excess RNA polymerase. Also, RNA polymerase at high concentration interacts with the -10 regions of all three promoters to form complexes which are not transcriptionally active. One can detect differences between these promoters only in carefully quantitated studies of transcription (Y. Ho and M.R., unpublished data; B. Hoopes and M. McClure, personal communication).

The 21 pRE and P22 pRE promoters

The experience with the pRE, pI and paQ promoters would lead one to predict that any 5'-TTGCN_6TTGC-3' sequence, separated from an appropriate -10 region sequence by the proper spacing, would be efficiently transcribed in the presence of cII protein. Such sequences exist at the pRE promoters of the lambdoid phages 21 and P22 (Fig. 4). (The P22 sequence is 5'-TTGCN_6TTGT-3', which is like the λctr-1 pRE$^+$ variant of λ pRE; see the pRE PROMOTER section above.) The phages λ, 21 and P22 all have identical gene configurations in the region of the lambda cII gene, and all make cII-like proteins which activate their respective pRE promoters. Moreover, the pRE promoters of the three phages overlap the NH_2-terminal regions of their cII-like genes in exactly the same manner. Although these

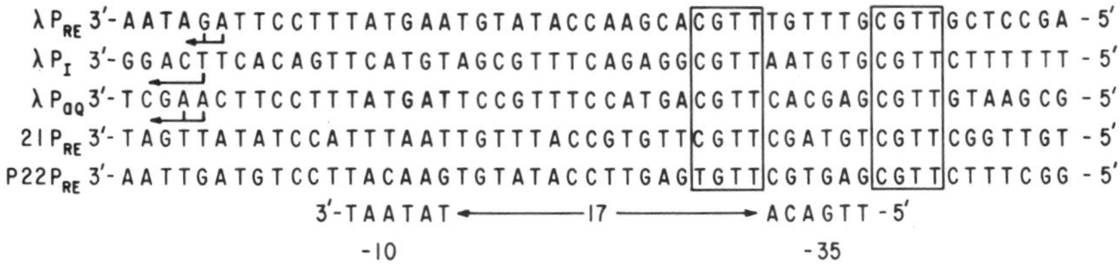

```
 λPRE 3'-AATAGATTCCTTTATGAATGTATACCAAGCA CGTT TGTTTG CGTT GCTCCGA -5'
  λPI 3'-GGACTTCACAGTTCATGTAGCGTTTCAGAGG CGTT AATGTG CGTT CTTTTTT -5'
 λPaQ 3'-TCGAACTTCCTTTATGATTCCGTTTCCATGA CGTT CACGAG CGTT GTAAGCG -5'
21PRE 3'-TAGTTATATCCATTTAATTGTTTACCGTGTT CGTT CGATGT CGTT CGGTTGT -5'
P22PRE 3'-AATTGATGTCCTTACAAGTGTATACCTTGAG TGTT CGTGAG CGTT CTTTCGG -5'
                 3'-TAATAT <------- 17 -------> ACAGTT -5'
                    -10                          -35
```

Figure 4. Nucleotide sequences of λ pRE (24,27,29), λ pI (1,4,15,26), λ paQ (13,16), 21 pRE (28), and P22 pRE (A. Poteete and R. Sauer, personal communication). The consensus sequences at the -10 and -35 regions (9,23) are also included. The TTGC repeat sequences are set off in boxes.

three promoters are widely divergent in sequence outside of the TTGC repeat sequences, they are less divergent than the λ pRE, pI and paQ promoters. The λ pRE, 21 pRE and P22 pRE promoters share the same bases in 24 of 50 positions (between -45 and +5 of the promoters), compared with only 14 homologous residues shared by the λ pRE, pI and paQ promoters (Fig. 4).

The relative activities of the cII-like proteins of λ, 21 and P22 with the pRE promoters of these three phages have been tested *in vivo* in all possible combinations (8,21), using a system in which cII protein from a derepressed defective prophage activates pRE transcription on a multicopy plasmid. Although cross-reactions are observed in all cases, the extents of cross-reaction are surprisingly small (<15% of the values observed with the homologous activator systems).

Cross-reactions have also been observed *in vitro* (Y. Ho, S. Keilty and M.R., unpublished observations). When activation is carried out in the presence of limiting amounts of RNA polymerase and activator, then the *in vitro* results are comparable with those obtained *in vivo*. However, promoter discrimination by the various activators is reduced dramatically in the presence of excess activator protein and RNA polymerase. Apparently, the various activators have the potential of exhibiting significant cross-reactivity, but, in fact, show high specificity due to their intracellular concentrations and the limiting available levels of RNA polymerase in the cell.

Recent studies (Y. Ho, S. Keilty and M. R. unpublished observations) demonstrate directly that the cII analogues of phages λ and P22 do not recognize all $TTGCN_6TTGC$ repeat sequences equally. Hence, promoter discrimination can in part be due to differences in activator recognition, presumably imposed by the sequence context surrounding the TTGC repeat sequences. However, in the case of λ pRE recognition, the λ and P22 activators exhibit similar affinities for the TTGC repeat recognition sequence, yet promoter discrimination is still observed. It was demonstrated that pRE promoter discrimination by these two activators is determined primarily by the subsequent recognition of RNA polymerase for the activator-DNA complex and not by the initial affinity of the activator. That is, RNA polymerase recognizes the λ activator-λpRE complex far better than it recognizes the P22 activator-λ pRE complex. This discrimination can be eliminated by simply increasing the concentration of RNA polymerase to achieve polymerase excess. Thus, promoter discrimination does not necessarily result from differences in activator recognition of the DNA signal, but can occur at the subsequent step of RNA polymerase recognition of the activator-DNA complex. This novel mode of signal discrimination and recognition specificity is quite distinct from that exhibited by other DNA binding proteins (e.g., repressors) where specificity results entirely from differential binding of the protein to different DNA recognition sequences.

How, then, does sequence determine promoter discrimination? Initial observations have been made *in vivo* (D.W. and M. Mahoney, unpublished observations). First, the fact that the P22 pRE repeat sequence is 5'-$TTGCN_6TTGT$-3' rather than 5'-$TTGCN_6TTGC$-3' does not account for the poor activation of λ pRE by the P22 cII-like protein (35). This result is easily understandable from the *in vitro* results described above, in which it was shown that discrimination of the λ pRE promoter for its own activator protein over the P22 protein occurs primarily at the level of the subsequent interaction of the cII protein-DNA complex with RNA polymerase. One would predict that a single base change in a region primarily responsible for the initial binding of the activator protein to DNA would not affect this discrimination. Second, λ*dya*-5 pRE is more active than $λ^+$ pRE when assayed with any cII-like protein (35). This is understandable since the *dya*-5 mutation affects an RNA polymerase contact, but not an activator protein contact (see THE pRE PROMOTER, above). Third, the *dya*-8 (Glu27→Lys) mutation causes the λ cII protein to become much more highly specific for the λ pRE promoter as compared to P22 pRE (D.W. and M. Mahoney, unpublished observations). The *dya*-8 mutation is an alteration in the helix-2 region of the cII protein, namely, the region which makes contact with the phosophodiester backbone of the DNA, and probably results in a protein which binds more tightly to DNA. The helix-2 region of the *dya*-8 cII protein has precisely the same charge and the same charge distribution as the helix-2 region of the P22 activator protein. The observed effect, which is exactly opposite to that predicted by the most simpleminded model, points to the complexity of promoter discrimination, which, as discussed above, depends upon both the primary interaction of the activator protein with DNA, and subsequent recognition by RNA polymerase.

The Symmetry Problem

The recognition site of cII protein is different from those of many other DNA binding proteins in that the recognition sequence is a direct repeat sequence rather than an inverted repeat. Proteins which recognize inverted repeat sequences are all oligomeric (dimers or tetramers) and their subunits are arranged so as to exhibit a 180° dyad axis of symmetry. In models of protein-DNA interaction, this dyad axis in the protein reflects the dyad axis in the DNA recognition sequence. How does cII protein, also an oligomer (tetramer) composed of identical subunits, recognize a direct repeat sequence? Although it is possible that the subunits are not arranged with a 180° dyad axis, it is hard to imagine how the subunits could be arranged so as to have a direct repeat axis. One possibility would be the use of two adjacent monomers of the tetramer, but a sharp bend in the DNA would be required in order to form a complex. On the much more likely presumption that the subunits are arranged with a two-fold symmetry axis, we can imagine three possible solutions to the symmetry problem. The first postulates a hidden inverted repeat in the DNA, 5'-TTGCAA-3' (5), in which the "AA" shows great sequence divergence. (This is equivalent to extending each box in Figure 4 two bases to the left.) Everyone is free to make their own conclusions from the sequences shown in Figure 4, but we do not see the "AA". Moreover, a central tenant of this hypothesis is the inverted repeat centered around the "GC" portion of the sequence. But a mutation in one of these "GC" sequences to "GT" (the <u>ctr</u>-1 mutation) has very little influence on <u>p</u>RE activity. The second solution postulates an internal symmetry in the helix-3 region of the cII protein, involving a Lys-Ser-x-x-Ser-Arg sequence. In this hypothesis the Lys and Arg, both positively charged amino acids, are postulated to play equivalent roles in binding to the two halves of the recognition sequence. Arguing against this model are the observations that Lys→Arg and Arg→Lys mutations in the postulated recognition sequence both have cII⁻ phenotypes (M. Mahoney and D.W., unpublished observations), and that inversion of one sequence by chemical synthesis (to yield the sequence 5'-TTGCN_6GCAA-3') eliminates binding by cII protein (M.R., unpublished observation). The third solution postulates that each of the two TTGC sequences has a distinct mode of interaction with cII protein. Completion of X-ray crystallographic studies of cII protein structure promises to yield a quantum leap in our understanding of how the protein recognizes its DNA binding site, and we hope will lead to solution of the symmetry problem.

ACKNOWLEDGEMENTS

Work from D.W.'s laboratory was supported by U. S. Public Health Service grant GM-28370. We thank Mrs. K. Schuff for her help in preparing the manuscript.

REFERENCES

1. Abraham, J., D. Mascarenhas, R. Fischer, M. Benedik, A. Campbell, and H. Echols: DNA sequence of the regulatory region for

the integration gene of bacteriophage λ. Proc Natl Acad Sci USA 77:2477, 1980.
2. Chung, S., and H. Echols: Positive regulation of integrative recombination by the cII and cIII genes of bacteriophage λ. Virology 79:312, 1977.
3. Court, D., S. Adhya, H. Nash, and L. Enquist: The phage λ integration protein (Int) is subject to control by the cII and cIII gene products. In: DNA Insertion Elements, Plasmids, and Episomes (A.I. Bukhari, J.A. Shapiro and S.L. Adhya, eds.). Cold Spring Harbor, New York, p. 389, 1977.
4. Davies, R.W.: DNA sequence of the Int xis pI region of bacteriophage lambda: overlap of the int and xis genes. Nucleic Acids Res 8:1765, 1980.
5. Ebright, R.H.: Proposed amino acid-base pair contacts for thirteen sequence-specific DNA binding proteins. In: Protein Structure, Folding, and Design (D. Oxender, Ed.). Alan R. Liss, Inc., New York, in the press.
6. Echols, H. and L. Green: Establishment and maintenance of repression by bacteriophage lambda: the role of the cI, cII and cIII proteins. Proc Natl Acad Sci USA 68:2190, 1971.
7. Eliason, J., M. Weiss, and M. Ptashne: NH_2-terminal arm of phage λ repressor contributes energy and specificity to repressor binding. Proc Natl Acad Sci USA 82:2339, 1985.
8. Fien, K., A. Turck, I. Kang, S. Keilty, D.L. Wulff, K. McKenney, and M. Rosenberg: CII-dependent activation of the pRE promoter of coliphage lambda fused to the Escherichia coli galK gene. Gene 32:141, 1984.
9. Hawley, D.K., and W.R. McClure: Compilation and analysis of Escherichia coli promoter DNA sequences. Nucleic Acids Res 11:2237, 1983.
10. Herskowitz, I., and D. Hagen: The lysis-lysogeny decision of phage λ: explicit programming and responsiveness. Ann Rev Genet 14:399, 1980.
11. Ho, Y.S., M. Lewis, and M. Rosenberg: Purification and properties of a transcriptional activator. J Biol Chem 257:9128, 1982.
12. Ho, Y., and M. Rosenberg: Characterization of the phage λ regulatory protein cII. Ann Microbiol (Inst Pasteur) 133A:215, 1982.
13. Ho, Y.S., and M. Rosenberg: A new lysogenic promoter on phage lambda coordinately activated by cII protein. J Biol Chem in press, 1985.
14. Ho, Y.S., D.L. Wulff, and M. Rosenberg: Bacteriophage λprotein cII binds promoters on the opposite face of the DNA helix from RNA polymerase. Nature 304:703, 1983.
15. Hoess, R.H., C. Foeller, K. Bidwell, and A. Landy: Site-specific recombination functions of bacteriophage λ: DNA sequence of the regulatory regions and overlapping structural genes for Int and Xis. Proc Natl Acad Sci USA 77:2482, 1980.
16. Hoopes, B.C., and W.R. McClure: A cII-dependent promoter is located within the Q gene of bacteriophage λ. Proc Natl Acad Sci USA 82:3134, 1985.

17. Katzir, N., A. Oppenheim, M. Belfort, and A.B. Oppenheim: Activation of the lambda int gene by the cII and cIII gene products. Virology 74:324, 1976.
18. Jones, M.O., and I. Herskowitz: Mutants of bacteriophage λ which do not require the cIII gene for efficient lysogenization. Virology 88:199, 1978.
19. Pabo, C.O., W. Krovatin, A. Jeffrey, and R.T. Sauer: The N-terminal arms of λ repressor wrap around the operator DNA. Nature 298:441, 1982.
20. Pabo, C.O., and R.T. Sauer: Protein-DNA recognition. Annu Rev Biochem 53:293, 1984.
21. Place, N., K. Fien, M.E. Mahoney, D.L. Wulff, Y.S. Ho, C. Debouck, M.C. Shih, and G.N. Gussin: Mutations that alter the DNA binding site for bacteriophage lambda cII protein and affect the translation efficiency of the cII gene. J Mol Biol 180:865, 1984.
22. Reichardt, L., and A. Kaiser: Control of λ repressor synthesis. Proc Natl Acad Sci USA 68:2185, 1971.
23. Rosenberg, M., and D. Court: Regulatory sequences involved in the promotion and termination of RNA transcription. Annu Rev Genet 13:319, 1979.
24. Rosenberg, M., D. Court, H. Shimatake, C. Brady, and D. Wulff: The relationship between function and DNA sequence in an intercistronic regulatory region in phage lambda. Nature 272:424, 1978.
25. Sauer, R.T., R.R. Yocum, R.F. Doolittle, M. Lewis and C.O. Pabo: Homology among DNA-binding proteins suggests use of a conserved super-secondary structure. Nature 298:447, 1982.
26. Schmeissner, U., D. Court, K. McKenney, and M. Rosenberg: Positively activated transcription of λ integrase gene initiates with UTP in vivo. Nature 292:173, 1981.
27. Schmeissner, U., D. Court, H. Shimatake, and M. Rosenberg: Promoter for the establishment of repressor synthesis in bacteriophage λ. Proc Natl Acad Sci USA 77:3191, 1980.
28. Schwarz, E.: Sequenzanalyse der DNA lambdoider Bakteriophagen: Gene and Signalstrukturen der Transkriptionskontrolle and der DNA-Replikation. Ph.D. thesis. U. Freiburg, 1980.
29. Schwarz, E., G. Scherer, G. Hobom, and H. Kossel: Nucleotide Sequence of cro, cII and part of the O Gene in phage λ DNA. Nature 272:410, 1978.
30. Shih, M.-C., and G.N. Gussin: Differential effects of mutations on discrete steps in transcription initiation at the λ p_{RE} promoter. Cell 34:941, 1983.
31. Shih, M.-C., and G.N. Gussin: Kinetic analysis of mutations affecting the cII activation site at the pRE promoter of bacteriophage λ. Proc Natl Acad Sci USA 81:6432, 1984.
32. Shimatake, H., and M. Rosenberg: Purified λ regulatory protein cII positively activates promoters for lysogenic development. Nature 292:128, 1981.
33. Shine, J., and L. Delgarno: The 3'-terminal sequence of Escherichia coli 16S ribosomal RNA: complementarity to nonsense triplets and ribosome binding sites. Proc Natl Acad Sci USA 71:1341, 1974.

34. Wulff, D.L., M. Beher, S. Izumi, J. Beck, M. Mahoney, H. Shimatake, C. Brady, D. Court, and M. Rosenberg: Structure and function of the cy control region of bacteriophage λ. J Mol Biol 138:209, 1980.
35. Wulff, D.L., and M.E. Mahoney: Cross-specificities of functionally identical DNA-binding proteins from different lambdoid bacteriophages. In: Sequence Specificity in Transcription and Translation, UCLA Symposia on Molecular and Cellular Biology, New Series, Volume 30 (R. Calendar and L. Gold, eds.). Alan R. Liss, Inc., New York, in the press, 1985.
36. Wulff, D.L., Mahoney, M., Shatzman, A., and M. Rosenberg: Mutational analysis of a regulatory region in bacteriophage λ that has overlapping signals for the initiation of transcription and translation. Proc Natl Acad Sci USA 81:555, 1984.
37. Wulff, D.L., and M. Rosenberg: The establishment of repressor synthesis. In: Lambda II (J. Hendrix, J. Roberts, F. Stahl, and R. Weisberg, eds.). Cold Spring Harbor, New York, p. 53, 1983.
38. Youderian, P., S. Bouvier, and M.M. Susskind: Sequence determinants of promoter activity. Cell 30:843, 1982.

5. How the Cyclic AMP Receptor Protein of *Escherichia coli* Works

Susan Garges and Sanker Adhya

Developmental Genetics Section, Laboratory of Molecular Biology, National Cancer Institute, Bethesda, MD 20892

INTRODUCTION

Cyclic AMP (cAMP) was first discovered in animal cells to act as a kind of "second messenger" in hormonal control. Specific hormone (and drug) receptors are associated with adenylate cyclase in the cell membrane of animal cells. In some cases, interaction of the receptor with its ligand stimulates adenylate cyclase to make cAMP from ATP, whereas in other cases it inhibits the adenylate cyclase (1). Cyclic AMP in the cell activates a cAMP-dependent protein kinase which in turn causes a number of changes in the cell's physiology. Until recently, all of the changes mediated by cAMP in animal cells were thought to be caused by changes induced in the enzymic actions of various proteins mediated by cAMP-dependent protein kinase (2). Recently, however, evidence is emerging that a number of cAMP-induced proteins may be regulated at the level of transcription (3-9). This is very significant in light of the fact that cAMP in *Escherichia coli* apparently functions only to affect transcription (10-13). Thus, although there indeed are differences between how cAMP functions in prokaryotes and eukaryotes, there may be more similarities than previously thought. The study of cAMP and how it works in *E. coli* has provided not only an interesting system in which to study gene regulation, protein structure, protein-nucleic acid interactions, and protein-protein interactions but also a model system with applications to some of these topics in higher organisms.

In *E. coli*, cAMP is known to act only in conjunction with a protein, the cAMP receptor protein (called CRP here; others have called it CAP for catabolite gene activator protein, but since the protein does more than gene activation, and the only commonality among the functions is binding to cAMP, we prefer the term CRP). This article will concern itself with how CRP works with cAMP in *E. coli*. First, we will discuss the role of CRP·cAMP in *E. coli*, in general. Then we will describe how, at the molecular level, these effects are actually accomplished.

WHAT CRP·cAMP DOES

Almost four decades ago, it was noted that *E. coli* grown in the presence of glucose could not utilize lactose (14). This "glucose effect" was also known as catabolite repression: glucose prevented or repressed the synthesis of the enzymes necessary for lactose utilization. Years later, it was found that cAMP addition could overcome this catabolite repression, and expression from the lactose operon could be restored (15). Through biochemical and genetic studies, it was found that cAMP, with a protein that became known as CRP, was necessary to activate transcription of the lactose operon (16). The glucose effect was due to glucose inhibiting adenylate cyclase, resulting in low intracellular levels of cAMP.

Primarily through the use of genetics, the role of CRP·cAMP has been elucidated. Mutants of *E. coli* were isolated that lack cAMP because of a defect in the *cya* gene which encodes adenylate cyclase (17,18) and that lack a functional CRP because of a defect in the *crp* gene encoding CRP (19,20). These two types of mutants are indistinguishable except that the former type (*cya*⁻) can be corrected by exogenous cAMP.

Comparing the cellular physiology of these mutants to wild type, it has been shown that cells lacking a functional CRP·cAMP complex cannot utilize certain carbohydrates such as lactose, arabinose, maltose, and mannitol (11). These mutants are more resistant than wild type to mutagens such as methylmethanesulfonate (21), ultraviolet light and gamma rays (21,22), antibiotics such as nalidixic acid and ampicillin, phages such as Lambda and T6 (21), and high temperature (21, Garges and Adhya, unpublished results). *cya*⁻ or *crp*⁻ strains show altered patterns of synthesis of many gene products such as those from *ompA*, *deo*, *tna*, *lac*, and *gal* (for reviews see 10-13). These pleiotropic effects of the *cya* or *crp* mutations are caused, in those cases that have been examined in detail, by the lack of the CRP·cAMP complex causing a change in the transcriptional regulation of that gene. It is important to note that in some cases, such as the lactose operon, the CRP·cAMP complex is necessary for activation of transcription, whereas in others, such as *ompA*, it is necessary for inhibition, direct or indirect, of transcription.

There may be a pattern to regulation by CRP·cAMP in *E. coli*. Those operons whose gene products would be considered to have catabolic functions, such as *lac*, are positively regulated by CRP·cAMP. Those, whose gene products participate in anabolic functions, are negatively regulated by CRP·cAMP. One such example is *ompA* encoding a major outer membrane protein. CRP·cAMP is important in allowing the cell to utilize a variety of substances found in the environment, when easily metabolizable substances such as glucose are not available. When conditions are better, cAMP levels decrease, and the cell can "concentrate" on building up its reserves of structural and other components. One possible inconsistency with this rather simplistic classification of what is negatively and what is positively regulated by CRP·cAMP may be that strains lacking a functional CRP·cAMP complex would seemingly appear to do so well in the presence of perturbing factors such as heat, mutagens, etc., implying that CRP·cAMP would actually impede a cell's ability to survive under harsh conditions. Perhaps that cya⁻ and crp⁻ strains grow slowly compared to wild type, and that the ability to obtain nutrition is far more important than the ability to survive infrequent bouts with inhibitory factors produce pressure to keep intact cya and crp genes.

HOW CRP·cAMP WORKS:

As stated previously, cAMP always works in conjunction with CRP. How the complex is formed and how it works to activate or inhibit transcription is still not known in its entirety, but a good deal of information has come from both biochemical and genetic studies. We will go through a step-by-step description of how CRP·cAMP is currently believed to work.

1. cAMP binds to CRP

Without cAMP, CRP exists as a dimer in a conformation that will bind to DNA, but non-specifically. Upon cAMP addition, the dimer undergoes a conformational change such that it becomes sensitive to certain proteases to which it was insensitive without cAMP (23,24). The conformational change has also been detected by fluorescence studies (25,26) and by X-ray scattering (27). In this cAMP-induced conformation, the protein becomes a specific DNA-binding protein, that is, one that recognizes certain specific sequences on the DNA. How this allosteric change comes about is a subject that is currently under investigation. Biochemical studies showed that cAMP binds to the larger aminoterminal domain of the protein (28). This is corroborated by the crystal structure of the CRP·cAMP complex which has been solved to 2.9 Å resolution by Steitz and his coworkers (29). The crystal structure shows the cAMP molecule bound within a pocket created by a series of anti-parallel β-sheets (see Fig. 1). The DNA-binding portion of CRP is in the carboxyterminal domain, quite distant from the cAMP binding region. To try to define what changes are imparted by cAMP in one domain that are ultimately felt at the other end of the protein, we have isolated and characterized a number of CRP mutants which function independently of cAMP. Our mutants have changes in amino acids in the D α-helix near the hinge region connecting the two domains of the protein (30). The substituted amino acids have longer or bulkier side groups, which would cause an increased distance by van der Waals repulsion between the C and D α-helices in the case of two of the mutants or an increased distance between the D α-helix and the DNA-binding F α-helix (see Fig. 2). Aiba et al. have recently reported similar cAMP-independent CRP mutants located in the D α-helix (31). In addition, they, as well as K. Chapman and M. Ptashne (personal communication), have found cAMP-independent CRP mutants with

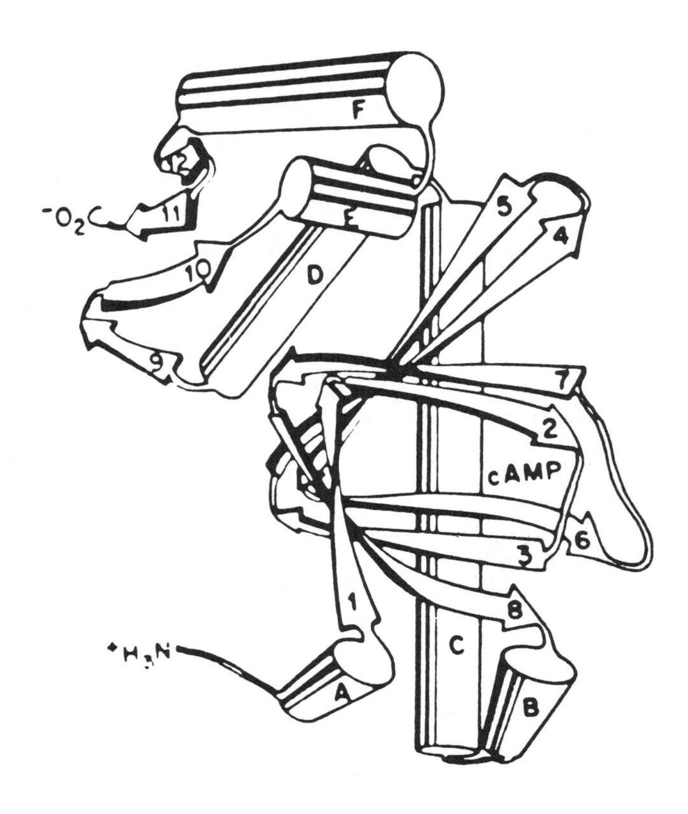

Fig. 1. Schematic drawing of a CRP monomer (from ref. 29). The regions that are α-helices are represented as cylinders lettered A-F. The regions in β-conformation are represented as arrows 1-12. The larger aminoterminal domain consists of α-helices A-C and β-sheets 1-8. The smaller carboxyterminal domain consists of the D - F α-helices and β-sheets 9-12. The two domains are connected covalently by a tetrapeptide segment between the C and D α-helices.

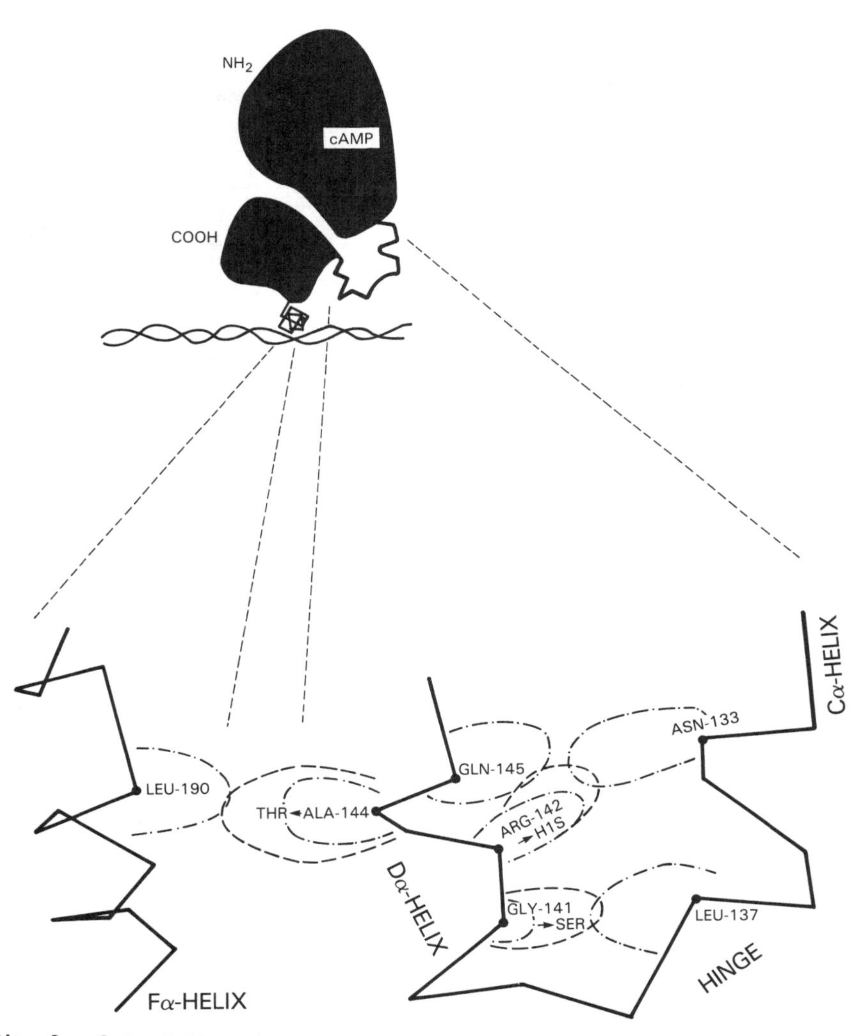

Fig. 2. Interactions between amino acids involved in cAMP-induced allosteric transition in CRP (from ref. 30). At the top, a CRP monomer is shown schematically, interacting with DNA. The hinge region, between the C and D α-helices, is shown enlarged at the bottom right, along with part of the C and D α-helices. The axis of the F α-helix that binds to the DNA is perpendicular to the page in the three dimensional structure of CRP. A planar version of a portion of the F α-helix is also shown enlarged. The bold solid line represents the α-carbon backbone of CRP. (-•-) Schematically represents the extent of the van der Waals radii of the side chains of the wild type amino acids. (---) Represents the same for the side chains of the substituted amino acids.

changes in the β-sheets 4 and 5. We believe these particular mutants may also change the distance or the angle between the cAMP-binding aminoterminal domain and the DNA-binding carboxyterminal domain of the protein.

Currently, the structure of CRP in the absence of cAMP is not known. However, based on the location of cAMP within CRP, the fact that the hinge angle is different in each subunit of the dimer (29), and the locations of mutations that allow CRP to function independently of cAMP, we present the following model as to how cAMP causes an allosteric change in CRP, causing it to become a specific DNA-binding protein:

(i) CRP without cAMP exists in a closed conformation.
(ii) Cyclic AMP binds within the aminoterminal domain, making contact with the roll of anti-parallel β-sheets, with residues wihin β-sheets 4 and 5, and with residues in the C α-helix. The cAMP contact causes a shift in these particular residues relative to other parts of the protein.
(iii) The intramolecular shifts within the protein cause an opening of the protein. Changes in the C α-helix are transmitted to the D α- helix, and the hinge between them opens up. The D α-helix, in turn, pushes against the DNA-binding F α-helix, making DNA contacts more favorable. Simultaneously, the opening of the protein is facilitated by a similar change transmitted through β-sheets 4 and 5 to the E α-helix, and then to the F α-helix.

Determination of the structure of CRP without cAMP and some of the cAMP-independent CRPs will verify the authenticity of the allosteric transition scheme defined above from mutational studies.

2. CRP·cAMP binds to DNA

Once the CRP·cAMP complex is formed, it will bind specifically to DNA to a site which has the consensus sequence AA-TGTGA--T---TCA-ATPu (32). Note that the consensus sequence is approximately symmetrical. CRP binds as a dimer; presumably each subunit contacts one half of the binding site. CRP shares with many other DNA binding proteins the helix-turn-helix motif (33). This two-helix motif (in the case of CRP, they are the E and F α-helices) appears to be important in DNA site recognition and binding.

Weber and Steitz have built a model of CRP·cAMP bound to DNA (34). In their model, the two F α-helices of the dimer lie diagonally across the DNA, with the amino end of each F α-helix interacting in successive major grooves. Indeed, Ebright et al. (32) have examined a mutant of CRP which is altered in its specificity. The mutation in position 181 at the amino end of the F α-helix allows a mutant DNA sequence to be recognized.

By model building, one can account for four amino acids making contact with the GTGA of the consensus sequence, but specific hydrogen bond interactions cannot be made for more than eight base pairs, which is shorter than the twenty base pairs to which DNAse footprinting experiments show CRP·cAMP to bind. However, Weber and Steitz (34) have found that by curving or bending the DNA, additional hydrogen bonds and salt links can be made between the DNA and the protein, extending the interaction site up to twenty bases.

Using changes in electrophoretic mobility, Wu and Crothers (35) have shown that CRP·cAMP bound to DNA causes a bend in the DNA a few bases away from the symmetric center of the lac CRP binding site. Whether this bending would be the same as that suggested by Weber and Steitz remains to be seen. Currently, DNA-CRP·cAMP co-crystals are being structurally analyzed.

3. CRP·cAMP bound to DNA can cause transcription activation

CRP·cAMP can bind to sites and activate transcription of certain operons such as lac. The initiation of transcription can be divided into two steps: (i) the formation of a "closed" complex between RNA polymerase and the promoter, and (ii) the isomerization step, where an "open" complex is formed (36). McClure and his coworkers (37, 38) have developed an abortive initiation assay whereby the two steps in transcription initiation can be quantitated separately in vitro. Malan et al. (39), studying lac, using the abortive initiation assay, found that CRP·cAMP enhances the rate of the first step, i.e., the closed complex formation.

There are two basic models which are not mutually exclusive to explain how CRP·cAMP increases the rate of closed complex formation. In the first model, CRP·cAMP binding to the DNA may cause a change in the DNA so that RNA polymerase can bind more readily at the promoter. Since, as mentioned earlier, CRP·cAMP causes a bend in DNA to which it is bound, it is possible that this bending may alter the structure of the DNA some distance away from a CRP binding site. The effect at a distance is an important consideration because the CRP binding sites in operons where CRP·cAMP activates transcription can be located as far as 100 base pairs from the start of transcription. In the second model, CRP·cAMP bound to DNA interacts with the RNA polymerase, effecting a change to promote RNA polymerase binding. Presumably, CRP·cAMP bound to DNA could interact with the RNA polymerase. In the cases where the CRP binding site and the promoter are separated by a large distance, the DNA between the CRP·cAMP and RNA polymerase could loop to bring the two proteins adjacent. Consistent with this concept, other regulatory proteins have been inferred to contact each other after binding to DNA sites separated by more than a hundred base pairs (40,41). Ptashne and colleagues (personal communication) have also searched for a mutant CRP that could still bind to DNA yet was unable to activate lac transcription. By the first model this CRP mutant would lose the ability to bend or alter the structure of the DNA, and, by the second model, would lose the ability to make the proper contacts with RNA polymerase. The mutant is currently being analyzed.

4. CRP·cAMP bound to DNA can cause repression

CRP·cAMP bound to specific sites can inhibit the transcription of certain operons such as crp (42), cya (43), galP2 (44), ompA (45), psiE and psiO (46). In the cases where studied in vitro, e.g., cya and crp, binding of CRP·cAMP to the CRP binding site does not prevent RNA polymerase from binding. As shown by DNAse footprinting studies for cya and crp, RNA polymerase still binds when CRP·cAMP is present, arguing against the possibility that CRP·cAMP acts as a repressor like the lac repressor, preventing RNA polymerase from binding. It is possible in these two cases, at least, that there may be some interaction between CRP·cAMP and RNA polymerase which precludes either proper RNA polymerase binding at the promoter or adequate transcription initiation. The cya CRP binding site overlaps the -10 of the promoter, whereas in crp, the site is twenty base pairs downstream of the transcription initiation site.

In the cases of galP2 and ompA, the CRP binding site overlaps with the -35 region of the respective promoter (47,45). In these two cases, repression by CRP·cAMP may just inhibit RNA polymerase binding, but DNAse footprinting experiments to address this have not been done.

FUTURE WORK WITH CRP·cAMP

Although much is known about how CRP·cAMP works, a number of points remain to be studied: (i) How does cAMP cause the allosteric change? Currently, we are using genetic studies to test our model of how cAMP brings about the conformational change, but clearly, structural studies of CRP without cAMP as well as of the cAMP-independent mutants are necessary. (ii) The mechanism by which CRP·cAMP activates or represses transcription, especially from a distance, needs to be elucidated. Is it through RNA polymerase contact, or through changing the DNA structure, or both? Genetic studies combined with biochemistry should answer this important question which may provide insight into how other gene regulatory elements such as enhancers and silencers which can work from a long distance function. It is possible that enhancers and silencers work by DNA site-specific recognition by proteins which may make contact with RNA polymerase, as suggested by the second CRP model.

REFERENCES

1. Lefkowitz, R., J. Stadel, and M. Caron: Adenylate cyclase-coupled beta-adrenergic receptors: structure and mechanisms of activation and desensitization. Ann Rev Biochem 52:159, 1983.

2. Kuo, J. and P. Greengard: Cyclic nucleotide dependent protein kinases. IV. Widespread occurrence of adenosine 3',5'-monophosphate-dependent protein kinase in various tissues and phyla of the animal kingdom. Proc Natl Acad Sci 64:1349, 1969.

3. Maurer, R: Transcriptional regulation of the prolactin gene by ergocryptine and cyclic AMP. Nature 294:94, 1981.

4. Lamers, W., R. Hanson, and H. Meisner: cAMP stimulates transcription of the gene for cytosolic phosphoenol-pyruvate carboxykinase in rat liver nuclei. Proc Natl Acad Sci 79:5137, 1982.

5. Murdoch, G., M. Rosenfeld, and R. Evans: Eukaryotic transcriptional regulation and chromatin-associated protein phosphorylation by cyclic AMP. Science 218:1315, 1982.

6. Jungmann, R., D. Kelley, M. Miles, and D. Milkowski: Cyclic AMP regulation of lactate dehydrogenase. J Biol Chem 258:5312, 1983.

7. Evans, M. and G. McKnight: Regulation of the ovalbumin gene: effects of insulin, adenosine 3',5'-monophosphate, and estrogen. Endocrinol 115:368, 1984.

8. Hashimoto, S., W. Schmid, and G. Schutz: Transcriptional activation of the rat liver tyrosine aminotransferase gene by cAMP. Proc Natl Acad Sci 81:6637, 1984.

9. Sasaki, K., T. Cripe, S. Koch, T. Andreone, D. Petersen, E. Beale, and D. Granner: Multihormonal regulation of phosphoenolpyruvate carboxykinase gene transcription. J Biol Chem 259:15242, 1984.

10. Pastan, I. and S. Adhya: Cyclic adenosine 3',5'-monophosphate in Escherichia coli. Bacteriol Rev 40:527, 1976.

11. Adhya, S. and S. Garges: How cyclic AMP and its receptor protein act in Escherichia coli. Cell 29:287, 1982.

12. Ullmann, A. and A. Danchin: Role of cyclic AMP in bacteria. Adv Cycl Nucl Res 15:1, 1983.

13. de Crombrugghe, B., S. Busby, and H. Buc: Cyclic AMP receptor protein: role in transcription activation. Science 224:831, 1984.

14. Monod, J: The phenomenon of enzymatic adaptation. Growth 11:223, 1947.

15. Perlman, R. and I. Pastan: Cyclic 3',5'-AMP: stimulation of β-galactosidase and tryptophanase induction in E. Coli. Biochem Biophys Res Commun 30:656, 1968.

16. Anderson, W., A. Schneider, M. Emmer, R. Perlman, and I. Pastan: Purification and properties of the cyclic adenosine 3',5'-monophosphate-receptor protein which mediates cyclic adenosine 3',5'-monophosphate dependent gene transcription in Escherichia coli. J Biol Chem 246:5929, 1971.

17. Perlman, R. and I. Pastan: Pleiotropic deficiency of carbohydrate utilization in an adenyl cyclase deficient mutant of Escherichia coli. Biochem Biophys Res Commun 37:151, 1969.

18. Schwartz, D. and J. Beckwith: Mutants missing a factor necessary for the expression of catabolite-sensitive operons. In J. Beckwith and D. Zipser (eds.) The Lactose Operon. Cold Spring Harbor, NY, 1970.
19. Emmer, M., B. de Crombrugghe, I. Pastan, and R. Perlman: Cyclic AMP receptor protein of E. coli: its role in the synthesis of inducible enzymes. Proc Natl Acad Sci 66:480, 1970.
20. Zubay, G., D. Schwartz, and J. Beckwith: Mechanism of activation of catabolite-sensitive genes: a positive control system. Proc Natl Acad Sci 66:104, 1970.
21. Kumar, S: Properties of adenyl cyclase and cyclic adenosine receptor protein-deficient mutants of Escherichia coli. J Bacteriol 125:545, 1976.
22. Swenson, P., J. Joshi, and R. Schenley: Regulation of cessation of respiration and killing by cAMP and CRP after U. V. irradiation of Escherichia coli. Mol Gen Genet 159:125, 1978.
23. Krakow, J. and I. Pastan: Cyclic adenosine monophosphate receptor: loss of cAMP dependent DNA binding after proteolysis in the presence of cyclic adenosine monophosphate. Proc Natl Acad Sci 70:2529, 1973.
24. Eilen E., C. Pampeno, and J. Krakow: Production and properties of the α core derived from the cyclic adenosine monophosphate receptor protein of Escherichia coli. Biochem 17:2469, 1978.
25. Wu, F., K. Nath, and C. Wu: Conformational transitions of cyclic adenosine monophosphate receptor protein of Escherichia coli. A fluorescent probe study. Biochem 13:2567, 1974.
26. Wu, C. and F. Wu: Conformational transitions of cyclic adenosine monophosphate receptor protein of Escherichia coli. A temperature jump study. Biochem 13:2573, 1974.
27. Kumar, S., N. Murthy, and J. Krakow: Ligand-induced change in the radius of gyration of cAMP receptor protein from E. coli. FEBS Lett 109:121, 1980.
28. Aiba, H., and J. Krakow: Isolation and characterization of the amino and carboxy proximal fragments of the adenosine cyclic 3',5'-phosphate receptor protein of Escherichia coli. Biochem 20:4774, 1981.
29. McKay, D., I. Weber, and T. Steitz: Structure of catabolite gene activator protein at 2.9 Å resolution. J Biol Chem 257:9518, 1982.
30. Garges, S. and S. Adhya: Sites of allosteric shift in the structure of the cyclic AMP receptor protein. Cell 41:745, 1985.
31. Aiba, H., T. Nakamura, H. Mitani, and H. Mori: Mutations that alter the allosteric nature of cAMP receptor protein of Escherichia coli. EMBO J 4:3329, 1985.
32. Ebright, R., P. Cossart, B. Gicquel-Sanzey, and J. Beckwith: Mutations that alter the DNA-sequence specificity of the catabolite gene activator protein of E. coli. Nature 311:232, 1984.
33. Steitz, T., D. Ohlendorf, D. McKay, W. Anderson, and B. Matthews: Structural similarity in the DNA-binding domains of catabolite gene activator and cro repressor proteins. Proc Natl Acad Sci 79:3097, 1982.
34. Steitz, T. and I. Weber: Structure of catabolite gene activator protein. In F. Jurnak and A. McPherson (eds.) Biological Macromolecules and Assemblies. John Wiley and Sons, NY, 1985.
35. Wu. H. and D. Crothers: The locus of sequence-directed and protein-induced DNA bending. Nature 305:509, 1984.

36. Chamberlin M: The selectivity of transcription. Ann Rev Biochem 43: 721, 1974.
37. McClure, W: Rate-limiting steps in RNA chain initiation. Proc Natl Acad Sci 77:5634, 1980.
38. Johnston, D. and W. McClure: Abortive initiation of in vitro RNA synthesis on bacteriophage λ DNA. In R. Losick and M. Chamberlin (eds.) RNA Polymerase. Cold Spring Harbor, NY, 1976.
39. Malan, T., A. Kolb, H. Buc, and W. McClure: Mechanism of CRP-cAMP activation of lac operon transcription initiation. I. Activation of the P1 promoter. J Mol Biol 180:881, 1984.
40. Majumdar, A. and S. Adhya: Demonstration of two operator elements in gal: in vitro repressor binding studies. Proc Natl Acad Sci 81:6100, 1984.
41. Dunn, T., S. Hahn, S. Ogden, and R. Schleif: An araBAD operator at -280 base pairs that is required for P_{BAD} repression: addition of DNA helical turns between the operator and promoter cyclically hinders repression. Proc Natl Acad Sci 81: 5017, 1984.
42. Aiba, H.: Autoregulation of the Escherichia coli crp gene: CRP is a transcriptional repressor for its own gene. Cell 32:141, 1983.
43. Aiba, H: Transcription of the Escherichia coli adenylate cyclase gene is negatively regulated by cAMP-cAMP receptor protein. J Biol Chem 260:3063, 1985.
44. Musso, R., R. DiLauro, S. Adhya, and B. de Crombrugghe: Dual control for transcription of the galactose operon by cyclic AMP and its receptor protein at two interspersed promoters. Cell 12:847, 1977.
45. Movva, R., P. Green, K. Nakamura, and M. Inouye: Interaction of cAMP receptor protein with the ompA gene, a gene for a major outer membrane protein of Escherichia coli. FEBS Lett 128:186, 1981.
46. Wanner, B: Overlapping and separate controls on the phosphate regulon in Escherichia coli K12. J Molec Biol 166:283, 1983.
47. Taniguchi, T., M. O'Neill, and B. de Crombrugghe: Interaction site of Escherichia coli cyclic AMP receptor protein and DNA of galactose operon promoters.
Proc Natl Acad Sci 76:5090, 1979.

6. Prokaryotic Chromatin: Site-Selective and Genome-Specific DNA Binding by a Virus-Coded Type II DNA-Binding Protein

Jonathan R. Greene, Krzysztof Appelt, and E. Peter Geiduschek

Department of Biology, University of California at San Diego, La Jolla, CA 92093

PROKARYOTIC CHROMATIN

The DNA of bacteria, which is confined to a condensed and segregated, but not membrane-enclosed, structure, the nucleoid, is associated with numerous proteins. Some of these are enzymes and enzyme complexes executing the replication, repair, recombination and transcription functions of the genetic material; others are abundant, basic and, as far as is known, non-enzymatic (1-5). The entire collection of components is entitled to be called "chromatin". Indeed, Griffith found DNA in a beads-on-a-string structure, resembling polynucleosomes, in gently lysed bacteria (6). The abundant basic proteins of Escherichia coli chromatin go by a variety of names (4, 5): Histone-like protein I (HLPI) or bacterial histone I (BHI) is a 17 kdal. protein and is known to be encoded by the fir A gene (7). A protein, called H (8), whose amino acid composition resembles that of histone H2A, is thought to be a heterodimer of 27-28 kdal. subunits. The most abundant of these proteins, called HLPII, HU, NS, or DBP(DNA-binding protein)II is a heterodimer of 9 kdal. subunits (HU_a and HU_b or NS1 and NS2). The gene coding for one of the E. coli HU subunits has recently been cloned (9). There are estimated to be 15-60,000 molecules of the HU/DBPII/HLPII dimer (4, 7), 15,000 molecules of the H protein dimer (8), and 15-25,000 copies of HLPI (11) per cell. Of all these proteins, and of several others that can be detected in the E. coli nucleoid, HU/DBPII has the highest affinity for DNA and is the least readily lost from nucleoids during preparation (5).

There is no convincing indication that these basic DNA-binding proteins stably associate with each other as the histones do in forming the protein core of the nucleosome. Thus, the term "histone-like" is unsuitable in the sense that the structural basis of the DNA-binding properties of these bacterial proteins and of the nucleosome core is likely to differ in important ways.

THE TYPE II DNA-BINDING PROTEINS

Of all these proteins, only DBPII has been studied extensively in other bacteria than E. coli. It is ubiquitous in the eubacteria and highly conserved with regard to sequence. E. coli and, possibly, Synechocystis spp.(11, 12) have a heterodimeric DBPII; the majority of the hitherto investigated bacteria form a homodimeric protein. A reasonably closely related protein is even made in the archaebacterium, Thermoplasma acidophilus (13). However, when DNA-binding proteins of a representive of the sulfur-dependent archaebacteria were examined, no DBPII homologue was found (14). It is, thus, not yet established that the DBPII are ubiquitous in the archaebacteria (see 15 for descriptions of archaebacterial taxonomy); this point might bear on the evolution of the prokaryotes.

The three-dimensional structure of the Bacillus stearothermophilus DBPII has recently been solved (16). The protein forms a homodimer in which the momoners are related by two-fold symmetry. Each monomer folds into a globular core of three α helical segments and three strands of antiparallel β sheet. Two of the β strands extend into a loop. Model building suggests that in binding to DBPII dimer, DNA is cradled between those two extended, symmetry-related loops, bringing the basic amino acids of the protein's binding domain into close proximity to the phosphate-sugar backbone of the B form of double stranded DNA (Figures 1 and 2).

Further model building suggests that lateral interactions between dimeric, wedge-shaped DBPII units bound to DNA are capable of generating rosette-like structures (16) on which a screw sense can be imposed by interaction with DNA, to form negatively supercoiled, nucleosome-like objects (17). Each bacterial DBPII dimer would bind to approximately one turn of DNA B helix. While the E. coli DBPII is thought to bind non-specifically, no decisive test to exclude selective DNA binding has been performed. For example, since this protein is known to be a component of the enzyme system initiating replication at the E. coli chromosomal origin of replication (18), there is a possibility of site-specific or preferential binding somewhere along that DNA segment, although the alternative possibility, that only general binding by DBPII to this region is required for replicon initiation, is not excluded.

E. coli also contain two other proteins that are related to the DBPII: the α and β subunits of the heterodimeric integration host factor, IHF (19, 20). This protein participates in site-specific recombination between lambdoid phage genomes and the bacterial chromosome and binds preferentially to certain sites in the phage DNA. IHF is also implicated in regulating the activity of several bacterial genes (21-26), and in the packaging of lambdoid phage DNA (25, 26), but the modes of action remain to be determined. IHF is perhaps an order of magnitude less abundant than DBPII, to judge by purification schedules (24).

TF1: A VIRUS-SPECIFIC TYPE II DNA-BINDING PROTEIN

SP01, one of the large B. subtilis bacteriophages, in whose 140 kbp. of DNA 5'-hydroxymethyluracil (hmUra) entirely replaces thymine, codes for a DBPII. This protein is called TF1 (standing for SPO1 transcription factor 1), because it was first identified and purified through its ability to selectively inhibit the transcription of hmUra-containing SPO1 DNA in vitro (27). Like the bacterial proteins that it resembles, TF1 forms a dimer (28); unlike the other DBPII, it binds with greater affinity to hmUra-containing DNA than to T-containing DNA (29). The cloning and sequencing of the TF1 gene (30) allows its taxonomic relationship to the eubacterial DBPII to be seen (Figure 1). TF1 has 99 amino acids rather than the 90 of the E. coli and B. stearothermophilus proteins; the extra amino acids are located at the C-terminus. Sixty-four of the 90 amino acids in the B. stearothermophilus DBPII are either identical in TF1, conserved with regard to hydrophilic or hydrophobic character, or identically replaced in at least one other eubacterial DBPII. Moreover, all matches between TF1 and B. stearothermophilus DBPII are in perfect register and the greatest homology with the bacterial DBPII is at positions in the sequence that are highly conserved among this family of proteins as a whole: residues 1-11, 43-50, and 60-65. Thus it is highly probable that TF1 and the bacterial homologue have identical or very similar overall structures. Nevertheless, TF1 is more distantly related to the eubacterial DBPII than they are to each other, suggesting a relatively ancient divergence of its viral gene (30).

TF1 BINDS PREFERENTIALLY TO SPECIFIC SITES IN hmUra-CONTAINING SPO1 DNA

Preferential binding of TF1 to certain sites in SPO1 DNA has been analyzed by footprinting methods. The complex properties of these binding sites imply that the DNA interactions of TF1 differ entirely from the simple RNA polymerase-promoter or repressor-operator relationship. At low TF1 concentrations, specific binding sites, whose lengths range from 40 to 80 bp, are occupied (38, 40). A plausible model of DNA binding by the bacterial DBPII (see above) suggests that a protein dimer interacts with approximately 10 bp. of DNA, and we therefore hypothesize that these binding sites (designated as "core" sites) must hold multiple molecules of TF1 dimer. Since the lengths of core binding sites at different locations on the SPO1 genome vary so greatly, one can imagine different numbers of TF1 molecules binding to them. As the TF1 concentration in equilibrium with DNA is increased, the zone of protection from DNase I expands on either side of the core binding sites (38,40). This suggests that protein binding spreads from each core by lateral accretion, probably involving protein-protein interactions. In view of the complexity of this binding, a non-saturating mixture of TF1 with a preferred-site-containing fragment of SPO1 DNA must contain multiple molecular species. The individual complexes of such mixtures have been at least partly resolved by polyacrylamide gel electrophoresis and separately analyzed by footprinting, with results that are consistent with the preceding summary (39). The general decrease of DNase I sensitivity of entire probes, regardless of changes of digestion pattern, accompanying increased TF1 concentration implies that considerable non-specific binding of TF1 to DNA also occurs. It has been shown that, when TF1 binds to a DNA fragment containing two of these preferred sites and two others of only slightly lesser strength, it constrains negative supercoils (40).

A number of the best-analyzed TF1 binding sites overlap promoters of the major _B. subtilis_ RNA polymerase holoenzyme. (These are the promoters that direct transcription during the first stage of viral development.) That is consistent with the original observation that TF1 inhibits transcription of SPO1 DNA by bacterial RNA polymerases _in vitro_ (27, 38). However, preferred binding sites are not confined to early viral promoters nor even to the segment of the SPO1 chromosome containing the early genes (38, 40). Several preferred TF1 binding sites in SPO1 DNA have been sequenced (39), three overlapping early promoters and two not. The binding sites are degenerate, i.e. without perfect sequence homology, but do share certain common structural features. The three binding sites overlapping early promoters all have two short, closely spaced blocks of alternating purine-pyrimidine residues, situated between extremely A-hmUra-rich blocks with dyad symmetry, one of which also has a highly asymmetric purine distribution. The two non-promoter preferred TF1 binding sites share the alternating purine-pyrimidine elements next to highly A-hmUra-rich stretches of DNA. Spacings between these structural features are not precisely conserved among the binding sites. The individual contributions that these elements make to the protein binding, and exactly how each of the sites achieves its higher affinity for TF1, are not yet clear, but the character of the sequences and their lack of precise homology suggest that local deformations of the DNA helix may play a part in the binding.

The hmUra-specificity of TF1 binding was first surmised on the basis of its ability to selectively inhibit transcription of a variety of hmUra-containing phage DNAs _in vitro_. In more definitive, recent experiments, it has been shown, first, that TF1 binds selectively to a preferred binding site in hmUra-containing DNA, but not to the same sequence in T-containing DNA and, secondly, that the _B. stearothermophilus_ DBPII does not bind selectively to the TF1-preferred binding site in hmUra-containing _or_ T-containing DNA (39).

SPECULATIONS ABOUT THE STRUCTURE, FUNCTION AND EVOLUTION OF TF1

From the structural point of view, the fact that TF1 binds to specific sites on SPO1 DNA is interesting because its three-dimensional structure is likely to be very similar to that of the bacterial DBPII, which, as far as is known, binds nonspecifically to DNA. The selective and specific binding of TF1 to hmUra-containing DNA, but not to the same sequences in T-containing DNA, must be determined by structural properties of the protein and its cognate DNA. An examination of the aligned amino acid sequences of TF1 and the _B. stearothermophilus_ DBPII (Figure 1) suggests that two important features in the TF1 primary structure might be responsible for its selectivity and specificity of interaction. (1) TF1 diverges from the bacterial DBPII in the proposed DNA-binding domain (Figure 1). In the _B. stearothermophilus_ protein, at least one of the arginine residues of the DNA-binding domain interacts with the phosphate groups of DNA (41). In TF1, two arginines of the DNA-binding domain are substituted with valine (residue 53) and phenylalanine (residue 61), reducing the electrostatic contribution to non-specific DNA-binding. The location of Phe-61 in the three-dimensional structure of the bacterial protein would put it in close proximity to the DNA helix as proposed in model building studies (16). This location of Phe make it capable of facilitating the unstacking of base pairs in a locally deformed DNA helix and might be a major factor in generating hmUra-selectivity for TF1. (2) TF1 has nine more amino acids at its C-terminus than the bacterial DBPII. In the previously proposed model of DBPII-DNA interaction (16), the C-terminal α-helix does not participate in binding. In TF1, these nine additional amino acids might form a turn followed by another short α-helix whch can be placed in close proximity to the DNA helix. In fact, amino acids 93-99 display clear α-helical preference (42). Sequence preference and specificity for hmUra could be contributed by these segments of TF1 dimer through hydrogen bonding interactions involving the hydroxyl group of hmUra which projects into the major groove of the DNA double helix. This part of TF1's interaction with DNA would resemble that proposed for the *cI* and *cro* repressors and CAP activator (43). The above ideas will be tested with the help of site-specific mutagenesis of the cloned TF1 gene combined with structural studies.

HmUra-containing DNA is less stable than T-containing DNA of identical G+C content (44). Protein-induced local deformations of DNA structure are therefore likely to be energetically less costly, especially in proximity to hmUra-rich blocks. In T-containing DNA local sequence affects the detailed structure of DNA B helices, the most obvious consequence of which is the generation of variations in the widths of the grooves in which protein-DNA interactions take place, and curvature of the DNA helix backbone (45-49). Local sequence might also affect the ease of unstacking of DNA (for example, the wedge angle of $\frac{A}{T}\frac{A}{T}$ base pairs is relatively great; 50). Since TF1 binds preferentially to DNA sequences in close proximity to, or between, A-hmUra rich blocks, the deformability of DNA helices might be an important determinant of preferential binding.

TF1 is made at a very high rate during a part of the infectious cycle of phage SPO1, accumulating to at least 10^5 (dimer) molecules/cell, but it is not packaged with DNA in the phage head (28). Little is yet known about the function of this virus-coded DBPII in the phage development. Its role may relate to the fact that, in contrast to the T phages of *E. coli*, the virulent SPO1 phage does not degrade the host cell's DNA, yet does shut off almost all host protein synthesis. Thus, the virus may be limited in acquiring host cell chromatin-forming proteins for its own, rapidly replicating DNA, and the TF1 gene may fill that particular void. Its hmUra-selectivity segregates TF1 on viral chromatin. Depending on its specific protein-protein interaction properties (of which nothing is yet known) TF1 might thereby help to segregate viral from bacterial chromatin in the infected cell.

In closing, we want to note that the chromatin of prokaryotes has been relatively little studied from either a molecular or a genetic perspective. The identification and isolation of genes coding for major chromatin proteins and the structural analysis of those proteins is likely to open up some interesting molecular biology.

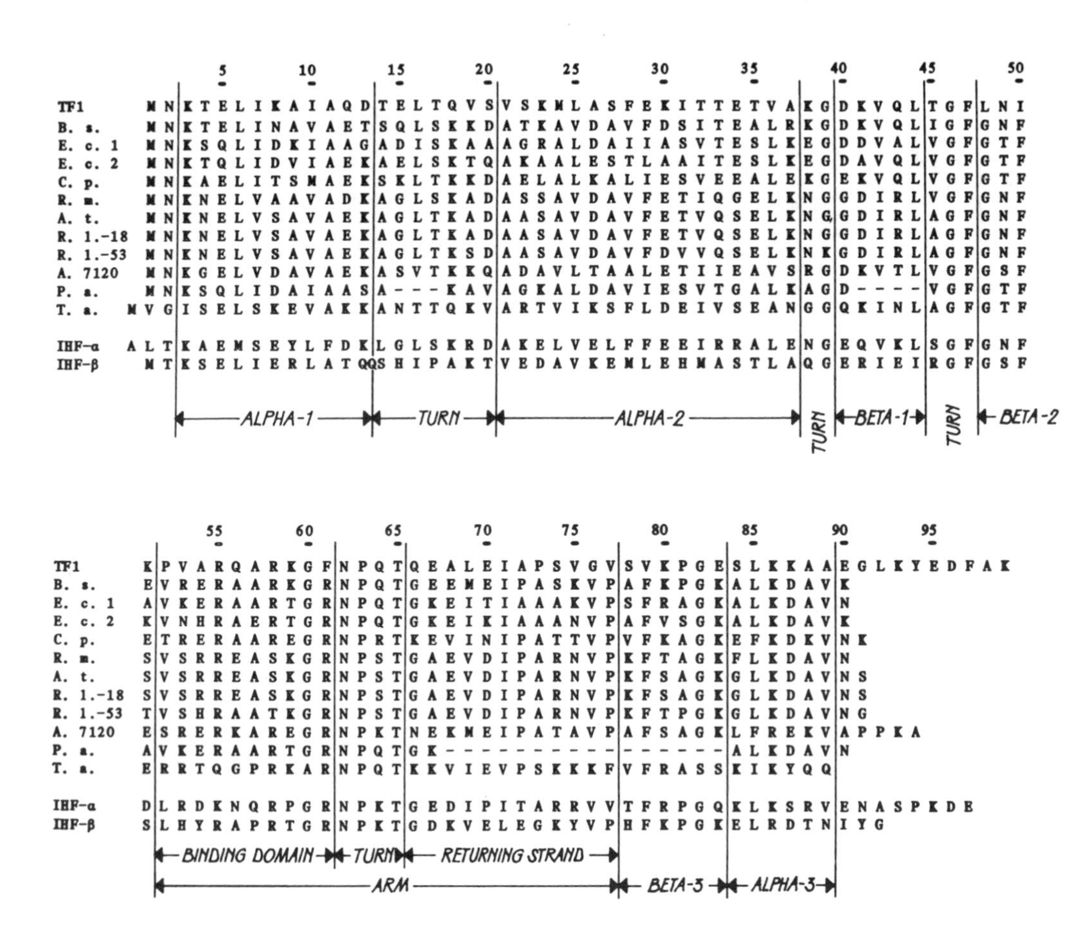
5 10 15 20 25 30 35 40 45 50
TF1 M N K T E L I K A I A Q D T E L T Q V S V S K M L A S F E K I T T E T V A K G D K V Q L T G F L N I
B. s. M N K T E L I N A V A E T S Q L S K K D A T K A V D A V F D S I T E A L R K G D K V Q L I G F G N F
E. c. 1 M N K S Q L I D K I A A G A D I S K A A A G R A L D A I I A S V T E S L K E G D D V A L V G F G T F
E. c. 2 M N K T Q L I D V I A E K A E L S K T Q A K A A L E S T L A A I T E S L K E G D A V Q L V G F G T F
C. p. M N K A E L I T S M A E K S K L T K K D A E L A L K A L I E S V E E A L E K G E K V Q L V G F G T F
R. m. M N K N E L V A A V A D K A G L S K A D A S S A V D A V F E T I Q G E L K N G G D I R L V G F G N F
A. t. M N K N E L V S A V A E K A G L T K A D A A S A V D A V F E T V Q S E L K N G G D I R L A G F G N F
R. 1.-18 M N K N E L V S A V A E K A G L T K A D A A S A V D A V F E T V Q S E L K N G G D I R L A G F G N F
R. 1.-53 M N K N E L V S A V A E K A G L T K S D A A S A V D A V F D V V Q S E L K N K G D I R L A G F G N F
A. 7120 M N K G E L V D A V A E K A S V T K K Q A D A V L T A A L E T I I E A V S R G D K V T L V G F G S F
P. a. M N K S Q L I D A I A A S A - - - K A V A G K A L D A V I E S V T G A L K A G D - - - - V G F G T F
T. a. M V G I S E L S K E V A K K A N T T Q K V A R T V I K S F L D E I V S E A N G G Q K I N L A G F G T F
IHF-α A L T K A E M S E Y L F D K L G L S K R D A K E L V E L F F E E I R R A L E N G E Q V K L S G F G N F
IHF-β M T K S E L I E R L A T Q Q S H I P A K T V E D A V K E M L E H M A S T L A Q G E R I E I R G F G S F
ALPHA-1
TURN
ALPHA-2
TURN
BETA-1
TURN
BETA-2
55 60 65 70 75 80 85 90 95
TF1 K P V A R Q A R K G F N P Q T Q E A L E I A P S V G V S V K P G E S L K K A A E G L K Y E D F A K
B. s. E V R E R A A R K G R N P Q T G E E M E I P A S K V P A F K P G K A L K D A V K
E. c. 1 A V K E R A A R T G R N P Q T G K E I T I A A A K V P S F R A G K A L K D A V N
E. c. 2 K V N H R A E R T G R N P Q T G K E I K I A A A N V P A F V S G K A L K D A V K
C. p. E T R E R A A R E G R N P R T K E V I N I P A T T V P V F K A G K E F K D K V N K
R. m. S V S R R E A S K G R N P S T G A E V D I P A R N V P K F T A G K F L K D A V N
A. t. S V S R R E A S K G R N P S T G A E V D I P A R N V P K F S A G K G L K D A V N S
R. 1.-18 S V S R R E A S K G R N P S T G A E V D I P A R N V P K F S A G K G L K D A V N S
R. 1.-53 T V S H R A A T K G R N P S T G A E V D I P A R N V P K F T P G K G L K D A V N G
A. 7120 E S R E R K A R E G R N P K T N E K M E I P A T A V P A F S A G K L F R E K V A P P K A
P. a. A V K E R A A R T G R N P Q T G K - - - - - - - - - - - - - - - - A L K D A V N
T. a. E R R T Q G P R K A R N P Q T K K V I E V P S K K K F V F R A S S K I K Y Q Q
IHF-α D L R D K N Q R P G R N P K T G E D I P I T A R R V V T F R P G Q K L K S R V E N A S P K D E
IHF-β S L H Y R A P R T G R N P K T G D K V E L E G K Y V P H F K P G K E L R D T N I Y G
BINDING DOMAIN
TURN
RETURNING STRAND
ARM
BETA-3
ALPHA-3

Figure 1. Comparison of TF1 with Type II DNA-binding (HU) proteins.

The amino acid sequences of TF1, the eubacterial DBPIIs from B. stearothermophilus (B.s.), E. coli (E.c.1 and E.c.2), Clostridium pasteurianum (C.p.), Rhizobium melilloti (R.m.), Agrobacterium tumefaciens (A.t.), two strains of Rhizobium leguminosarum, (R.l.-18) and (R.l.-53), Anabaena 7120 (A.7120), Pseudomonas aeruginosa (P.a.), the archaebacterial analog protein from Thermoplasma acidophilus (T.a.) (30-37), and the two subunits of the E. coli Integration Host Factor (IHF) (19, 20) are presented in the IUPAC one-letter code. The DBPII from E. coli is a heterodimer (11) as is IHF, while the other DBPII's are homodimers. The dashes refer to missing information from the incomplete sequence of the Pseudomonas aerugirosa protein. The secondary structure elements of the B. stearothermophilus protein (16) are indicated at the bottom of the figure. "ALPHA" represents α-helical segments, "BETA" represents strands of β-sheet and "ARM" represents the loop of antiparallel β-strands which forms an arm-like domain. Participation in DNA binding has been proposed for one strand of the loop, indicated as the "BINDING DOMAIN" (16).

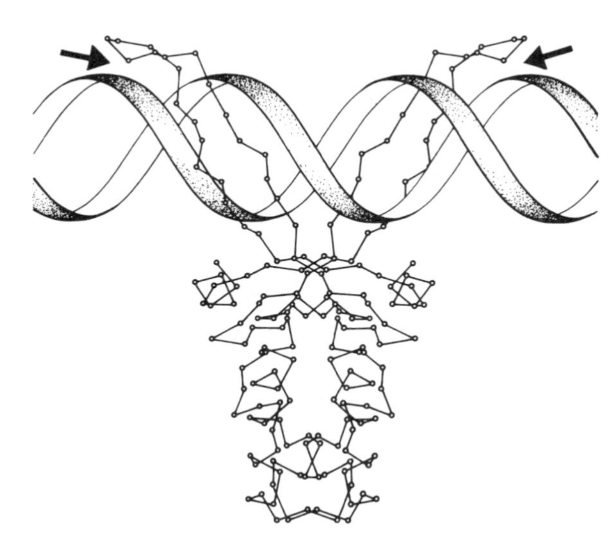

Figure 2. A Proposal for the Interaction between B. stearothermophilus DBPII-dimer and the double helix of B-DNA.

The arms follow the DNA minor groove rather than the previously suggested alternative in which the arms follow the major groove (16). In the proposed model of TF1 interaction with hmUra-containing DNA (see text), the two Phe-61 which might facilitate unstacking of DNA base pairs, are located at the top of each arm (arrows), and the nine additional C-terminal amino acids are in C-terminal helices (not shown) that might interact with slightly bent DNA.

BIBLIOGRAPHY

1. Rouvière-Yaniv, J., F. Gros (1977) In Organisation and Expression of the Eucaryotic Genome. E.M. Bradbury and K. Javaherian, eds., New York: Academic Press, pp. 211-231.
2. Kornberg, A. (1982) Supplement to DNA Replication, W.H. Freeman and Company, San Francisco.
3. Varshavsky, A.J., S.A. Nedospasov, V.V. Bakayev, T.G. Bakayeva, G.P. Georgiev (1977) Nucl. Acids Res. 4, 2724-2745.
4. Pettijohn, D.E. (1982) Cell 30, 667-669.
5. Yamazaki, K., A. Nagata, Y. Kano and F. Imamoto (1984) Mol. Gen. Genet. 196, 217-224.
6. Griffith, J.D. (1976) Proc. Natl. Acad. Sci. USA 73, 563-567.
7. Lathe, R., H. Buc, J.-P. Lecocq and E.K.F. Bautz (1980) Proc. Natl. Acad. Sci. USA 77, 3548-3552.
8. Hübscher, U., H. Lutz and A. Kornberg (1980) Proc. Natl. Acad. Sci. USA 77, 5097-5101.
9. Kano, Y. and S. Yoshino, M. Wada, K. Yokoyama, M. Nobuhara and F. Imamoto (1985) Mol. Gen. Genetics 201, 360-362.
10. Lathe, R. and J.-P. Lecocq (1977) Mol. Gen. Genet. 154, 53-60.
11. Rouvière-Yaniv, J. and N.O. Kjeldgaard (1979) FEBS Lett. 106, 297-300.
12. Aitken, A. and J. Rouvière-Yaniv (1979) Biochem. Biophys. Res. Comm. 91, 461-467.
13. DeLange, R.J., L.C. Williams and D.G. Searcy (1981) J. Biol. Chem. 256, 905-911.
14. Green, G.R., D.G. Searcy and R.J. DeLange (1983) Biochim. Biophys. Acta 741, 251-257.
15. Schleifer, K.H. and Stackebrandt, E., eds. (1985) Evolution of Prokaryotes (Academic Press, New York).
16. Tanaka, I., K. Appelt, J. Dijk, S.W. White and K.S. Wilson (1984) Nature 310, 376-381.
17. Rouvière-Yaniv, J., M. Yaniv and J.-E. Germond (1979) Cell 17, 265-274.
18. Dixon, N.E. and A. Kornberg (1984) Proc. Natl. Acad. Sci. USA 81, 424-428.
19. Flamm, E.L. and R.A. Weisberg (1985) J. Mol. Biol., 183, 117-128.
20. Miller, H.I. (1985).Cold Spring Harbor Symposia. 49, 691-699.
21. Craig, N.L and H.A. Nash (1984) Cell 39, 707-716.
22. Friden, P., Voelkel, K., Sternglanz, R. and Freundlich, M. (1986) J. Mol. Biol. In press.
23. Szekely, E., H.I. Miller and M.I. Simon (1986) J. Bact., in press.
24. Nash, H. and C. Robertson (1981) J. Biol. Chem. 256, 9246-9253.
25. Mozola, M.A. and D.I. Friedman (1985) Virology 140, 313-327.
26. Mozola, M.A., D.C. Carver and D.I. Friedman (1985) Virology 140, 328-341.
27. Wilson, D.L. and E.P. Geiduschek (1969) Proc. Nat. Acad. Sci. USA 62, 514-520.
28. Johnson, G.G. and E.P. Geiduschek (1972) J. Biol. Chem. 247, 3571-3578.
29. Johnson, G.G. and E.P. Geiduschek (1977) Biochem. 16, 1473-1485.
30. Greene, J.R., S.M. Brennan, D.J. Andrew, C.C. Thompson, S.H. Richards, R.L. Heinrikson, and E.P. Geiduschek (1984) Proc. Natl. Acad. Sci., 81, 7031-7035.

31. Mende, L., B. Timm and A.R. Subramanian (1978) FEBS Lett. 96, 395-398.
32. Kimura, M. and K.S. Wilson (1983) J. Biol. Chem. 258, 4007-4011.
33. Hawkins, A.R. and J.C. Wooton (1981) FEBS Lett. 130, 275-278.
34. Imber, R., H. Bachinger and T.A. Bickle (1982) Eur. J. Biochem. 122, 627-632.
35. Laine, B., D. Belaiche, H. Khanaka, and P. Sautière (1983) Eur. J. Biochem. 131, 325-331. press.
36. Nagaraja, R. and R. Haselkorn (1986), personal communication.
37. Khanaka, H., B. Laine, P. Sautière and J. Guillaume (1985) Eur. J. Biochem. 147, 343-349.
38. Greene, J.R. and E.P. Geiduschek (1985) EMBO J. 4, 1345-1349.
39. Greene, J.R. and E.P. Geiduschek (1986), manuscripts submitted for publication.
40. Greene J.R. and E.P. Geiduschek (1985) In Sequence Specificity in Transcription and Translation (R. Calendar and L. Gold, eds, Plenum Press, New York) p. 255-269.
41. Lammi, M., M. Paci and C.O. Gualerzi (1984) FEBS Lett. 170, 99-104.
42. Chou, P.Y. and G.D. Fasman (1974) Biochemistry 13, 222-230.
43. Takeda, Y., D.H. Ohlendorf and B.W. Mathews (1983) Science 221, 1020-1026.
44. Okubo, S., B.S. Strauss and M. Stodolsky (1964) Virology 24, 552-562.
45. Dickerson, R.E. (1983) J. Mol. Biol. 166, 419-441.
46. Drew, H.R. and A.A. Travers (1984) Cell 37, 491-502.
47. Calladine, C.R. (1982) J. Mol. Biol. 161, 343-352.
48. Trifonov, E.N. and J.L. Sussman (1980) Proc. Natl. Acad. Sci. 77, 3816-3820.
49. Marini, J.C., S.D. Levene, D.M. Crothers and P.T. Englund (1982) Proc. Natl. Acad. Sci. 79, 7664-7668.
50. Ulanovsky, L., M. Bodner, E.N. Trifonov and M. Choder (1986) Proc. Natl. Acad. Sci. 83, in press.

7. N4 Virion RNA Polymerase-Promoter Interaction

Alexandra Glucksmann and Lucia B. Rothman-Denes

Departments of Molecular Genetics and Cell Biology and of Biophysics and Theoretical Biology, University of Chicago, 920 East 58th Street, Chicago, IL 60637

INTRODUCTION

Transcription of DNA-containing bacteriophages typically requires the activity of the host DNA-dependent RNA polymerase (1). Regulation of phage gene expression during the growth cycle occurs through modification of its activity by phage-coded proteins at the level of initiation (i.e. repressors, activators or sigma factors) (2,3,4) or termination (i.e. antiterminators) (5,6). In other cases, however, the host RNA polymerase is responsible only for transcription of the phage early genes (i.e. T7-like phages) (7). One of the products of this transcription codes for a phage-coded enzyme which is responsible for transcription of the phage-late genes (8).

Bacteriophage N4, which contains 72 kb of linear, double-stranded DNA (9) is unique among DNA containing bacteriophages in that transcription of the early region of its genome is carried out by a virion-encapsulated, phage-coded RNA polymerase (10). In this paper, we summarize our knowledge of the properties of this enzyme and its interaction with its template.

Properties of N4 Virion RNA Polymerase

Phage-specific RNA synthesis in the absence of active *E. coli* RNA polymerase lead us to postulate the existence of an N4 virion-encapsulated RNA polymerase (11). A rifampicin-resistant, RNA polymerizing activity was detected in extracts of N4 virions (10) and N4-infected cells (12). The enzyme is purified from N4 virions, where it is present in one to two copies (13). The RNA polymerase is active as a 320,000 molecular weight monomer (13). Attempts to generate a smaller active domain through controlled proteolysis have failed (14). The purified polymerase activity is resistant to rifampicin and streptolidigin, in agreement with the resistance of early N4 RNA synthesis to these drugs (11,13). Through the isolation and characterization of of temperature-sensitive mutants, we have demonstrated that the activity is required for early N4 RNA synthesis (10) and N4 DNA replication (15).

N4 virion RNA Polymerase Promoters

In vitro, the N4 virion RNA polymerase shows a distinctive and peculiar template specificity. It is incapable of transcribing native N4 DNA or any other double-stranded DNA. Unlike any other RNA polymerases, however, the enzyme shows a striking specificity towards denatured N4 DNA (16). RNA polymerases, in general are unable to discriminate between denatured DNAs; indeed, these templates are transcribed poorly and without specificity. The ability of the N4 virion RNA polymerase to preferentially transcribe denatured N4 DNA lead us to determine the sites of transcription initiation on this template. To our surprise, we found that the enzyme initiates transcription at specific sites on denatured N4 DNA and, most importantly, these sites coincide with the *in vivo* sites of initiation of early N4 RNA synthesis (17). The sequence of the three sites utilized by the enzyme on the N4 genome are presented in Figure 1.

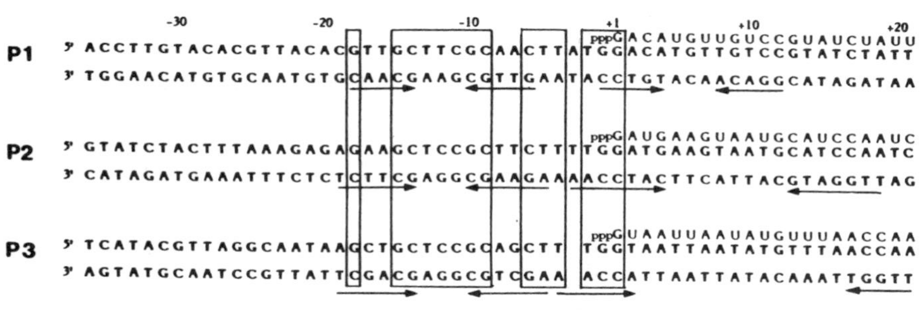

Figure 1. N4 virion RNA Polymerase sites of transcription initiation. Conserved sequences are boxed. Inverted repeats are designated by arrows.

The sequences, aligned with respect to the site of transcription initiation (+1), share extensive sequence homology from positions -18 to +1. A G-C rich heptamer is centered at -12. Moreover, two sets of inverted repeats are present in all three transcription initiation sites. One set is centered around the conserved G-C rich heptamer and encompasses both conserved and non-conserved sequences. The second set of inverted repeats encompasses the site of transcription initiation. These sequences are very different from other known prokaryotic promoters (i.e. eubacterial-type or T7 RNA polymerase-like) (18,8).

Since only three promoters for N4 virion RNA polymerase are present in the 72 kb pair N4 genome, we have recently investigated the characteristics of N4 virion RNA polymerase on heterologous templates (19). We found that the enzyme initiates at specific sites on heterologous templates (i.e. T7, pBR322, ØX174, SV40 DNAs) but with very low efficiency (2% or less of a wild type N4 promoter). Inspection of the DNA sequences surrounding these sites of transcription initiation reveals some sequence homology to the N4 promoters. In some cases, the G-C rich heptamer is present. Indeed, these sequences can be considered wild type promoters which have undergone substitutions, deletions and/or insertions.

Host Factors Required for N4 Early RNA Synthesis

As mentioned above, purified N4 virion RNA polymerase transcribes only single-stranded N4 DNA, no activity is detected on a native template. One could expect that alterations of the enzyme or template might occur upon injection which would render them competent for transcription. This consideration lead us to search for host factors that allowed transcription of the N4 double-stranded genome. Early N4 RNA synthesis is blocked by inhibitors of *E. coli* DNA gyrase such as coumermycin or nalidixic acid upon infection of drug-sensitive cells (16). No such inhibition is observed upon infection of cells resistant to these drugs suggesting that supercoiling of the genome is required for transcription by the N4 virion RNA polymerase. The enzyme is inactive, however, on supercoiled plasmids carrying any of the N4 virion RNA polymerase promoters (P. Markiewicz, unpublished results). Since the enzyme has no detectable affinity for double-stranded DNA (17), transient opening of the helix facilitated by supercoiling might not be sufficient for productive RNA polymerase-promoter interaction. Since single-stranded DNA binding proteins have the ability to shift the nucleic acid helix to coil conformational equilibrium by binding single-stranded DNA (20), we examined the effect of mutations of the *E. coli* single-stranded DNA binding (SSB) protein on N4 early transcription. Cells carrying the *ssb-1* mutation (21) do not support early N4 RNA synthesis suggesting that this protein is required for early transcription *in vivo* (22).

In vitro Transcription of a Double-stranded Template by N4 Virion RNA Polymerase Requires Supercoiling and E. coli Single-stranded DNA Binding Protein

The effect of *E. coli* SSB protein on N4 virion RNA polymerase activity on a supercoiled plasmid containing promoters P1 and P2 is shown in Figure 2. No activity is detected in the absence of SSB protein. Addition of SSB protein activates transcription reaching a maximum at a ratio of SSB/DNA (w/w) of 1. Specific transcripts, due to the presence of a virion RNA Polymerase transcription terminator, are detectable. SSB does not activate transcription when the plasmid is linear or when it is relaxed or nicked (not shown). Moreover, SSB-1 protein, at low concentrations, is defective in transcriptional activation (not shown) in agreement with the properties of the mutant protein.

Although SSB protein is not required for transcription of a single-stranded template, it is capable of activating transcription at high RNA polymerase to DNA ratios (not shown).

If SSB protein activates transcription by stabilizing single-stranded regions of the double-stranded template, one would expect that other single-stranded DNA binding proteins could substitute for SSB. This is not the case, no other single-stranded DNA binding protein tested

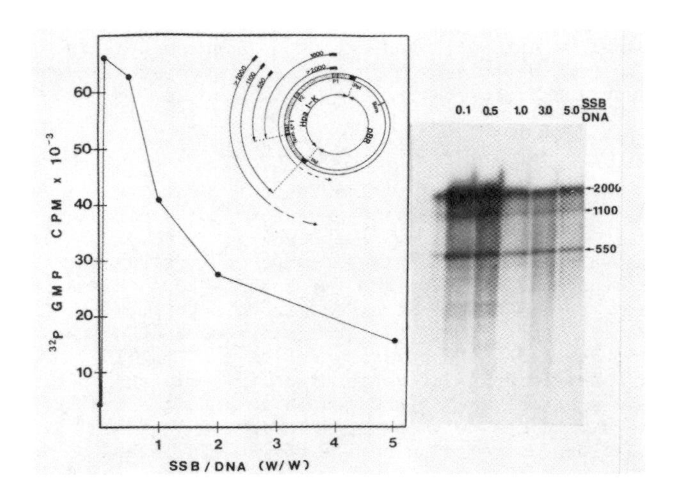

Figure 2. SSB protein allows N4 virion RNA Polymerase to transcribe supercoiled template. Supercoiled pBRK plasmid, diagrammed at the top, (0.5 μg) was preincubated with SSB protein at various protein/DNA ratios in transcription assay mixture for 5 min at 37°. Polymerase and nucleotides were added in the Standard assay (13). Total acid-precipitable radiolabeled material was measured (Left panel) and individual RNA species were resolved on an 8M urea-8% polyacrylamide gel (Right panel).

(i.e. T4 gene 32 protein, fd gene 5 protein, T7 gene 2.5 product, N4 single-stranded DNA binding protein) can activate transcription (not shown). These results suggest that activation is due to a specific structure of the template induced by the manner in which SSB protein binds to DNA, to specific protein-protein interactions between SSB protein and virion RNA polymerase, or to a combination of both possibilities.

Interactions of Virion RNA Polymerase and SSB Protein With a Single-stranded Template

In order to elucidate the architecture of the N4 virion RNA polymerase-initiation complex, we have recently studied the interaction between RNA polymerase, SSB protein, and single-stranded template by footprinting analysis. The P1 promoter was cloned into the HincII site of M13mp7 which is flanked by symmetrical cloning sites. Reannealing of the symmetrical cloning sites on the viral DNA containing the P1 template strand and subsequent digestion by EcoR1 or BamH1 releases the single-stranded insert containing the site of transcription initiation (23). This DNA is either used as a template for transcription or 5' end-labeled and used for footprinting analysis. Figure 3 shows the results of footprinting analysis using DNAse I as a probe. Two salient features are worth mention. First, addition of SSB protein to the promoter

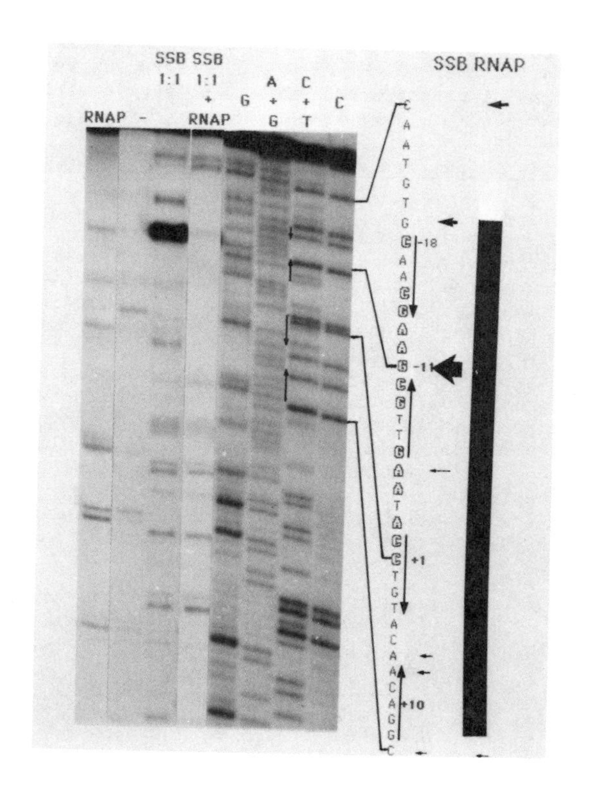

Figure 3. DNAse I Footprinting of SSB protein and N4 virion RNA Polymerase interaction with a single-stranded DNA template. A 136 nucleotide fragment generated as described in the text was 5' end labeled and treated with DNAse I in the presence of SSB protein and/or N4 virion RNA Polymerase. The products were analyzed on a 8M urea-12% polyacrylamide gel along with Maxam-Gilbert base specific reactions on the same fragment.

containing fragment induces the appearance of a DNAse I hypersensitive site, located in between the two upstream inverted repeats at position -12. Second, virion RNA polymerase protects the promoter region from DNAse I cleavage.

The site-specific pattern of DNAse I cleavage induced by SSB protein binding is surprising since SSB protein is a non-specific, single-stranded DNA binding protein.

Two possible explanations can be advanced for such behavior: 1) phasing of SSB protein binding by the ends of the DNA fragment; 2) phasing of SSB protein binding by a structure present on the DNA. Using an 18 nucleotide longer fragment, we have detected DNAse I hypersensitivity at the same site. This result suggests that some other feature of the DNA is the determinant of the DNAse I cleavage pattern. Moreover, no other single-stranded DNA binding protein elicits DNAse I hypersensitivity at that site (not shown).

Mutational Analysis of N4 Virion RNA Polymerase-SSB Protein-template Interactions

We had previously questioned the significance of inverted repeats at the N4 virion RNA polymerase sites of transcription initiation since promoter sequences are thought to be asymmetric reflecting the asymmetry of the transcriptional process. When present at promoters, inverted repeats show strong sequence conservation and indicate the binding site of activators or repressors which possess dyad symmetry (24). In contrast, the inverted repeats at the N4 promoters do not show overwhelming sequence conservation leading us to suggest that they might play a structural role in RNA polymerase-promoter recognition. We have recently tested this hypothesis by generating, through oligo-directed, site-specific mutagenesis, sequence changes at the non-conserved positions of the upstream inverted repeats. One such example is presented in Figure 4 where we have tested for the appearance of the SSB-induced DNAse I hypersensitive site when the upstream inverted repeat has been disrupted.

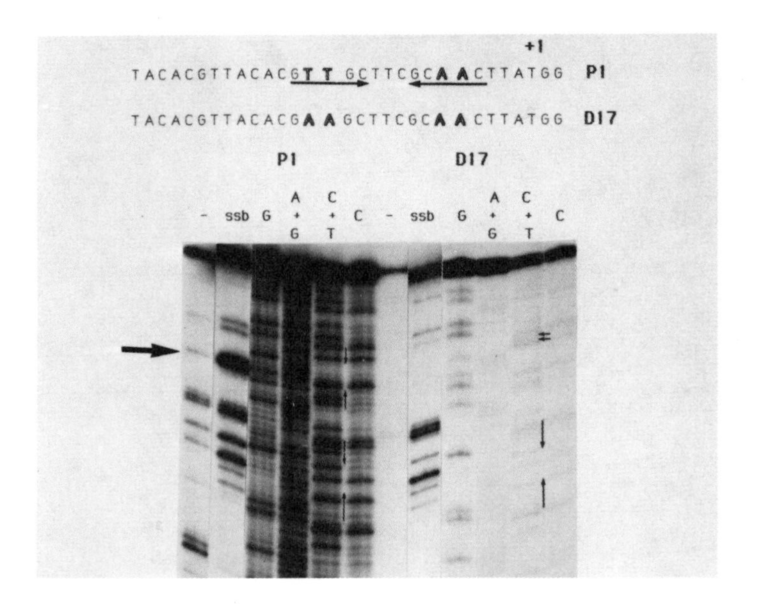

Figure 4. DNAse I footprinting of SSB-single-stranded DNA promoter interaction. Wild type P1 promoter containing DNA and D17, a mutant generated by oligo-directed, site specific mutagenesis (27), were

incubated with SSB protein and treated as described in the legend to Figure 3.

It is evident that, when the upstream inverted repeat is disrupted, the hypersensitive site disappears suggesting that the inverted repeat provides a signal to SSB protein binding.

To determine the effect of disruption of the inverted repeats on virion RNA Polymerase activity, we cloned the wild type and mutant promoters upstream of a transcription termination signal. Recognition of the initiation and termination signals leads to the synthesis of a specific RNA. Figure 5 shows that, in contrast to the wild type promoter (P1), "9" shows drastically decreased activity. Restoration of the inverted repeats, with sequences that are not present at the N4 promoters (compare to Figure 1), yields a new active promoter. These results strongly suggest that the upstream inverted repeats are important for promoter activity.

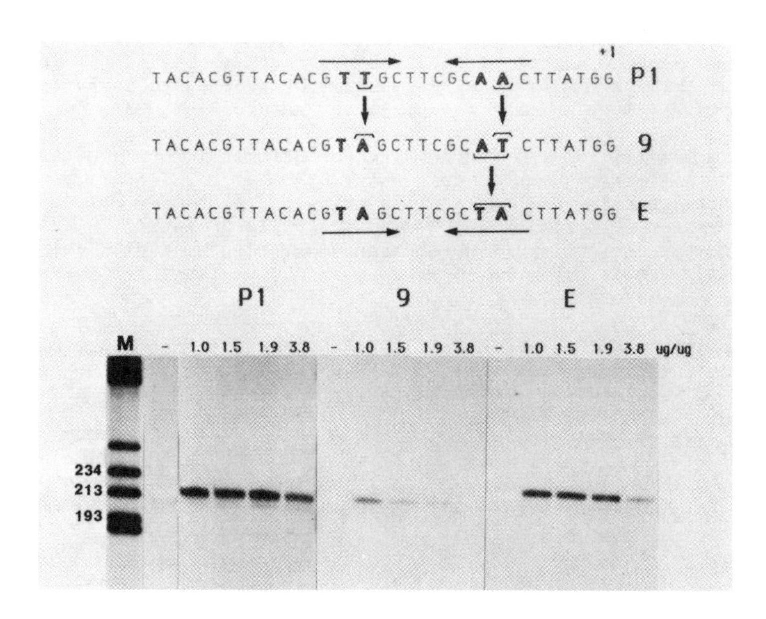

Figure 5. The upstream inverted repeats are necessary for N4 virion RNA polymerase activity. Promoter P1 or its derivatives, generated by oligo-directed, site-specific mutagenesis, were cloned in pEMBL9 upstream of an N4 transcriptional terminator. Correct initiation and termination yields a nucleotide RNA. The supercoiled plasmids were used as templates for N4 virion RNA polymerase as described in the legend to Figure 2.

Conclusions

N4 virion RNA polymerase is unique among DNA-dependent RNA polymerases in its capacity to recognize specific sequences for transcription initiation on single-stranded DNA. At this point we can only speculate on the implications of this special activity. We propose that this activity reflects part of the recognition reaction on a double-stranded template. N4 virion RNA polymerase promoters share not only sequence homology but also the presence of two sets of inverted repeats. Preliminary evidence suggests that the conserved sequences are required for activity on both single-stranded and on supercoiled, SSB-activated templates. In contrast, the set of inverted repeats present upstream of the site of transcription initiation, while required for activity on supercoiled, SSB-activated DNA, are not essential for activity on single-stranded templates.

We suggest that supercoiling of the DNA is required for SSB binding. This binding would be directed to the promoter sequences by the formation of cruciform structures at the upstream inverted repeats. The SSB-activated promoter provides the "structure" required for N4 virion RNA polymerase recognition of specific sequences to yield a productive initiation complex. Present experiments are directed to investigate the structure of the initiation complex on supercoiled DNA.

We had previously proposed the involvement of protein-protein (SSB-virion RNA polymerase) and protein-DNA (SSB-supercoiled promoter template) interactions to explain the activation of supercoiled template by SSB protein (25). If, indeed, only SSB protein is capable of interacting properly with promoter sequences (Figure 4 and unpublished experiments), there might be no need to invoke virion RNA polymerase-SSB interactions.

SSB protein is required for *E. coli* DNA replication and recombination (26). Its requirement for early N4 RNA synthesis provides the first instance of an involvement of SSB protein in transcription.

ACKNOWLEDGEMENTS

This work was supported by USPHS grant AI 12575 to L.B.R.-D. A.G. was a trainee supported by NIH grant 5T32 AI07099.

REFERENCES

1. Rabussay, D. and E. P. Geiduschek: Regulation of gene action in the development of lytic bacteriophages in Comprehensive Virology 8: 1 (H. Fraenkel-Conrat and R. Wagner, eds.) Plenum Press, N.Y. 1977.
2. McClure, W. R.: Mechanism and control of transcription initiation in prokaryotes. Ann. Rev. Biochemistry 54:171, 1985.
3. Gussin, G. N., A. D. Johnson, C. Pabo, and R. T. Sauer: Repressor and Cro protein: structure, function and role in lysogenization In Lambda II: 99 (Hendrix, R., Roberts, J.W., Stahl, F. and Weisberg, R. A., eds.) Cold Spring Harbor, New York, 1983.

4. Losick, R. and J. Pero: Cascade of sigma factors. Cell 25:582, 1981.
5. Friedman, D. I. and M. Gottesman: Lytic mode of lambda development, In Lambda II:21 (Hendrix, R., Roberts, J. W., Stahl, F. and Weisberg, R. A., eds.) Cold Spring Harbor, New York, 1983.
6. Grayhack, E. and J. Roberts: The phage lambda Q gene product: activity of a transcription antiterminator in vitro. Cell 30:367, 1983.
7. Minkley, E. G. and D. Pribnow: Transcription of the early region of bacteriophage T7: selective initiation with dinucleotides. J. Mol. Biol. 77:255, 1973.
8. Chamberlin, M. J. and T. Ryan: Bacteriophage DNA-dependent RNA Polymerases, In The Enzymes 15:87 (Boyer, P., ed.) Academic Press, N. Y., 1982.
9. Zivin, R., C. Malone, and L. B. Rothman-Denes: Physical map of coliphage N4 DNA. Virology 104, 205, 1980.
10. Falco, S. C., K. Vander Laan, and L. B. Rothman-Denes: Virion-associated RNA Polymerase required for bacteriophage N4 development. Proc. Natl. Acad. Sci. U.S.A. 74:520, 1977.
11. Rothman-Denes, L. B. and G. C. Schito: Novel transcribing activities in N4-infected E. coli. Virology 60:65, 1974.
12. Falco, S. C. and L. B. Rothman-Denes: Bacteriophage N4-induced transcribing activities in E. coli I. Detection and characterization in cell extracts. Virology 95:454, 1979.
13. Falco, S. C., W. A. Zehring, and L. B. Rothman-Denes: DNA-dependent RNA Polymerase from bacteriophage 4 virions. Purification and characterization. J. Biol. Chem. 255:4339, 1980.
14. Markiewicz, P. Ph.D. Thesis, The University of Chicago, 1984.
15. Guinta, D. G., J. Stambouly, S. C. Falco, J. K. Rist, and L. B. Rothman-Denes: Host and phage-coded functions required for N4 DNA replication. Virology 150:33, 1986.
16. Falco, S. C., R. Zivin, and L. B. Rothman-Denes: Novel template requirements of N4 virion RNA Polymerase. Proc. Natl. Acad. Sci. U.S.A. 75: 3220, 1978.
17. Haynes, L. L. and L. B. Rothman-Denes: N4 virion RNA Polymerase sites of transcription initiation. Cell 41:597, 1985.
18. Hawley, D. and W. McClure: Compilation and analysis of E. coli promoter DNA sequences. Nucl. Acid Res. 11:1137, 1983.
19. Markiewicz, P., A. Glucksmann and L. B. Rothman-Denes, submitted.
20. Kowalczykowski, S., D. Bear, and P. H. Von Hippel: Single-stranded DNA binding proteins, In The Enzymes 14:374 (Boyer, P., ed.). Academic Press, N. Y., 1983.
21. Glassberg, J., R. Meyer, and A. Kornberg: Mutant single-strand binding protein of E. coli: genetic and physiological characterization. J. Bact. 140:14, 1979.
22. Markiewicz, P., C. Malone. J. Chase, and L. B. Rothman-Denes, submitted.
23. Been, M. D. and J. J. Champoux: Cutting of M13mp7 phage DNA and excision of cloned single-stranded sequences by restriction endonucleases, In Methods in Enzymology 101:90. Recombinant DNA, part C (WU, R., Grossman, L., and Moldave, K., eds.) Academic Press, 1983.
24. Miller, J. H. and W. S. Reznikoff, eds., The Operon, Cold Spring Harbor Laboratory, CSH, 1978.

25. Rothman-Denes, L. B., L. L. Haynes, P. Markiewicz, A. Glucksmann, C. Malone, and J. W. Chase: Bacteriophage N4 virion RNA Polymerase promoters, In Sequence Specificity in Transcription and Translation 41, Alan R. Liss, Inc., New York, 1985.
26. Chase, J. W. and K. R. Williams: Single-stranded DNA binding proteins required for DNA replication. Ann. Rev. Biochem., in press, 1986.
27. Kunkel, T. A.: Rapid and efficient site-specific mutagenesis without phenotypic selection. Proc. Natl. Acad. Sci. U.S.A. 82:488, 1985.

8. Bacteriophage Lambda N System: Interaction with the Host NusA Protein

Alan T. Schauer,[a] *Eric R. Olson,*[b] *and David I. Friedman*[c]

[a]Biological Laboratories, Harvard University, 16 Divinity Avenue, Cambridge, MA 02138; [b]Molecular Biology, Upjohn Company, Kalamazoo, MI 49001; [c]Department of Microbiology and Immunology, University of Michigan Medical School, Ann Arbor, MI 48109

INTRODUCTION

Coliphage λ gene expression is temporally regulated by two transcription antitermination systems (for review, see 13). The first of these systems to operate during phage development is controlled by the λ *N* gene product. Upon infection (or induction of a prophage), a program of gene expression begins with the initiation of transcription by *E. coli* RNA polymerase at two promoters, *p*L and *p*R (see Figure 1). In order for the full complement of lytic and lysogenic genes to be expressed, RNA polymerase must transcribe through termination sites *t*L1 in the left operon, and *t*R1, *t*R2 and *t*R3 in the right operon. The interactions among N protein, phage nucleic acid loci (*nut* and *boxA*) and host proteins (Nus) that enable RNA polymerase to transcribe through most termination sites is the subject of this chapter. We first briefly summarize the current understanding of components that are known to be involved in this antitermination system. We then review recent experiments from our laboratory that are part of a continuing genetic analysis of the tripartite N-NusA-*boxA* interaction. Our recent studies suggest that the strength of the overall p*N*-modification reaction, resulting in an alteration of RNA polymerase activity, can be viewed as a sum of a set of smaller subreactions. Thus, a change caused by a mutation in one interaction that leads to a failure to antiterminate can be overcome, not only by a compensating alteration in its protein or nucleic acid target, but also by improving any one of the other subreactions. Such improvements can arise by either mutational changes or an increase in the amount of one component.

The first gene to be expressed in the left operon, *N*, encodes a 12kd basic protein (p*N*) that is essential for transcription antitermination (33,10). λ mutants that grow lytically independent of p*N* include a class (λ*nin*) that are deleted for the *t*R2 and *t*R3 transcription terminators (3,8). Although physical evidence is not yet available, genetic experiments have shown that the sites of p*N* action are the *nut* regions (discussed below). One interesting aspect of the p*N*-*nut* interaction is its specificity. Transcription initiating at promoters in other phage does not become termination-resistant in the presence of λ p*N* (16,1,9). Phages that are closely related to λ have their own *N* genes and cognate *nut* sites (16,9,20). For example, λ p*N* does not work with the *nut* sites from the related phage 21. The molecular mechanism(s) by which the *N* gene product and the *nut* sites render RNA polymerase termination-resistant is not known, and the hypothesis that p*N* binds to *nut* region RNA or DNA has not been tested. Nevertheless, it is well established that p*N* binds to the *E. coli* NusA protein and RNA polymerase (see below).

The sites of p*N* action are in the *nutL* and *nutR* regions of the left and right operons respectively (see Figure 1). Initially, the *nutL* locus was defined by a mutation in a 17 base-pair region of dyad symmetry that resulted in defective antitermination (35). Sequence analysis of the right operon revealed a homologous segment that was named *nutR* (34). Because other homologies exist between the left and right operons (and among other lambdoid phage), the *nut region* is defined as the larger fragment that includes the other homologies in addition to the dyad. Deletion analysis and subcloning of the *nut* region downstream of bacterial promoters showed that the *nut* region is both necessary and sufficient to render RNA polymerase termination-resistant in the presence of p*N* (4,6).

The other internal segments of homology among *nut* regions are called *boxA* and *boxC*. The eight base-pair *boxA* sequence (CGCTCTTA in λ) is found just upstream of the *nut* dyad in all lambdoid phage *nut* regions that have been sequenced to date (13,14,39). Two types of mutation in *boxA* have been isolated: 1) the *boxA1* allele that enables λ to use the NusA protein from *Salmonella* (14), and 2) other changes in *boxA* that reduce N-mediated antitermination, demonstrating that an intact *boxA* sequence is required for the N reaction, at least at higher temperatures (30,32). The *boxA1* class of mutation was the basis of an argument that *boxA* is a target site for the host NusA protein (described below). The *boxC* segment of homology is located on the opposite side of the dyad from *boxA*. It is not yet known what function, if any, it serves.

Several *E. coli* proteins are also required for p*N* action. These were identified by the isolation of *nus* mutations that block λ growth (see 13 for review). Table 1 summarizes the major characteristics of the *nus* loci; three of the mutations lie in genes that have previously been shown to be involved in transcription or translation. All *nus* mutations share the following phenotypes: 1) they restrict p*N* action rather than p*N* synthesis; 2) they are recessive, suggesting that they result in a loss of function; and, 3) they are more restrictive for λ growth at high temperature. All of the Nus proteins have been shown to be required *in vitro*, although the complexity of the p*N* reaction has precluded the reconstruction of the reaction in a purified *in vitro* system (5,17). *nusA* was the first identified and is the best understood *nus* gene.

The *nusA* gene was identified by the isolation and mapping of the *nusA1* mutation (11,12). This mutation blocks λ growth at 42° because the N system is inactive. The *nusA* gene encodes an acidic 55kd protein (as determined from sequence; 23) which runs at 69kd on polyacrylamide gels (18). In a variety of transcriptional systems, NusA enhances the pausing of RNA polymerase as well as transcription termination and antitermination (19,24,7,26,38,36,31,37). Conditionally lethal *nusA* mutations have been isolated, demonstrating the essential nature of NusA activity for cell viability (29, Schauer et al., manuscript submitted).

The initial evidence suggesting that NusA and N proteins interact with each other was the isolation of λ*punA1* (15). λ*punA1* carries a mutation in the *N* gene that allows λ growth on the *nusA1* host at the nonpermissive (42°) temperature. Direct proof of this interaction came from *in vitro* experiments (18). NusA exhibits specific binding to RNA polymerase core enzyme *in vitro* (18).

The construction of a viable *E. coli-Salmonella* hybrid host showed that the *Salmonella nusA* gene, when expressed in *E. coli*, can supply any NusA activity that is essential for *E. coli* growth (2,12). This hybrid bacterium carries a small *Salmonella* DNA substitution in the *E. coli* chromosome; it expresses the wild type *Salmonella nusA* gene ($nusA^S$) instead of the *E. coli* gene. However, the $NusA^S$ protein does not provide the activity required for λ p*N* action. As in the case of the *nusA1* mutant, the hybrid bacterium (hereafter called the $NusA^S$ host) blocks the growth of N-dependent λ derivatives, but supports the growth of N-independent phages such as λ*nin*.

RESULTS

A NusA recognition site in the λ *nutR* region was initially identified by the isolation of frameshift mutations within the *cro* gene (which is located just upstream of *nutR*; see Figure 1; 30). It was shown that single base-pair deletions in *cro* interfere with λ growth because translation extends a few bases into *nutR* before encountering an in-frame stop signal. A direct effect of an altered *cro* product has been ruled out. This implied that the new in-frame *cro* stop codon in the mutant caused the ribosome to sterically block access of a factor to a nucleic acid site. Examination of possible sites at which steric interference could occur suggested that a seven base-pair sequence (*boxA*) might be recognized by a protein factor (14). This sequence was found in analogous positions in all other lambdoid phage, as well as in bacterial operons where NusA was known to participate in control of transcription. The idea that NusA protein was the factor recognizing *boxA* was tested by selecting for a *boxA* mutation.

The *boxA1* mutation was obtained by selecting for a λ mutant that could grow in the $NusA^S$ host (14). The two-stage selection procedure took into consideration the results of *in vivo* and *in vitro* studies of p*N* interaction with NusA. λ*punA1* was chosen as the starting parental phage because the *punA1 N* mutation, which allows phage growth in an *E. coli nusA1* mutant, was assumed to result in expression of a more active p*N*. The first step in the selection yielded a phage that made a tiny plaque on the $NusA^S$ host. This phage carries a second mutation in *N*: *punA133*.

The λ*punA1,133* phage was used in the second round of mutagenesis to generate a mutant that formed normal-sized plaques on the *nusA*^{S} host; the mutation responsible for this phenotype was mapped to the *nutR* region. DNA sequencing identified the responsible change in the *boxA* region: a single base-pair transversion resulting in the *boxA* sequence CGCTCTT*T* compared to the wild type CGCTCTT*A*. This change had been predicted based on the knowledge that other lambdoid phages that can normally utilize NusA^S (P22 and 21) have the three T's at the 3' end of their *boxA* sequences.

The *boxA* sequence with three T's could either be more specific for *Salmonella* NusA, or could represent a more optimal sequence for both NusA proteins in general. To distinguish between these alternatives, we constructed derivatives of λ that carry *boxA1* but which have a wild type *N* gene (Schauer et al., manuscript submitted). This permitted direct observation of the effects of *boxA1* in the absence of the *N* (*punA*) mutations. The *boxA1* mutation alone permits λ to grow well in the *nusA1* host under conditions where wild type λ fails to plate. On the other hand, it does not markedly improve the growth of λ in other *nus* mutants, such as *nusB5*. This suggests that CGCTCTTT is a more optimal signal than CGCTCTTA in general.

The idea that the three T's are more optimal but not specific for *Salmonella* NusA implies that *boxA*$^+$ might function wih NusA^S if NusA^S is overproduced. To test this, we used a host that overproduces NusA^S, and a λ derivative that carries *boxA*$^+$ and the *punA1* and *133 N* mutations (Schauer et al., manuscript submitted). These mutations convert the λ N into one that functions with NusA^S protein. Therefore, the block in growth of λ*punA1,133* in the NusA^S host is due solely to the presence of the wild type *boxA* sequence in the *nutR* region. The effect of excess NusA^S on λ growth was determined by measuring the burst of phage following infection. The burst of λ*punA1,133* is ten-fold greater in the NusA^S overproducer than in the normal NusA^S host. This ability of NusA^S to function with the wild type *boxA* means that the three T's in *boxA* do not produce a qualitatively different *boxA*, but rather a more optimal *boxA* sequence. This is consistent with the results described above for mutant *E. coli* NusA.

Mapping of the *punA133* mutation placed it in or near the *N* gene (14). To determine the precise nature of the change(s) in *N* that permit use of NusA^S, we sequenced the *N* gene from the λ*punA1,133* phage (Schauer et al., manuscript submitted). There were only two changes from wild type, with resulting amino-acid changes: 1) the *punA1* serine to arginine substitution that was present in the parental phage, and 2) a second change five codons promoter-proximal to *punA1* that causes an arginine replacement of lysine. Although we have so far been unable to separate the two mutations from each other, it is likely that both are required for interaction with the *Salmonella* NusA protein, because we have been unable to isolate single-step mutations in λ*N*$^+$*boxA1* that plaque on the NusA^S host. Reconstruction experiments demonstrated that only these mutations are required, along with the *boxA1* mutation, to permit phage λ growth in the NusA^S host.

Another example of how an increase in the amount of one component can overcome a normally incompatable reaction is our finding that overproduction of λ N protein allows a P22 *N*⁻ to grow. The cloning of the λ *N* gene on a high copy-number plasmid allowed us to explore the question of p*N* specificity (Schauer et al., manuscript submitted). Although previous studies had shown that λ could grow when supplied with the p*N* of phage P22 (gene *24* product), there was no indication that P22 could function with λ p*N* (20,21). Using a host containing a p*N*-overproducing plasmid, we found that a λ-P22 hybrid with an amber mutation in its *N* gene (*24am*) forms normal plaques. Thus, just as with the $NusA^S$-$boxA^+$ case described above, the N_λ-nut_{P22} interaction can be pushed to the point that antitermination is sufficient to allow good phage growth.

In comparisons of NusA activities between *E. coli* and *S. typhimurium*, only the difference in the ability to facilitate the action of λ p*N* appears to be significantly different; $NusA^S$ can function in *E. coli* growth and support the action of p*N* analogues of lambdoid phages 21 and P22. Our sequence analysis of $nusA^S$ and comparison with *E. coli nusA* (manuscript submitted) identifies three segments of heterogeneity between the two proteins in the amino two-thirds of the protein. We know from work with a truncated *E. coli* NusA protein that the carboxy-terminal one-third of the protein is dispensable; therefore the important regions for interactions both with N protein and with RNA polymerase are likely to be identified by some of these short stretches of heterogeneity between the two protein sequences. Computer analysis fails to detect any significant conformational difference in the structure of the two proteins, suggesting that these amino-acid heterogeneities represent actual sites of interaction rather than sites which alter tertiary protein structure. Experiments to localize the important contact residues in NusA are in progress.

An antitermination system in *E. coli* ribosomal RNA operons has been discovered; it requires a *nut*-like sequence and appears to rely on many of the same *E. coli* components required for the action of p*N* (28, 27,22,37,M. Cashel, personal communication). This fact, along with the demonstration of a role for NusA (and possibly *boxA*) in attenuation within the *trp* leader (40,7,25) and the finding that *nusA* is required for cell viability, indicates that information gained from analysis of the N system will be valuable for a wide range of reactions that control transcription and translation.

REFERENCES

1. Adhya, S., M. Gottesman, and B. de Crombrugghe: Release of Polarity in *Escerichia coli* by gene *N* of phage λ: termination and antitermination of transcription. Proc Natl Acad Sci USA 71: 2534, 1974.
2. Baron, L. S., E. Penido, I.R. Ryman, and S. Falkow: Behavior of coliphage λ in hybrids between *Escherichia coli* and *Salmonella*. J Bacteriol 102: 221, 1970.
3. Court, D., and K. Sato: Studies of novel transducing variants of lambda: dispensability of gnes *N* and *Q*. Virology 39: 348, 1969.
4. Dambly-Chaudiere, C., M. Gottesman, C. Debouck, and S. Adhya: Regulation of the *p*R operon of bacteriophage lambda. J Mol Appl Genet 2: 45, 1983.

5. Das, A. and K. Wolska: Transcription antitermination *in vitro* by lambda *N* gene product: requirement for a phage *nut* site and the products of host *nusA*, *nusB* and *nusE* genes. Cell 38: 165, 1984.
6. de Crombrugghe, B., S. Adhya, M. Gottesman, and I. Pastan: Effect of rho on transcription of bacterial operons. Nature New Biol 241: 260, 1979.
7. Farnham, P.J., J. Greenblatt, and T. Platt: Effects of NusA protein on transcription termination in the tryptophan operon of *Escherichia coli*. Cell 29: 945, 1982.
8. Fiandt, M., Z. Hradecna, H.A. Lozeron, and W. Szybalski: In The Bacteriophage Lambda, A. Hershey, ed., Cold Spring Harbor Laboratory, Cold Spring Harbor, N.Y., p. 329, 1971.
9. Franklin, N.C.: Altered reading of genetic signals fused to the N operon of bacteriophage λ: genetic evidence for the modivication of polymerase by the protein product of the N gene. J Mol Biol 89: 33, 1974.
10. Franklin, N.C. and G.N. Bennett: The N protein of bacteriophage lambda, defined by its DNA sequence, is highly basic. Gene 8: 107, 1979.
11. Friedman, D.I.: A bacterial mutant affecting λ development. In The Bacteriophage Lambda, A. D. Hershey, ed., Cold Spring Harbor Laboratory, Cold Spring Harbor, N.Y., p. 733, 1971.
12. Friedman, D.I. and L.S. Baron: Genetic characterization of a bacterial locus involved in the activity of the N function of phage λ. Virology 58: 141, 1974.
13. Friedman, D.I. and M. Gottesman: Lytic mode of λ development. In Lambda II, R. Hendrix et al., eds., Cold Spring Harbor Laboratory, Cold Spring Harbor, NY., p. 21, 1983.
14. Friedman, D.I., and E.R. Olson: Evidence that a nucleotide sequence, "boxA", is involved in the action of the NusA protein. Cell 34: 143, 1983.
15. Friedman, D.I. and R. Ponce-Campos: Differential effect of phage regulator functions on transcription from various promoters: evidence that the P22 gene and the λ gene N products distinguish three types of promoters. J Mol Biol 98: 537, 1975.
16. Friedman, D.I., G.S. Wilgus, and R.J. Mural: Gene N regulator function of phage λ*imm*21: evidence that a site of N action differes form a site of N recognition. J Mol Biol 81: 505, 1973.
17. Goda, Y. and J. Greenblatt: Efficient modification of *E. coli* RNA polymerase *in vitro* by the *N* gene transcription antitermination protein of bacteriophage lambda. Nucleic Acids Res 13: 2569, 1985.
18. Greenblatt, J. and J. Li: Interaction of the sigma factor and the *nusA* gene protein of *E. coli* with RNA polymerase in the initiation-termination cycle of transcription. Cell 24: 421, 1981.
19. Greenblatt, J., M. McLimont, and S. Hanly: Termination of transcription by *nusA* gene protein of *Escherichia coli*. Nature (London) 292: 215, 1981.
20. Hilliker, S. and D. Botstein: Specificity of genetic elements controlling regulation of early functions in temperate bacteriophages. J Mol Biol 106: 537, 1976.
21. Hilliker, S., M. Gottesman, and S. Adhya: The activity of *Salmonella* phage P22 gene 24 product in *Escherichia coli*. Virology 86: 37, 1978.

22. Holben, W.E., S.W. Prasad, and E.A. Morgan: Antitermination by both the promoter and the leader regions of an *Escherichia coli* ribosomal RNA operon. Proc Natl Acad Sci USA 82: 5073, 1985.
23. Ishii, S., M. Ihara, T. Maekawa, Y. Nakamura, H. Uchida, and F. Immamoto: The nucleotide sequence of *nusA* and its flanking region of *Escherichia coli*. Nucleic Acids Res 12: 3333, 1984.
24. Kingston, R.E. and M.J. Chamberlin: Pausing and attenuation of *in vitro* transcription in the *rrnB* operon of *Escherichia coli*. Cell 27: 523, 1981.
25. Landick, R. and C. Yanofsky: Stability of an RNA secondary structure affects *in vitro* transcription pausing in the *trp* operon leader region. J Biol Chem 259: 11550, 1984.
26. Lau, L., J.W. Roberts, and R. Wu: Transcription terminates at λt_{R1} in three clusters. Proc Natl Acad Sci USA 79: 6171, 1982.
27. Li, S.C., C.L. Squires, and C. Squires: Antitermination of *Escherichia coli* ribosomal RNA transcription is caused by a control region segment containing lambda *nut*-like sequences. Cell 38: 851, 1984.
28. Morgan, E.A.: Insertions of Tn*10* into an *Escherichia coli* ribosomal RNA operon are incompletely polar. Cell 21: 257, 1980.
29. Nakamura, Y. and S. Mizusawa: *In vivo* evidence that the *nusA* and *infB* genes of *Escherichia coli* are part of the same multi-gene operon which encodes at least four proteins. EMBO J 4: 527, 1985.
30. Olson, E.R., C.-S.C. Tomich, and D.I. Friedman: The NusA recognition site: alteration in its sequence or position relative to upstream translation interferes with the action of the N antitermination function of phage lambda. J Mol Biol 180: 1053, 1984.
31. Peacock, S., J.R. Lupski, G.N. Godson, and H. Weissbach: *In vitro* stimulation of *Escherichia coli* RNA polymerase sigma subunit synthesis by *nusA* protein. Gene 33: 227, 1985.
32. Peltz, S.W., A.L. Brown, N. Hasan, A.J. Podhajska, and W. Szybalski: Thermosensitivity of a DNA recognition site activity of a truncated *nutL* antiterminator of coliphage lambda. Science 228: 91, 1985.
33. Roberts, J.W.: Termination factor for RNA synthesis. Nature (London) 224: 1168, 1969.
34. Rosenberg, M., D. Court, H. Shimatake, C. Brady, and D.L. Wulff: The relationship between function and DNA sequence in an intercistronic regulatory region of phage λ. Nature (London) 272: 414, 1978.
35. Salstrom, J.S. and W. Szybalski: Coliphage λ*nutL*$^-$: a unique class of mutants defective in the site of gene N product utilization for antitermination of leftward transcription. J Mol Biol 124: 195, 1978.
36. Schmidt, M.C. and M.J. Chamberlin: Amplification and isolation of *E. coli nusA* protein and studies of its effects on *in vitro* RNA chain elongation. Biochemistry 23: 197, 1984.
37. Sharrock, R.A., R.L. Gourse, and M. Nomura: Defective antitermination of ribosomal RNA transcription and derepression of transfer RNA and RNA synthesis in the *nusB5* mutant of *Escherichia coli*. Proc Natl Acad Sci USA 82: 5275, 1985.
38. Simons, R.W. and N. Kleckner: Translational control of IS*10* transposition. Cell 34: 683, 1983.
39. Tanaka, S. and A. Matsushiro: Characterization and sequencing of the region containing gene *N*, the *nutL* site and *t*L1 of bacteriophage φ80. Gene 38: 119, 1985.

40. Ward, D.F. and M.E. Gottesman: The *nus* mutations affect transcription termination in *Escherichia coli*. Nature (London) 292: 212, 1981.

TABLE 1. *nus* genes, products, and mutations which block early gene expression in wild type bacteriophage λ, but not in λ *N*-independent mutants.

Gene	Map Location	Product	Size	Mutations
nusA	69 min.	NusA	55 kd	missense (*nusA1*) cold-sensitive (*nusA10*) nonsense "foreign" (*Salmonella nusA*)
nusB	11 min.	NusB	16 kd	missense (e.g. *nusB5*)
nusC	88 min.	β subunit RNA polymerase	140 kd	missense (*nusC60*)
nusD	84 min.	termination factor ρ	50 kd	missense (*hdf026*)
nusE	72 min.	ribosomal protein S10	12 kd	missense (*nusE71*)

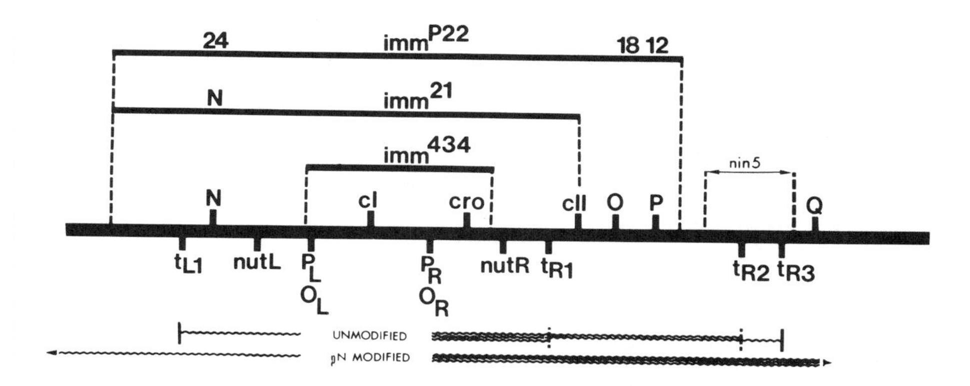

Figure 1. Genetic map of bacteriophage λ immunity region. The repressor gene, *cI*, lies between the two early promoters, *p*L and *p*R, and its protein product represses this transcription during lysogeny by binding to the operators *o*L and *o*R. During execution of the lytic program, transcripts terminate partially at *t*R1 and more completely at *t*L1, *t*R2 and *t*R3. Once N protein has achieved a sufficient concentration, it acts at the *nutL* and *nutR* regions, resulting in transcription that proceeds through the terminators and into *Q*, the gene encoding the late transcription antiterminator. p*Q* action permits transcripts to reach the distal morphogenetic genes. Also shown are the extents of substitution from the related lambdoid phages 434, 21 and P22 that create viable hybrids, and the *nin5* terminator deletion that relieves λ of N-dependency. The transcripts are depicted by the wavy lines at the bottom of the figure.

9. Nuclear Magnetic Resonance Techniques for Studies of Protein-DNA Interactions

Kurt Wüthrich

Institut für Molekularbiologie und Biophysik, ETH-Hönggerberg, CH-8093 Zurich, Switzerland

INTRODUCTION

Nuclear magnetic resonance (NMR) is presently the only experimental method besides diffraction studies with single crystals which can provide detailed information on the molecular conformation of biopolymers (1). Besides solution NMR (2), solid state NMR techniques (3) may also be employed for studies of biological macromolecules. The ability of NMR to be applicable to non-crystalline samples makes it particularly attractive as a complementary technique to single crystal X-ray studies. In the field of protein-DNA interactions the following features of NMR are of special interest: (i) Since spatial structure determinations by NMR are not dependent on the availability of a crystal structure, a meaningful comparison of the conformations in single crystals and in non-crystalline states can be obtained, and NMR can also be applied to molecules or multimolecular aggregates for which no single crystals are available. (ii) The solution conditions for NMR studies (e.g. pH, temperature, ionic strength, addition of different buffers) can usually be varied over a wide range. (iii) In addition to its use for studies of conformation and intermolecular interactions, NMR allows direct, quantitative measurements of the rates of certain infrequent, high activation energy motional processes and at least semiquantitative information on additional high frequency processes, and thus provides experimental access to the molecular dynamics of biopolymer molecules (4).

NMR studies of DNA and intermolecular interactions with DNA have made rapid progress since the recent introduction of improved techniques for DNA synthesis, which can supply oligo- and poly-nucleotides with tailor-made sequences in milligram quantities. NMR data on DNA-protein interactions are still scarce and the more sophisticated ones among the experiments described in the following are currently at the stage of initial practical applications. However, while this is not further considered in this brief survey, extensive NMR studies have for some time been pursued with complexes formed between DNA and low molecular weight compounds, such as drugs and small peptides (5). These may also serve as a useful source of reference for various aspects of NMR studies with DNA and proteins.

QUESTIONS FOR NMR

Quite generally, when dealing with multimolecular assemblies, obvious problems to be solved by experimental techniques relate to the structure, the thermodynamic stability and the kinetic properties. In the special case of NMR applied to proteins interacting with DNA, information on all three of these areas may be obtained (6,7). When working with relatively low molecular weight systems, NMR is capable of comparing the conformations of free protein and DNA with those in the complexes formed by these molecules. NMR is a suitable technique for delineating the contact regions in mutually interacting polymer chains, both on the level of the primary structure and the spatial structure. It can further be used to investigate the equilibrium between free and aggregated molecules in the system, and, depending on the time scales involved, NMR experiments may provide quantitative data on the internal mobility of the individual molecules in the complex as well as on the dynamics of the multimolecular aggregate.

NMR MANIFESTATIONS OF PROTEIN-DNA INTERACTIONS

Inspection of the literature on NMR studies of Protein-DNA interactions shows that from the point of view of spectroscopy, these can be classified into three groups. The first, and so far most widely used approach relies on observation of the effects of interactions with the DNA on NMR parameters in the protein, and vice versa. Most commonly, the measurements bear on the chemical shifts, either of the naturally present hydrogen or phosphorous nuclei, or of isotopes such as ^{13}C, ^{15}N or ^{19}F which are introduced into the system for the purpose of the NMR experiments (8-14). (Instead of the chemical shifts, variation of spin relaxation times or spin-spin couplings upon complexation could in principle also be measured). All these experiments have in common that they can only indicate a change in molecular conformation or, more generally, in the microenvironment of certain structural components in the observed molecule. It is usually difficult to attribute chemical shift changes to specific structural features or intermolecular interactions. In particular, they could be a consequence of long range effects from the intermolecular interactions on the conformation in molecular areas far from the site of interaction.

A second strategy relies on nuclear Overhauser enhancement (NOE) experiments, which can directly manifest close proximity of nuclear spins in different, interacting molecules (15-17). NOE measurements were only recently applied to protein-DNA complexes (18-20), but they represent a very promising approach for the future, since they can pinpoint intermolecular close contacts more directly than any other technique except single crystal X-ray methods.

A third group of NMR experiments uses observations on different accessibility of specified molecular segments in the free and complexed protein and/or DNA molecules. For example photochemically induced dynamic nuclear polarization (photo-CIDNP) probes the accessibility of aromatic rings in the biopolymers to dye molecules, which were added to the solution for this experiment (21,22). As an alternative, the accessibility of surface groups to paramagnetic shift reagents could be used (6,7). Compared to NOE measurements, all these experiment provide less direct evidence on site-specific intermolecular interactions since, at least in principle, different accessibility of certain groups of atoms could also result from long range effects on the conformation in molecular regions far from the intermolecular contacts.

THE PIVOTAL ROLE OF SEQUENCE-SPECIFIC RESONANCE ASSIGNMENTS

Independent of the choice of the NMR experiments used for protein-DNA systems, the amount of information obtained is dramatically increased when sequence-specific resonance assignments are available. To illustrate this point we consider the analysis of NOE measurements. As mentioned in the preceding section, NOE's can manifest short distances between nuclear spins located in different, interacting molecules. Without sequence-specific resonance assignments such data merely indicate that complex formation has occured, but when combined with resonance assignments they identify the sites of intermolecular contacts in the polymer chains. As is schematically shown in Fig. 1 for a polypeptide chain interacting with a polynucleotide chain, such studies can initially be conducted on the primary structure level. They may then be combined with the results from studies on the spatial structure so that, for example, close contacts of α-helical or β-sheet regions in the protein with the DNA can then be identified.

Since approximately 1981 methods have become available for obtaining nearly complete, sequence-specific resonance assignments in biopolymers, using exclusively NMR data and the chemically determined primary structure. ^{1}H NMR assignments were described for more than 10 small proteins and a selection of synthetic DNA fragments in duplex form. The assignment techniques rely on the use of two-dimensional (2D) NMR experiments, in particular 2D correlated spectroscopy (COSY) and 2D NOE spectroscopy (NOESY) (2).

For proteins, ^{1}H assignments can rely entirely on homonuclear ^{1}H NMR. COSY is used to identify the ^{1}H spin systems (6) of the individual amino acid residues, and NOESY provides sequential assignments (1,23-25).

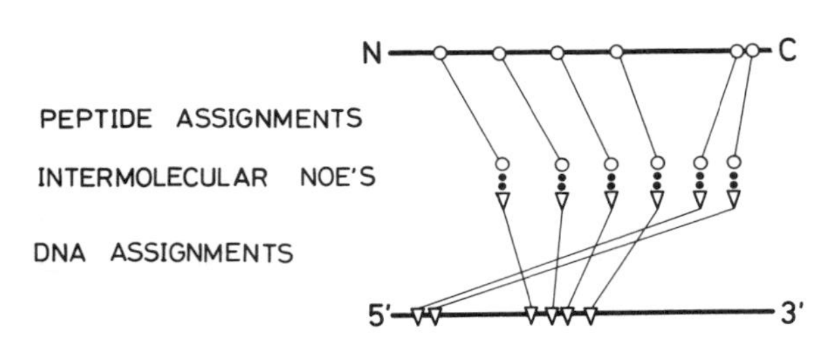

Fig. 1 Identification of intermolecular contacts between a polypeptide and a polynucleotide by intermolecular NOE's when sequence-specific NMR assignments are available for both molecules. Circles and triangles represent nuclear spins in the polypeptide and the polynucleotide, respectively.

For DNA fragments, two different strategies have been developed. The first is largely analogous to the approach used for proteins, and uses ^{1}H NOESY experiments for establishing sequential connectivities between the non-labile protons in neighboring nucleotides (26-29). Additional NOE experiments can yield the assignments for the labile protons in the hydrogen bonds of the base pairs (30). The second strategy uses ^{1}H COSY for identification of the deoxyribose ^{1}H spin systems and heteronuclear ^{1}H-^{31}P COSY for obtaining the sequential assignments (31-33). The homonuclear ^{1}H NMR procedure for resonance assignments in DNA duplexes has so far been more widely applied. However, with regard to studies of protein-DNA interactions, it is an attractive feature of the heteronuclear approach that sequence-specific assignments are also obtained for the phosphate groups, since these could be observed in the complexes without interference from nuclear spins in the protein. In this context the use of isotope labelling for assigning the phosphate resonances is also an interesting alternative (34,35).

AN ILLUSTRATION: DNA-INTERACTIONS WITH E.COLI LAC REPRESSOR

The *E.coli lac* repressor was extensively investigated by NMR, and consultation of the original literature on this system presents probably the best illustration of recent NMR applications for protein-DNA interactions. The following is a brief survey of these studies, supplemented by a description of some experiments done in our laboratory.

E. coli lac repressor is a tetrameric protein of molecular weight 152,400. As such it would be far too big for detailed studies by

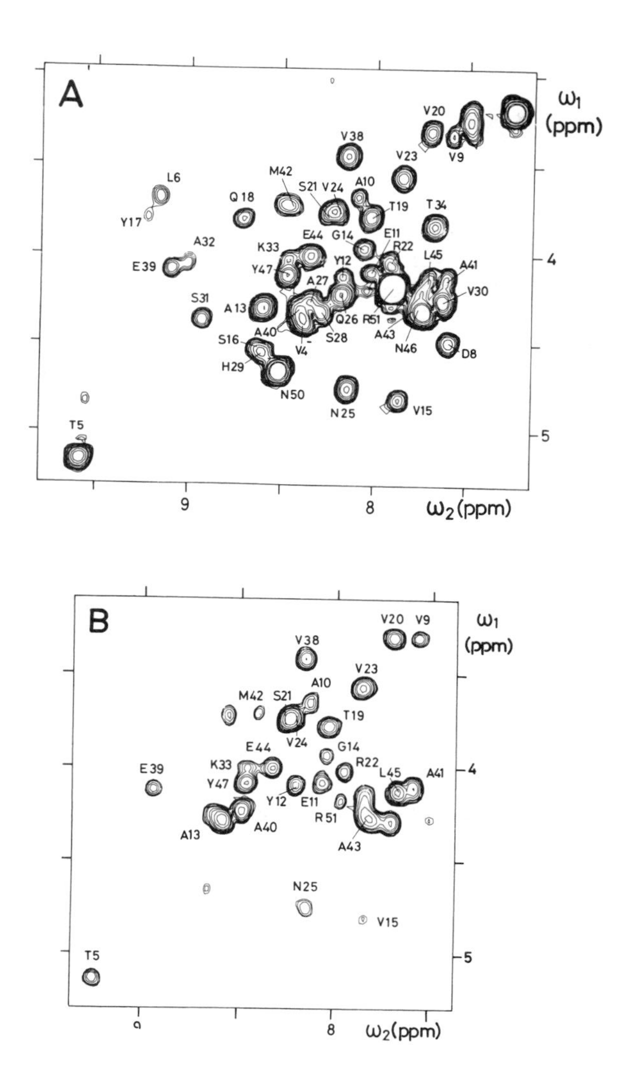

Fig. 2 Fingerprint region containing the NH-αH cross peaks in ^{1}H COSY spectra of the E.Coli lac repressor DNA-binding domain 1-51 in H_2O (A) and in D_2O (B). Both spectra were recorded at 500 MHz and are in the absolute value mode. (Reproduced from ref. 25 and 38).

solution NMR techniques. However, the DNA-binding domain formed in each subunit by the 51 or 59 N-terminal amino acids retains its spatial structure and the specific DNA binding capacity when isolated from the intact repressor by enzymatic cleavage (36,37), and extensive NMR studies were conducted with this polypeptide fragment. These include investigations of interactions with DNA based on observation of chemical shift changes (8,14,20), NOE's between protons in the protein and the DNA (14) and studies of surface accessibility of aromatic residues in the protein by CIDNP (21). Furthermore, for the isolated protein sequence-specific ^{1}H NMR assignments were obtained (25) and the secondary structure was determined (38). The following was a crucial experiment relating to the secondary structure.

The "fingerprint region" in ^{1}H COSY spectra of proteins (23) contains the cross peaks connecting the backbone amide- and α-protons in the individual amino acid residues. Each amino acid contributes one such peak (except that glycines may give two peaks, and proline has none). Fig. 2A shows the complete fingerprint of lac repressor 1-51 recorded in H_2O solution, with sequence-specific assignments for the individual peaks.

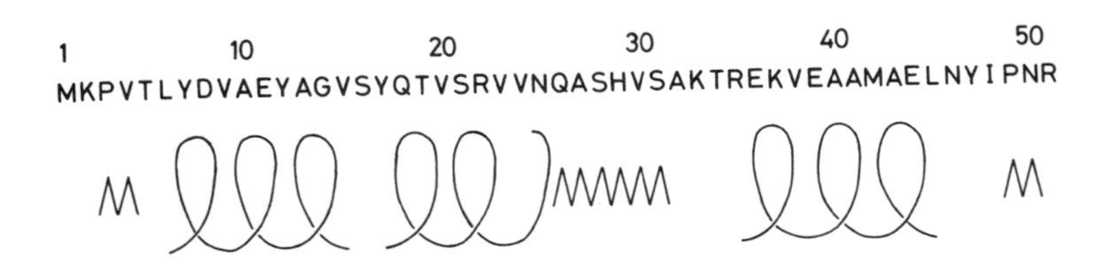

Fig. 3 Amino acid sequence of E. Coli lac repressor DNA binding domain 1-51 with indication of the locations of helical and extended polypeptide segments. (From ref. 40).

In a freshly prepared D_2O solution only those peaks can be observed which correspond to residues with slowly exchanging amide protons. On this basis the data of Fig. 2B were used to identify the hydrogen bonds in the helical secondary structure segments in lac repressor 1-51 (Fig. 3). Subsequently, the spatial arrangement of the three helices was determined from intramolecular distance constraints imposed by NOE experiments (39,40) and work on refinements of the molecular conformation is in progress (41). Obviously, the availability of these structural data can present a basis for detailed studies of the protein-DNA contacts in the complexes formed with synthetic operator DNA fragments.

ACKNOWLEDGEMENTS

The study on E. Coli lac repressor was a joint project with Prof. R. Kaptein, University of Groningen. Financial support of our Research projects by the Schweizerischer Nationalfonds (project 3.284.82).

REFERENCES

1 Wüthrich, K., G. Wider, G. Wagner and W. Braun: J. Mol. Biol. 155: 311, 1982.
2 Wider, G., S. Macura, Anil Kumar, R.R. Ernst and K. Wüthrich: J. Magn. Reson. 56:207, 1984.
3 Cross, T.A. and S.J. Opella: J. Mol. Biol. 182:367, 1985.
4 Wagner, G.: Quart. Rev. Biophys. 16:1, 1983.
5 Patel, D.H.: Acc. Chem. Res. 12:118, 1979.
6 Wüthrich, K.: NMR in Biological Research: Peptides and Proteins. North Holland Publishing Company, Amsterdam, 1976.
7 Jardetzky, O. and G.C.K. Roberts: NMR in Molecular Biology. Academic Press, New York, 1981.
8 Buck, F., K. D. Hahn, W. Zemann, H. Rüterjans, J.R. Sadler, K. Beyreuther, R. Kaptein, R. Scheek and W.E. Hull: Eur. J. Biochem. 132:321, 1983.
9 Prigodich, R.V., J. Casa-Finet, K.R. Williams, W. Konigsberg and J.E. Coleman: Biochemistry 23:522, 1984.
10 Kime, M.J.: FEBS Lett. 175:259, 1984.
11 Alma, N.C.M., B.J.M. Harmsen, J.H. van Boom, G. van der Marel and C.W. Hilbers: Eur. J. Biochem. 122:319, 1982.
12 Hahn, U., R. Desai-Hahn and H. Rüterjans: Eur. J. Biochem. 146:705, 1985.
13 Mirau, P.A., R.H. Shafer and T.L. James: Biochemistry 21:615, 1982.
14 Nick, H., K. Arndt, F. Boschelli, M.A. Jarema, M. Lillis, J. Sadler, M. Caruthers and P. Lu: Proc. Natl. Acad. Sci. USA 79:218, 1982.
15 Noggle, J.H. and R.E. Schirmer: The Nuclear Overhauser Effect. Academic Press, New York, 1971.
16 Wagner, G. and K. Wüthrich: J. Magn. Reson. 33:675, 1979.
17 Anil Kumar, G. Wagner, R.R. Ernst and K. Wüthrich, J. Amer. Chem. Soc. 103:3654, 1981.
18 Alma, N.C.M., B.J.M. Harmsen, J.H. van Boom, G. van der Marel and C.W. Hilbers: Biochemistry 22:2102, 1983.
19 Kime, M.J. and P.B. Moore: Biochemistry 22:2622, 1983.

20 Scheek, R.M., E.R.P. Zuiderweg, K.J.M. Klappe, J.H. van Boom, R. Kaptein, H. Rüterjans and K. Beyreuther: Biochemistry 22:228, 1983.
21 Buck, R., H. Rüterjans, R. Kaptein and K. Beyreuther: Proc. Natl. Acad. Sci. USA 77:6145, 1980.
22 Shirakawa, M., S.J. Lee, H. Akutsu, Y. Kyogoku, K. Kitano, M. Shin, E. Ohtsuka and M. Ikehara: FEBS Lett. 181:286, 1985.
23 Wagner, G. and K. Wüthrich: J. Mol. Biol. 155:347, 1982.
24 Billeter, M., W. Braun and K. Wüthrich: J. Mol. Biol. 155:321, 1982.
25 Zuiderweg, E.R.P., R. Kaptein and K. Wüthrich: Eur. J. Biochem. 137:279, 1983.
26 Hare, D.R., D.E. Wemmer, S. H. Chou, G. Drobny and B.R. Reid: J. Mol. Biol. 171:319, 1983.
27 Scheek, R.M., R. Boelens, N. Russo, J.H. van Boom and R. Kaptein: Biochemistry 23:1371, 1984.
28 Weiss, M.A., D.J. Patel, R.T. Sauer and M. Karplus: Proc. Natl. Acad. Sci. USA 81:130, 1984.
29 Clore, G.M. and A.M. Gronenborn: Eur. J. Biochem. 141:119, 1984.
30 Chou, S. H., D.E. Wemmer, D.R. Hare and B.R. Reid: Biochemistry 23:2257, 1984.
31 Pardi, A., R. Walker, H. Rapoport, G. Wider and K. Wüthrich: J. Amer. Chem. Soc. 105:1652, 1983.
32 Marion, D. and G. Lancelot: Biochem. Biophys. Res. Commun. 124:774, 1984.
33 Frey, M.H., W. Leupin, O. W. Sørensen, W.A. Denny, R.R. Ernst and K. Wüthrich: Biopolymers, in press, 1985.
34 Connolly, B.A. and F. Eckstein: Biochemistry 23:5523, 1984.
35 Lai, K. D.O. Skah, E. De Rose and D.G. Gorenstein: Biochem. Biophys. Res. Commun. 121:1021, 1984.
36 Miller, J.H., C. Coulondre, M. Hofer, U. Schmeissner, H. Sommer, A. Schmitz and P. Lu: J. Mol. Biol. 131:191, 1979.
37 Wade-Jardetzky, N., R.P. Bray, W.W. Conover, O. Jardetzky, N. Geisler and K. Weber, J. Mol. Biol. 128:259, 1979.
38 Zuiderweg, E.R.P., R. Kaptein and K. Wüthrich: Proc. Natl. Acad. Sci. USA 80:5837, 1983.
39 Zuiderweg, E.R.P., M. Billeter, R. Boelens, R.M. Scheek, K. Wüthrich and R. Kaptein: FEBS Lett. 174:243, 1984.
40 Zuiderweg, E.R.P., M. Billeter, R. Kaptein, R. Boelens, R.M. Scheek and K. Wüthrich: Progress in Bioorganic Chemistry and Molecular Biology, Yu. A. Ovchinnikov ed., Elsevier Science Publishers B.V., 1984.
41 Kaptein, R., E.R.P. Zuiderweg, R.M. Scheek, R. Boelens and W.F. van Gunsteren: J. Mol. Biol. 182:179, 1985.

Part II. *Eukaryotic Systems*

10. The Major Protein Components of hnRNP Complexes

G. Leser and T. E. Martin

Department of Molecular Genetics and Cell Biology, University of Chicago, 1103 East 57th Street, Chicago, IL 60637

INTRODUCTION

Recently there have been significant advances in the understanding of the sequence of events involved in the maturation of nuclear RNA molecules resulting in the appearance of messenger RNA in the cytoplasm (1). However, comparatively little is known about the detail of these processes *in vivo* and the protein:nucleic acid complexes presumably involved. Most relevant of recent findings to the present review are the results of cell-free systems which suggest that the formation of specific RNP complexes as a necessary prerequisite for the processing of exogenous model mRNA precursors (2,3). Much earlier electron microscope studies of transcriptionally active chromatin spread by the Miller technique suggested that as hnRNA is synthesized, still attached to the DNA template, it becomes associated with proteins which result in a particulate substructure of ca 20 nm diameter giving rise to a "beads on-a-string" appearance in the unfolded form (4,5). The particulate nature of these complexes serve to package and compact the hnRNA possibly facilitating its removal from the template. It is likely that they also maintain the hnRNA in a conformation necessary for subsequent maturational events. Beyer *et al*. (6) have suggested that the associations of some protein complexes along the nascent molecule observed by microscopy could be sequence specific.

Biochemical studies have shown that most hnRNA can be isolated as RNA:protein complexes (7,8,9). These RNP complexes have particulate substructure, revealed by limited RNase digestion as a relatively homogeneous peak sedimenting at approximately 30-40S in sucrose density gradients. These 20 nm RNA:protein complexes appear to involve a majority of the sequences in the precursor mRNA chain (10,11). There has been considerable controversy over the identity of the proteins associated directly with hnRNA (12), but in some cases the presence of certain species now seems well established. We have summarized some major components in Table 1. Not included in this summary are the protein constituents of snRNP (13), which can be shown to be associated with large hnRNP structures, but which are released upon nicking of the hnRNA chain to yield 30-40S subcomplexes (results not shown). In addition, it is to be expected that minor enzymatic or sequence specific proteins will subsequently be identified in these complexes.

A relatively simple set of "core" polypeptides is involved in the particulate substructure of hnRNP (9,14,15). Our laboratory examined 30S particles from mouse tumor cells and found 4-6 polypeptide classes ranging in molecular weights from 33,000 to 40,000 (9,16,17). Beyer et al. (14) examined the core proteins associated with HeLa cell hnRNP and observed three closely spaced doublets migrating between 32 and 44 kDa. These authors termed the pair of lowest molecular weight the A proteins (A1 and A2), the middle doublet B1 and B2, and the largest two C1 and C2. In mouse tumor cells the so-called C proteins were not abundant in purified 30S particles (9,16,17). The polypeptides which were present in highly purified 30S complexes (equivalent to A and B proteins) were shown to be sufficient for the in vitro reconstitution of RNP structures morphologically indistinguishable from native particles (18) and we term these the hnRNP core group proteins (chrp). This suggests that C group polypeptides and some higher molecular weight proteins (discussed below) that co-purify with RNP complexes might be associated with the particles but are not integral to the structure itself, possibly interacting with the portion of hnRNA between or on the surface of the 30S particles.

As a group, hnRNP core proteins appear to be highly conserved in size and amino acid composition among vetebrates (9,14,16,17,19). We have obtained and characterized polyclonal and monoclonal antibodies specific for the core group which cross-react with hnRNP proteins from a variety of organisms, and have employed these specific immunological probes to analyze the complexity of core polypeptides and examine their distribution in cells and on chromosomes (20,21,22,23).

hnRNP Core Proteins (chrp) are a Family of Related Polypeptides

The hnRNP core proteins have similar amino acid compositions, have a high glycine content, and contain the unusual modified amino acid N_G,N_G-dimethylarginine (9,14,15,16,19). Their isoelectric points are in the range pH 8-9. They also have in common the unfortunate trait of blocked N-termini thus complicating a direct analysis of the polypeptide sequence (16).

The proteins associated with hnRNP complexes from human HeLa and mouse Taper ascites cells displayed on Coomassie stained two dimension gels (Fig. 1) show a complexity difficult to describe using the A,B, and C nomenclature introduced by Beyer et al. (14). For instance in HeLa the core protein complement in this range is comprised of 3 proteins at 33 kDa, 3 at 34 kDa, 3-4 at 35 kDa, 3 at 37 kDa, 2 at 38 kDa, and 2 at 40 kDa. To what extent these spots represent charge isomers of a fundamental set of proteins is unknown. Furthermore the number of isomers could be affected by the physiological state of the cell thus further complicating matters. The precise nature of these modifications is unknown (see chapter by Holoubek in this volume). However, in light of the past biochemical evidence, the possibility of phosphorylation must be considered (9,24). The analysis of the relationships between individual chrp's will require more stringent criteria such as peptide mapping (24) or shared antigenic determinants (23,25). There are other proteins that reproducibly co-purify with hnRNP complexes and in some cases are present in preparations from

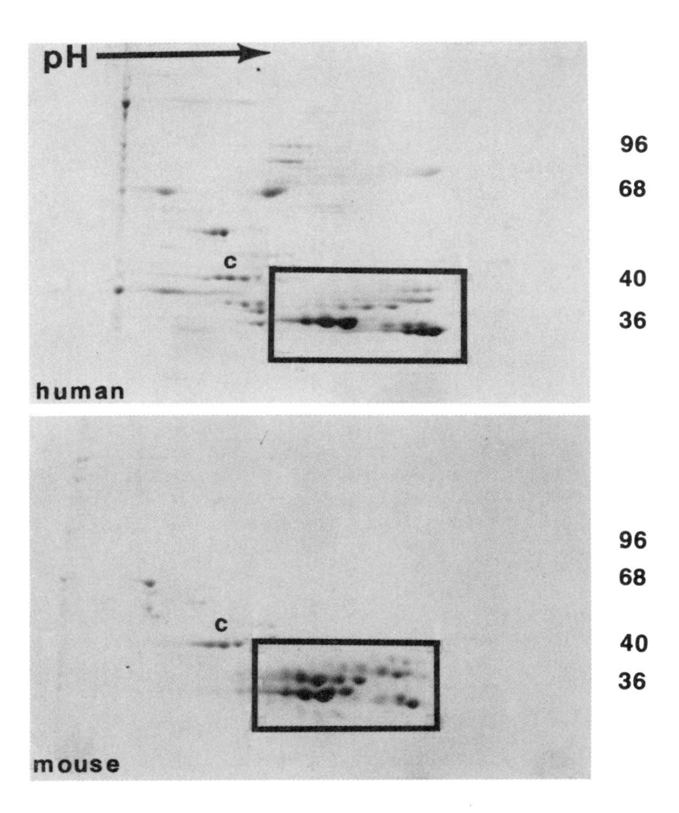

Figure 1. Proteins of isolated 30-40S hnRNP particles analyzed by two dimensional electrophoresis. hnRNP particles were prepared (12,25) from human HeLa cells and mouse ascites cells. The 30-40S fractions were electrophoresed on non-equilibrium pH gradient SDS gels and stained with Coomassie blue. The proteins detected on immunoblots with chrp specific antibodies are enclosed in boxes (refer Fig. 3). Other antigenically distinct proteins are also seen. The majority of these are relatively more acidic and include the so-called C proteins (C) and other higher molecular weight proteins associated with hnRNP, among them a prominent 68 KDa polypeptide.

different species. Most apparent are a group of proteins (denoted by C on Fig. 1) that presumably correspond to the "C group" proteins on the basis of their being relatively more acidic and migrating between 42 and 45 kDa. It has been previously suggested that hnRNP complexes are comprised of a large number of proteins, most being less predominant than the chrps and some that may be only loosely associated. The stained gels show several higher molecular weight proteins. Among them are species of 68 kDa which could include the polypeptide detected by cross-linking and single dimension SDS gels (25).

The relationships suggested by the limited chemical data are reinforced by our study of monoclonal antibodies specific for chrps (23). Of approximately 20 clones characterized, most react with all chrp polypeptides, only two clones have been characterized that recognize predominantly a subset of the group. This implies that a high degree of conservation exists among the individual polypeptide species. When core proteins are subjected to mild proteolytic digestion with either trypsin or _Staph. aureus_ V8 protease, the fragments generated retain their antigenicity and migrate between 20 and 28 kDa in a pattern analogous to undigested polypeptides (23). The uniform digestion and the common location of antigenic determinants recognized by several monoclonal antibodies further suggests an underlying structural similarity exists among these proteins.

Our monoclonal antibodies react solely with the basic core proteins or subsets of this group (23). Dreyfuss and co-workers (25) have obtained immunoglobulins that only recognize the more acidic C group proteins, suggesting that these polypeptides form an antigenically distinct group. There is in fact no cogent reason at present to group A and B polypeptides into pairs, nor to expect that the precise relationship of electrophoretic mobilities of this clustered group will be maintained in all eukaryotic species, indeed the electrophoretic patterns of avian and amphibian proteins differ significantly from mammals (9, 23). Since together, these basic polypeptides are sufficient _in vitro_ to form the 30S hnRNP substructure we prefer simply to group them under the term hnRNP core proteins (chrp).

The availability of specific antibodies has permitted the analysis of the total chrp complement of cells or tissues circumventing the potential artifacts inherent to hnRNP purification. We have subjected total cellular protein from HeLa cells (Fig. 2a) and mouse Taper ascites cells (Fig. 2b) to non-equilibrium pH gradient electrophoresis with a second dimension in SDS. These gels were immunoblotted (23) with a monoclonal antibody specific for all the basic core proteins. For both cell types 14-17 proteins migrating in a moderately basic pH range were detected. The core proteins detected in HeLa cells are represented diagramatically in Fig. 2c. Each protein band visible in single dimension SDS gels is resolved into multiple polypeptide species. The patterns are highly suggestive of multiple modifications of a fundamental set (5-6) of core proteins which are themselves related. Some minor differences in the migration of chrp polypeptides of human and mouse are apparent. In particular, the three spots at 37 kDa in mouse migrate in a relatively more acidic range when compared to human, also the two proteins found at approximately 32 kDa in mouse hepatoma are absent from

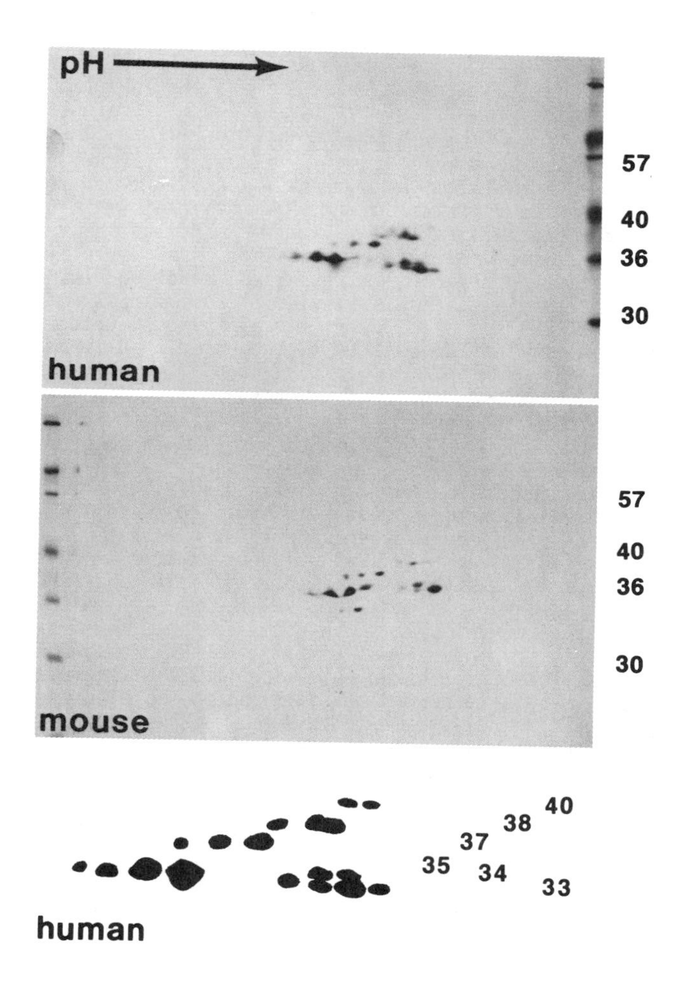

Figure 2. hnRNP core proteins (chrp) detected by immunoblots of total human HeLa and mouse Taper ascites cell proteins resolved on non-equilibrium pH gradient/SDS 2-dimensional gels. The blots were reacted with a monoclonal antibody shown to specifically react with the basic chrp group (A and B proteins). A diagram of the HeLa cell array of spots is shown.

HeLa. However, in general both species exhibit similarities in number and arrangement of these polypeptide constellations. It remains to be determined if these polypeptide relationships are a reflection of an extensive gene family, or represent the diverse products of post-translational modification of mRNA processing resulting from one or several genes.

hnRNP Core Proteins Associated with Transcriptionally Active Chromatin

The chrp polypeptides bind newly transcribed RNA in its nascent state and presumably remain complexed throughout subsequent maturational processes, being removed as mature mRNA is transported to the cytoplasm. Amphibian oocytes have provided a system where RNA synthesis can be visualized at the level of both the light and electron microscope. Carol Okamura of our laboratory used specific polyclonal antisera to show by indirect immunofluorescent staining that the nuclear distribution of chrps in an amphibian cell is analogous to that in mammalian cells (21,22). These antibodies were then used to localize the hnRNP polypeptides to the ribonucleoprotein matrix surrounding the transcriptionally active DNA that loops out from the axis of the chromosome (21, 22).

Although the chrp specific monoclonal antibodies were formed against chick hnRNP subcomplexes they cross-react with most vertebrate species. The chrp complement of *Xenopus* was identified by immunoblot analysis of single dimension SDS gels as five polypeptide classes migrating between 36 and 41 kDa, slightly larger overall than in mammalian species (23). We have used monoclonal antibodies specific for all or a subset (33 and 34 kDa species in mammals) to stain lampbrush chromosomes isolated from the newt, *Notopthalamus viridescens*. The reaction is specific for the RNP fibers associated with the loops, areas of chromatin lacking transcriptionally active DNA remain unstained (results not shown). In addition to confirming with monoclonal antibody probes the direct association of the core proteins with hnRNA still attached to the DNA template, we found no difference at this level of resolution in the distribution of a subset of the proteins when compared to that of the entire core group, suggesting that there is no marked difference in hnRNP core structure between different transcription units in this system.

Ultrastructural Distribution of Core Proteins in Interphase Mammalian Cells

Indirect immunofluorescent staining of mammalian cells with either chrp specific polyclonal or monoclonal antibodies has shown these polypeptides to be restricted to the nuclei of interphase cells but absent from the nucleolus (20,23). During cell division the proteins are dispersed throughout the cell; their re-entry into the daughter nuclei following mitosis occurs as RNA synthesis is restored (22). Significantly there are no differences visible at this level between the localization of all the chrp polypeptides and that of subsets of the group (23). The examination of the distribution of hnRNP proteins at the ultrastructural level promises a greater degree of resolution and the identification of any possible associations with nuclear substructures, locating nascent pre-mRNA and potential storage forms of the core proteins. For visualization of chrps we have used colloidal gold as a discrete and highly electron dense marker (26). The monoclonal antibodies were directly coupled to the gold particles and then used to stain thin sections of mouse pancreatic tissue fixed with periodate lysine paraformaldehyde (shown to maintain antigenicity by immunofluorescent staining)(23) prior to embedding in Epon resin. Fig. 3 shows the immunogold staining pattern of a monoclonal antibody specific for all chrp polypeptides of mouse. There is a high degree of labeling at the periphery of condensed chromatin regions (see arrows), areas where perichromatin fibers are found. Previous autoradiographic analysis at the electron microscope level has shown this to be a region of high transcriptional activity (27,28), reinforcing the concept of a rapid association of the chrp polypeptides with nascent hnRNA molecules. Immunogold labeling of a more diffuse

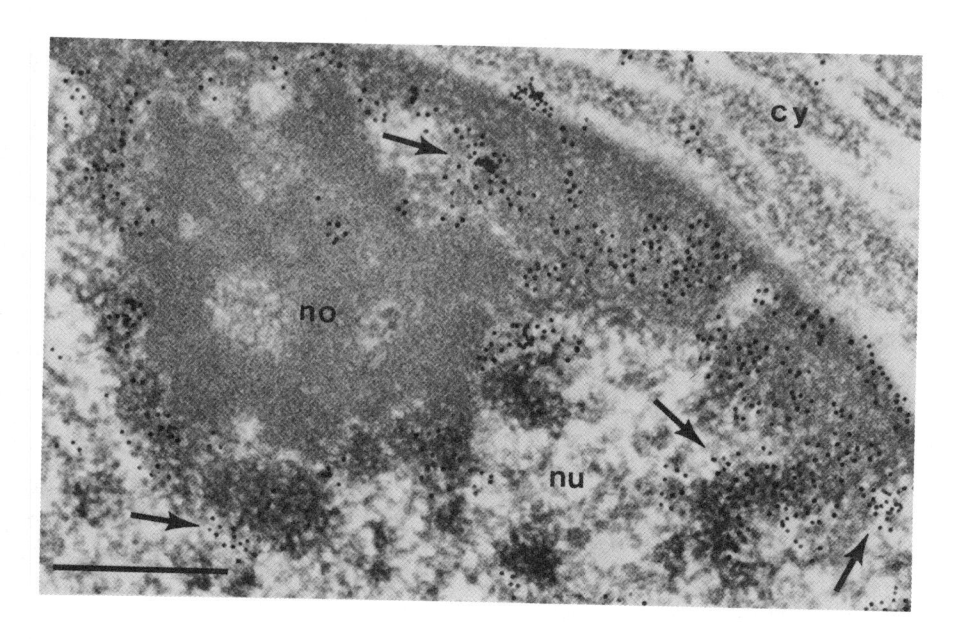

Figure 3. Ultrastructural distribution of chrp's in mouse pancreatic cells. Thin sections of Epon embedded cells fixed by the periodate-lysine-paraformaldehyde method, were reacted with a monoclonal antibody specific for all chrp proteins. Immunoglobulin binding was visualized by staining with goat anti-mouse IgG-IgM directly coupled to 10 nm colloidal gold. Gold granules are observed in clusters at the periphery of regions of chromatin (arrows), thought to be areas of transcriptional activity. The non-chromatin regions of the nucleus (nu) have low levels of diffuse staining while the cytoplasm (cy) and nucleolus (no) exhibit only non-specific background. Bar represents 0.5 μm.

pattern was observed in the interchromatin space, presumably representing core proteins free in the nucleus or those complexes containing post transcriptional hnRNA molecules. Interchromatin granules are rarely found associated with gold particles and then only a low degree of labeling has been observed (29); these regions are highly labeled by anti-snRNP antibodies. However, perichromatin granules are labeled with anti-chrp antibodies (29), suggesting the presence of hnRNP core proteins, perhaps in dormant state; there is some biochemical evidence for the presence of chrp in these granules (30). Conversely, perichromatin granules remain unstained by anti-snRNP antibodies (29). This suggests that neither interchromatin nor perichromatin granules represent sites of processing activity which is thought to involve both hnRNP and snRNP particles. The morphological manifestation of a pre-mRNA processing complex has yet to be found. However, we feel it is reasonable to expect that further high resolution immunocytochemical analysis of hnRNP and snRNP complexes both *in situ* and in nuclear spreads will provide useful information regarding possible sites of such activity.

ACKNOWLEDGMENTS

This research has been supported by the research grant CA 12550 and the University of Chicago Cancer Research Center facilities grant CA 14599 from the National Institutes of Health.

REFERENCES

1. Keller, W. The RNA lariat: A new ring to the splicing of mRNA presplicing of mRNA precursors. Cell 39: 423, 1984.
2. Padgett, R.A., S.F. Hardy, and P.A. Sharp. Splicing of adenovirus RNA in a cell free transcription system. Proc Natl. Acad Sci, USA 80: 5230, 1983.
3. Hernandez, N., and W. Keller. Splicing of in vitro synthesized messenger RNA precursors in HeLa cell extracts. Cell 35: 89, 1983.
4. Miller, O. L., Jr., and A. H. Bakken. Morphological studies of transcription. Acta Endocrinol Suppl 168: 155, 1972.
5. Angelier, N., and J. C. Lacroix. Complexes de transcription d'origines nucleolaire et chromosomique d'ovocytes de *Pleurodeles waltlii* et *P. poireti* (Amphibiens, Urodeles). Chromosoma 51:323, 1975.
6. Beyer, A. L., O. L. Miller, Jr., and S. L. McKnight. Ribonucleoprotein structure in nascent hnRNA is nonrandom and sequence-dependent. Cell 20: 75, 1980.
7. Samarina, O.P., E. M. Lukanidin, J. Molnar and G. P. Georgiev. Structural organization of nuclear complexes containing DNA-like RNA. J. Mol Biol 33: 251, 1968.
8. Martin, T. E., and B. J. McCarthy. Synthesis and turnover of RNA in the 30S nuclear ribonucleoprotein complexes of mouse ascites cells. Biochim Biophys Acta 277: 354, 1972.
9. Martin, T. E., P. Billings, A. Levey, S. Ozarslan, T. Quinlan, H. Swift, and L. Urbas. Some properties of RNA: protein complexes from the nucleus of eukaryotic cells. Cold Spring Harbor Symp.Quant. Biol 38: 921, 1973.
10. Kinniburgh, A.J., and T. E. Martin. Detection of mRNA sequences in nuclear 30S ribonucleoprotein subcomplexes. Proc Natl Acad Sci USA

73: 2725, 1976.

11. Pederson, T., and N. G. Davis. Messenger RNA processing and nuclear structure: Isolation of nuclear ribonucleoprotein particles containing β-globin messenger RNA precursors. J.Cell Biol 87: 47, 1980.
12. Martin, T. E., J. M. Pullman, and M. E. McMullen. Structure and function of nuclear and cytoplasmic ribonucleoprotein complexes. In D. M. Prescott and L. Goldstein (ed.), Cell Biology: a Comprehensive Treatise. Academic Press, Inc., New York, Vol. 4, 137, 1980.
13. Hinterberger, M., I. Pettersson, and I. A. Steitz. Isolation of small nuclear ribonucleoproteins containing U1, U2, U4, U5, and U6 RNAs. J.Biol Chem 258: 2604, 1983.
14. Beyer, A. L., M. E. Christensen, B. W. Walker and W. M. LeStourgeon. Identification and characterization of the packaging proteins of core 40S hnRNP particles. Cell 11: 127, 1977.
15. Karn, J., G. Vidali, L. C. Boffa, and V. G. Allfrey. Characterization of the non-histone nuclear proteins associated with rapidly labeled heterogeneous nuclear RNA. J Biol Chem 252: 7307, 1977.
16. Billings, P. B., and T. E. Martin. Proteins of nuclear ribonucleoprotein subcomplexes. Methods Cell Biol 17: 349, 1978.
17. Martin, T. E., P. B. Billings, J. M. Pullman, B. J. Stevens and A. J. Kinniburgh. Substructures of nuclear ribonucleoprotein complexes. Cold Spring Harbor Symp Quant Biol 42: 899, 1977.
18. Pullman, J. M. and T. E. Martin. Reconstitution of nucleoprotein complexes with mammalian heterogeneous nuclear ribonucleoprotein (hnRNP) core proteins. J Cell Biol 97: 99, 1983.
19. LeStourgeon, W. M., A. L. Beyer, M. E. Christensen, B. W. Walker, S. M. Poupore, and L. P. Daniels. The packaging proteins of core hnRNP particles and the maintenance of proliferative cell states. Cold Spring Harbor Symp Quant Biol 42: 885, 1977.
20. Jones, R. E., C. S. Okamura and T. E. Martin. Immunofluorescent localization of the proteins of nuclear ribonucleoprotein complexes. J Cell Biol 86: 235, 1980.
21. Martin, T. E., and C. S. Okamura. Immunocytochemistry of nuclear hnRNP complexes. In The Cell Nucleus (Busch, H., ed.), Academic Press, New York, Vol. IX: 119, 1981.
22. Martin, T. E., and C. S. Okamura. hnRNP protein distribution in various differentiated vertebrate cells. In: International Cell Biology 1980-81 (Schweiger, H.G., ed) Springer Verlag, Berlin. 77, 1981.
23. Leser, G. P., J. Escara-Wilke, and T. E. Martin. Monoclonal antibodies to heterogeneous nuclear RNA-protein complexes. J Biol Chem 259: 1827, 1984.
24. Wilke, H. E., H. Werr, D. Friedrich, H. H. Kiltz, and K. P. Schafer. The core proteins of 35S hnRNP complexes. Eur J Biochem 146: 71, 1985.
25. Choi, Y. D., and G. Dreyfuss. Isolation of heterogeneous nuclear RNA-ribonucleoprotein complex (hnRNP): A unique supramolecular assembly. Proc Natl Acad Sci USA 81: 7471, 1984.
26. Romano, E. L., C. Stolinski, and N. C. Hughes-Jones. An antiglobulin reagent labeled with colloidal gold for use in electron microscopy. Immunochem 11, 521, 1974.
27. Fakan, S., and W. Bernhard. Nuclear labeling after prolonged ^{3}H-uridine incorporation as visualized by high resolution autoradiography. Exp Cell Res 79: 431, 1973.

28. Puvion, E. and A. Viron. In situ structural and functional relationships between chromatin pattern and RNP structures involved in non-nucleolar chromatin transcription. J Ultrastruct Res 74: 351, 1981.
29. Fakan, S., G. P. Leser, and T. E. Martin. Ultrastructural distribution of nuclear ribonucleoproteins as visualized by immunocytochemistry on thin sections. J Cell Biol 98: 358, 1984.
30. Daskal, Y., L. Komaromy, and H. Busch. Isolation and partial characterization of perichromatin granules. A unique class of nuclear RNP particles. Exp Cell Res 126: 39, 1980.

TABLE I

Major hnRNP Polypeptides of HeLa Cells

chrp's (A+B)	33-40 kDa	basic pI
A family of 14-17 related species, sufficient for *in vitro* reconstitution of core particles, and have common antigenic determinants; partial peptide mapping and immunological analyses suggest that the chrp complement is generated by modification of a fundamental set of several parent polypeptides (18,20,23)		
C	41-45 kDa	acidic pI
2-4 species can be directly cross-linked to hnRNA,share antigenic determinants although distinct from those of chrp, and possibly serve to attach hnRNP complexes to nuclear substructure (25).		
68 kDa	ca. 68 kDa	
A single polypeptide that can be directly cross-linked to hnRNA, found associated with large hnRNP complexes (25).		
120 kDa	ca. 120 kDa	
Can be directly cross-linked to hnRNA, antigenically distinct (25).		

11. Interactions between the Mammalian Cell La Protein and Small Ribonucleic Acids

Jack D. Keene, Jasemine C. Chambers, and Barbara Martin

Department of Microbiology and Immunology and Department of Medicine, Division of Rheumatology and Immunology, Duke University Medical Center, Durham, NC 27710

INTRODUCTION

The mammalian cell protein referred to as La was originally defined by its reactivity with certain autoantibodies from patients with rheumatological conditions known as Sjogren's syndrome and systemic lupus erythematosus (35). Unlike other autoantigens such as Sm and RNP, which consist of multiple species of protein (17,23), La appears to be a single protein of approximately 50,000 MW (7). Smaller antigenic forms of La (9,24,36) as well as multiple species of the 50,000 MW protein have been observed, however (6,34). The protein has been isolated in various states of purity using both affinity methods (34,36) and standard chromatographic techniques (34). Recently, Chambers and Keene reported the isolation of complementary DNA clones encoding the human La protein (3). La antigen has been reported to be present in most mammalian species as well as in protozoa (5) but not in insects (20).

Interest in the biological function of La protein emerged with the findings of Steitz and co-workers (18,27) that La is associated with precursor transcripts of small cell RNAs such as pre-transfer RNA and pre-5S RNA (Table 1). In addition, small RNAs synthesized by Epstein-Barr virus (19) and by adenovirus (7,29) were found to be precipitable by La antisera. These viral RNAs are more stably bound to La protein than are the cell RNAs but they are also products of RNA polymerase III (pol III). Based upon these findings, Rinke and Steitz (27) proposed that La is a transcription factor for pol III (8,20,22,25).

Role of La Protein in Cell RNA Synthesis

Mammalian RNA polymerase III produces RNA transcripts of 70 to over 300 nucleotides in length that are generally localized in the cytoplasm and involved in mRNA translation (16,18,27). These include transfer RNA, 5S RNA, 7S RNA (4) and less well defined species such as 7-2, 7-3, 8-2 (10,26) and probably some forms of *in vitro* synthesized Alu transcripts (31). The virus encoded pol III products may carry out similar functions by interacting with their infected hosts. For example, the VA RNAs synthesized by adenoviruses are pol III products, are bound to La protein and have been shown to function in translational control during infection (30,33).

Transcription by pol III involves the recognition of promoter sequences that are contained internally within the gene coding region (16,32). Two active regions, the distal and proximal promoters, have been shown to interact with the polymerase and with pol III associated factors. Three specific pol III transcription factors have been defined and partially purified by Roeder and coworkers (16). The precise functions of TF III A, B and C are not totally known, but factor A binds to the internal promoter as well as to the 5S RNA transcript and, thus, may function in an autocatalytic fashion. TF-IIIA was recently reported to have DNA gyrase activity of type II topoisomerase (13). Attempts to determine whether the La protein is related to factors A, B or C have not shown a definitive relationship. Gottesfeld et al. (8) were able to inhibit the *in vitro* transcription of VA RNA, 5S RNA and an "identifier" pol III transcript using La antisera. Their studies indicated that a nonspecific DNA binding protein of 64 kd was associated with the La RNP complexes and that depletion of this protein was probably responsible for the loss of pol III activity. These investigators also found that a high salt eluate obtained by phosphocellulose chromatography of HeLa cell extracts was able to restore the anti-La depleted activity. Although it is possible that the 64 Kd La-associated factor is TF-IIIC, it is apparent that none of the proteins in this restoration fraction is identical to La protein (8). In addition, La protein does not appear to be TF-IIIC (R.Roeder, personal communication). In a similar study, Francoeur and Mathews (7) were not able to inhibit transcription of VA RNA *in vitro* after depletion of La protein. Thus, the role of La protein in pol III transcription and its relationship to pol III transcription factors remains unclear at the present time. In an intriguing study, Moore and Sharp found that La antisera can inhibit *in vitro* polyadenylation of mRNA (22). The significance of this finding is not clear, but a form of U6 RNA is known to bind La protein (25,28). The U6 RNA associates with U4 RNA (2) and Berget has proposed a role for U4 RNA in polyadenylation (1). Thus, although associations between La protein and RNA 3' termination events of both pol II and pol III have been observed, additional experimentation is needed to assess the functions of La protein in RNA synthesis.

Nature of Binding of La Protein to RNA

The property which all known RNAs that bind to La protein have in common is the presence of terminal uridylate residues at the 3' end of the RNA. The size range of RNA binding to La protein varies from 45 to over 300 nucleotides and, thus, size does not appear to be a critical factor in recognition. In addition, although secondary structure may influence binding to small RNAs as suggested for the VA RNAs (7), a common higher order structural feature has not been identified. The leader RNAs of VSV (12,15,37,38) and rabies virus (14) do not appear to have significant amounts of secondary structure and bind La protein with high efficiency. Considering the broad variation in the qualities of RNAs that bind La protein, it is possible that the presence of terminal uridylates is sufficient for recognition. Stefano (34) studied the binding of La protein to model transfer RNAs following purification by various methods including affinity chromatography over poly U Sepharose. He concluded that 3' terminal hydroxyl groups were preferred over terminal phosphates and that binding is saturated with the addition of more than 3 terminal uridylates. The dissociation constant of binding was estimated to be approximately 10^{-7}M. Furthermore, using reconstitution experiments and stability studies with NaCl and heparin, he found that model tRNAs with several various terminal nucleotide base adducts were capable of binding if sufficiently large amounts of the La protein were added. In addition, although there was a strong preference for 3' terminal Us, tetranucleotides of G, A, and C were also capable of some binding with decreasing affinity in that order. The purified La protein used in these studies showed heterogeneous properties that may have been due to the presence of RNA fragments and the recovery of La protein was only about 7%. Thus, the binding properties observed using these fractions may not have been representative of the total cell La protein. Heparin Sepharose was used to displace La bound RNAs, however, and to help assure that oligo U binding species of La protein were not preferentially selected.

Reconstitution of La protein with isolated cell RNAs has also been used by other investigators to study the binding of precursor species (4,6,7,21,25,27) and to determine the sequence recognition sites of binding (7,21,25). Using T1 oligonucleotides and *in vitro* reassociation, the VA RNA (7,21) and 4.5S RNA (25) were both shown to bind La protein at 3' terminal fragments. The role of the base paired 5' portion of these RNAs in La recognition is not clear from these studies, however. The 3' terminus of VA RNA, for example, can be covalently crosslinked to La protein using UV irradiation (21). Because structurally diverse RNAs bind La protein and do not all possess potential 3' and 5' terminal base pairing, it is unlikely that the 5' end is involved in recognition. These studies, however, suggest that sequences of up to 30 nucleotides from the 3' terminus may influence binding (7). In addition, the studies with model tRNAs (34) also leave open the possibility that sequences other than just the terminal uridylates influence recognition by La protein and suggest that the 3' U residues may function in the stabilization of attachment. It is also clear from these studies that addition of pCp to the 3' end interferes with binding but periodate oxidation of the 3' hydroxyl group does not interfere (25,34). In every case studied, the intact, nonfragmented RNAs consistently bound La protein more efficiently than oligonucleotide fragments. It is interesting to note, however, that Reddy et al. (25) were able to reconstitute the terminal oligonucleotide fragment of U6 RNA with La protein with only 40% efficiency while the intact U6 was bound with about 23 % efficiency.

Molecular Nature of the La Protein

The La protein is readily susceptible to proteolytic degradation and has been reported to have molecular weights varying from 25 kd to 68 kd (7,9,24,27,34,36). Most investigators currently agree that the protein is 45 to 50 kd in size with major degradation products of 40 kd and 25 kd (6,9). Immunofluorescence studies have shown that La is localized to the nucleus but also has a weak cytoplasmic signal (11,27). La protein leaks readily from the nucleus during biochemical fractionation, however (10,15). The La autoimmune specificity tends to occur together with the Ro specificity in patients with Sjogren's syndrome but Ro is a distinct protein of approximately 60 kd (7,11,36). Ro protein is bound to a subset of the La reactive small RNAs that are strictly cytoplasmic and that vary in sequence among mammalian species (11,39). Although the function of Ro RNPs is not known, their cytoplasmic locale suggests that they may play a role in the regulation of translation.

The isolation of cDNA encoding the La protein (3) has allowed determination of the genomic structure in human DNA by Southern blot analysis (Fig. 1A). DNA from peripheral blood lymphocytes of three individuals and from cultured HeLa cells showed no differences in restriction pattern following digestion with EcoR1. As estimated from the number and relative darkness of the DNA bands there are approximately 4 to 6 copies of the La gene. Northern blot analysis showed that the polyadenylated mRNA encoding the cloned cDNA was about 1.8 kb in size (Fig. 1B). DNA sequence data of several cDNA clones indicated that there was approximately 1.1 kb of coding sequence. Thus, one would expect on this basis to produce a protein of only 42 kd MW rather than 50 kd as estimated by migration on SDS gels. La protein may show aberrant migration due to post-translational modifications.

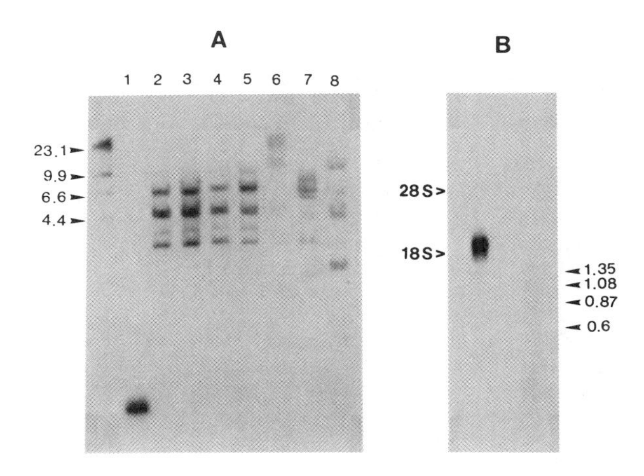

Figure 1A. Southern blot analysis of DNA from three normal humans (lanes 2-4) and from HeLa cells (lanes 5-8) using nick translated cDNA (La-6) encoding the amino half of the La protein (3). DNA markers are shown as kilobases. Lane 1, probe; lanes 2-5, EcoR1; lane 6, Bam H1; lane 7, Hind III; lane 8, Bgl II.
Figure 1B. Northern blot analysis of poly A selected messenger RNA from HeLa cells probed with La cDNA.

SUMMARY

The mammalian cell La protein is bound to cell and virus small RNAs that contain terminal uridylate residues. While the cellular bound RNAs are usually precursor transcripts synthesized by RNA polymerase III, the viral small RNAs do not appear to be precursors and are more stabily bound to La protein. In addition, La protein is associated with U1 and U6 small nuclear RNAs and this binding may also be mediated by the presence of terminal uridylates. The DNA containing viruses, Epstein-Barr virus and adenovirus, synthesize the La-bound EBER and the VA RNAs, respectively, using the cell RNA polymerase III. The negative strand containing RNA viruses, vesicular stomatitis virus and rabies virus, synthesize small leader RNAs using their own virion-bound RNA polymerases and are, thus, not products of RNA polymerase III. La protein has been proposed to function in the transcription, processing or transport of pol III transcripts. In addition, it has been suggested that La mediates the switch from a transcriptive mode to a replicative mode of RNA synthesis by the negative strand RNA viruses. Recombinant clones containing cDNA encoding the La protein have been isolated and Northern blot analysis indicates that the mRNA is about 1.8 kb and contains 1.1 kb of coding sequence. Analysis of the genomic structure of La indicates that there may be as many as 4 to 6 copies of the gene in human DNA (Fig. 1A). Only 2 or 3 of these gene fragments were reactive with cDNA representing the carboxyl half of La protein, however. Thus, some of the gene fragments analyzed by Southern blotting may represent processed pseudogenes. Data concerning the molecular nature of La protein-RNA interactions when taken together are compatible with the suggestion that the protein participates in the termination and release of RNA transcripts from their templates during RNA synthesis.

ACKNOWLEDGEMENTS

We appreciate the valuable contributions of Michael Kurilla, Jeffrey Wilusz and Helen Piwnica-Worms to the progress of this work. We thank Hilda Smith for help in preparing the manuscript.

REFERENCES

1. Berget, S.M.: Are U4 small nuclear ribonucleoproteins involved in polyadenylation? Nature 309:179,1984.
2. Bringmann, P., B. Appel, J. Rinke, R. Reuter, H. Theissen and R. Luhrmann: Evidence for the existence of snRNAs U4 and U6 in a single ribonucleoprotein complex and for their association by intermolecular base pairing. The EMBO J. 3:1357, 1984.
3. Chambers, J.C. and J.D. Keene: Isolation and analysis of cDNA clones expressing human lupus La protein. Proc. Natl. Acad. Sci. (USA) 82:2115, 1985.
4. Chambers, J.C., M.G. Kurilla and J.D. Keene: Association between the 7S RNA and the lupus La protein varies among cell types. J. Biol. Chem. 258:1438, 1983.
5. Francoeur, A.M., C.A. Gritzmacher, C.L. Peebles, R.T. Reese and E.M. Tan: Synthesis of small nuclear ribonucleoprotein particles by the malarial parasite Plasmodium falciparum. Proc. Natl. Acad. Sci. (USA) 82:3635, 1985.
6. Francoeur, A.M., E.K.L. Chan, J.I. Garrels and M.B. Mathews: Characterization and purification of lupus antigen La, an RNA-binding protein. J. Mol. Biol. 5:586, 1985.
7. Francoeur, A.M. and M.B.Mathews: Interaction between VA RNA and the lupus antigen La. Proc. Natl. Acad. Sci. (USA) 79:6772, 1982.
8. Gottesfeld, J.M., D.L. Andrews and S.O. Hoch: Association of an RNA polymerase III transcription factor with a ribonucleoprotein complex recognized by autoimmune sera. Nuc. Acids Res. 12:3185,1984.
9. Habets, W.J., J.H. den Brok, A.M.Th. Boerbooms, L.B.A. van de Putte and W.J. van Venrooij: Characterization of the SS-B (La) antigen in adenovirus-infected and uninfected HeLa cells. EMBO J. 2:1625, 1983.
10. Hashimoto, C. and J.A. Steitz: Sequential association of nucleolar 7-2 RNA with two different autoantigens. J. Biol. Chem. 258:1379, 1983.
11. Hendrick, J.P., S.L. Wolin, J. Rinke, M.R. Lerner and J.A. Steitz: Ro small cytoplasmic ribonucleoproteins are a subclass of La ribonucleoproteins: further characterization of the Ro and La small ribonucleoproteins from uninfected mammalian cells. Mol. Cell Biol. 1:1138, 1981.
12. Keene, J.D., M.G. Kurilla, J. Wilusz, and J.C. Chambers: Interactions between cellular La protein and leader RNAs. In, D.H.L. Bishop and R.W. Compans, eds., Negative Strand Viruses, Academic Press, Inc., New York, p. 103, 1984.

13. Kmiec, E.B. and A. Worcel: The positive transcription factor of the 5S RNA gene induces a 5S DNA-specific gyration in Xenopus oocyte extracts. Cell 41:945, 1985.
14. Kurilla, M.G., C.D. Cabradilla, B.P. Holloway and J.D. Keene: Nucleotide sequence and host La protein interactions of rabies virus leader RNA. J. Virol. 50:773, 1984.
15. Kurilla, M.G. and J.D. Keene: The leader RNA of vesicular stomatitis virus is bound by a cellular protein reactive with anti-La lupus antibodies. Cell 34:837, 1983.
16. Lassar, A.B., P.L. Martin, and R.G. Roeder: Transcription of class III genes: formation of preinitiation complexes. Science 222:740, 1983.
17. Lerner, M.R. and J.A. Steitz: Antibodies to small nuclear RNAs complexed with proteins are produced by patients with systemic lupus erythematosus. Proc. Natl. Acad. Sci. (USA) 76:5495, 1979.
18. Lerner, M.R., J.A. Boyle, J.A. Hardin and J.A. Steitz: Two novel classes of small ribonucleoproteins detected by antibodies associated with lupus erythematosus. Science 211:400, 1981a.
19. Lerner, M.R., N.C. Andrews, G. Miller and J.A. Steitz: Two small RNAs encoded by Epstein-Barr virus and complexed with protein are precipitated by antibodies from patients with systemic lupus erythematosus. Proc. Natl. Acad. Sci. (USA) 78:805, 1981b.
20. Madore, S.J., E.D. Wieben and T. Pederson: Eukaryotic small ribonucleoproteins: anti-La human autoantibodies react with Ul RNA-protein complexes. J. Biol. Chem. 259:1929, 1984.
21. Mathews, M.B. and A.M. Francoeur: La antigen recognizes and binds to the 3'-oligouridylate tail of a small RNA. Mol. Cell Biol. 4:1134, 1984.
22. Moore, C.L. and P.A. Sharp: Site-specific polyadenylation in a cell-free reaction. Cell 36:581, 1984.
23. Petterson, I., M. Hinterberger, T. Mimori, E. Gottlieb and J.A. Steitz: The structure of mammalian small nuclear ribonucleoproteins. Identification of multiple protein components reactive with anti-Ul ribonucloeprotein and anti-Sm autoantibodies. J. Biol. Chem. 259:5907, 1984.
24. Pizer, L.I., J.-S. Deng, R.M. Stenberg and E.M. Tan: Characterization of a phosphoprotein associated with the SS-B/La nuclear antigen in adenovirus-infected and uninfected KB cells. Mol. Cell. Biol. 3:1235, 1983.
25. Reddy, R., D. Henning, E. Tan and H. Busch: Identification of a La protein binding site in a RNA polymerase III tanscript (4.5S RNA). J. Biol. Chem. 258:8352, 1983.

26. Reddy, R., E.M. Tan, D. Henning, K. Nohga and H. Busch: Detection of a nucleolar 7-2 ribonucleoprotein and a cytoplasmic 8-2 ribonucleoprotein with autoantibodies from patients with scleroderma. J. Biol. Chem. 258:1383, 1983.

27. Rinke, J. and J.A. Steitz: Precursor molecules of both human 5S ribosomal RNA and tRNAs are bound by a cellular protein reactive with anti-La lupus antibodies. Cell 29:149, 1982.

28. Rinke, J. and J.A. Steitz: Association of the lupus antigen La with a subset of U6 snRNA molecules. Nuc. Acids Res. 13:2617, 1985.

29. Rosa, M.D., E. Gottlieb, M.R. Lerner and J.A. Steitz: Striking similarities are exhibited by two small Epstein-Barr virus-encoded ribonucleic acids and the adenovirus-associated ribonucleic acids VAI and VAII. Mol. Cell Biol. 1:785, 1981.

30. Schneider, R.J., B. Safer, S.M. Munemitsu, C.E. Samuel and T. Shenk: Adenovirus VAI RNA prevents phosphorylation of the eukaryotic initiation factor 2 subunit subsequent to infection. Proc. Natl. Acad. Sci. (USA) 82:4321, 1985.

31. Shen, C.-K.J. and T. Maniatis: The organization, structure, and _in vitro_ transcription of Alu family RNA polymerase III transcription units in the human alpha-like globin gene cluster: precipitation of in vitro transcripts by lupus anti-La antibodies. J. Mol. Appl. Gen. 1:343, 1982.

32. Shenk, T.: Transcriptional control regions: nucleotide sequence requirements for initiation by RNA polymerase II and III. Curr. Top. Microbiol. Immunol. 93:25, 1981.

33. Siekierka, J., T.M. Mariano, P.A. Reichel and M.B. Mathews: Translational control by adenovirus: lack of virus-associated RNA I during adenovirus infection results in phosphorylation of initiation factor eIF-2 and inhibition of protein synthesis. Proc. Natl. Acad. Sci. (USA) 82:1959, 1985.

34. Stefano, J.E.: Purified lupus antigen La recognizes an oligouridylate stretch common to the 3' termini of RNA polymerase III transcripts. Cell 36:145, 1984.

35. Tan, E.M.: Autoantibodies to nuclear antigens (ANA): their immunobiology and medicine, _in_ Advances in Immunology, vol. 33, H.Kunkel and F.Dixon, eds., Academic Press, Inc. p. 167, 1982.

36. Venables, P.J.W., P.R. Smith and R.N. Maini: Purification and characterization of the Sjogren's syndrome A and B antigens. Clin. Exp. Immunol. 54:731, 1983.

37. Wilusz, J. and J.D. Keene: Interactions of plus and minus strand leader RNAs of the New Jersey serotype of vesicular stomatitis virus with the cellular La protein. Virology 135:65, 1984.
38. Wilusz, J., M.G. Kurilla and J.D. Keene: A host protein (La) binds to a unique species of minus-sense leader RNA during replication of vesicular stomatitis virus. Proc. Natl. Acad. Sci. (USA) 80:5827,1983.
39. Wolin, S.L. and J.A. Steitz: Genes for two small cytoplasmic Ro RNAs are adjacent and appear to be single-copy in the human genome. Cell 32:735, 1983.

Table 1. Cellular and viral RNAs that are precipitable with La autoantibodies using mammalian cell extracts.

RNA	source	size (nucleotides)	polymerase
pre transfer	mammalian	80 - 100	III
hY1-hY5	human	80 - 110	III
mY1, mY2	mouse	80 - 110	III
4.5S	mouse	100	III
pre 5S	mammalian	130	III
7S, 7-2, 7-3	mammalian	300 - 400	III
VA I, II	adenovirus	160	III
EBER 1, 2	EBV	166, 172	III
leader RNAs	VSV rabies virus	45 - 50	virus specific
U_1 RNA	mammalian	165	II
U_6 RNA	mammalian	107	?

12. RNA-Protein Interactions in the Nuclear Ribonucleoprotein Particles

V. Holoubek

Department of Human Biological Chemistry and Genetics, University of Texas Medical Branch, Galveston, TX 77550

INTRODUCTION

In addition to control of gene expression at the transcriptional level, a second type of control acts on the level of posttranscriptional processing of the transcribed RNA. This was demonstrated as early as 1978, when it was shown that RNA sequences which encode specific embryonic proteins of sea urchins are still found in the nuclear polyadenylated RNA of adult animals, but not in their cytoplasmic mRNA (1). The dependence of the transport of specific mRNA into the cytoplasm on processing is best illustrated in analbuminemic rats which carry a mutation affecting processing of albumin mRNA. In these rats the precursors of this RNA are transcribed normally in the nuclei, but no albumin mRNA is found in the cytoplasm (2). In the cell nucleus the heterogeneous nuclear RNA (hnRNA), generally regarded to include the sequences of premessenger RNA (pre-mRNA), is associated with proteins and can be isolated from the nuclei in the form of ribonucleoprotein particles (hnRNP particles). Since the assembly of the hnRNP particles precedes pre-mRNA processing, many of the modifications required for the maturation of mRNA probably take place in the hnRNP particles. The presence of sequences of mRNA in the hnRNP particles has been demonstrated by many authors (for review see 3). The best support for the proposal that pre-mRNA is processed in the hnRNP particles was obtained by Pederson and Davis (4), who demonstrated the presence of both the 9S beta-globin mRNA, and its 15S precursor in the hnRNP particles from mouse Friend erythroleukemia cells. It is possible that the proteins in the hnRNP particles are only a packaging device for hnRNA and that the processing of premessenger RNA is not actually controlled by interactions between the RNA and these proteins. But the findings that many of the particle proteins are post-translationally modified, and the indications that the proteins are interacting with hnRNA in a sequence specific manner (for review see 3) support the opposite view, that the particle proteins play some regulatory role in the RNA processing.

DNA-protein interactions are the heart of most models of the regulation of gene transcription, as discussed extensively in other parts of this book. Much, but not all DNA transcription is believed to be controlled by interaction of proteins which interact as trans control elements. These bind specific DNA sequences which have the function of cis control elements. No involvement of specific RNA in the control of DNA transcription is envisioned in the majority of currently proposed models. Therefore, a possibly important difference in the regulation of RNA processing and in the control of DNA transcription should be mentioned. It is very probable that at least some stages of RNA processing require involvement of unique RNA. At present it cannot be excluded that in RNA processing a RNA and not a protein could act as the trans control element. If this is indeed the case, the proteins of the nuclear RNP particles would not determine which specific precursor of mRNA should be processed into mRNA. This conclusion does not preclude the interaction of the proteins with RNA in the hnRNP particles playing a role in the control of RNA processing, for example, by controlling the rate. Such an assumption is supported by the findings that the particle proteins are phosphorylated and the "core" proteins of the particles have some of their arginine residues modified by methylation. The regulation of hnRNA processing by the phosphorylation and dephosphorylation of the hnRNP particle proteins has been suggested mainly based on the findings that some of the enzymatic activities required for the phosphorylation and dephosphorylation of proteins are associated with the hnRNP particles (for review see 5). In our experiments we could demonstrate the simultaneous dephosphorylation of the hnRNP particle proteins and the distortion of RNA processing after the application of the hepatocarcinogen 3'methyl-4-dimethylaminoazobenzene (6). Experimental findings which support the conclusion that the methylation of hnRNP particle "core" proteins also can play a role in the modulation of RNA processing are discussed in this chapter.

HnRNP PARTICLES

As already mentioned, in eukaryotic cells the hnRNA is associated in the nuclei with specific proteins and can be isolated in the form of hnRNP particles. The structure of these particles and their protein composition is discussed in this book by T.E. Martin. A detailed review evaluating the controversy concerning the structure and composition of the hnRNP particles has been published (3). Therefore, the following description of the hnRNP particle structure and of the particle proteins is limited to conclusions which are absolutely necessary for the discussion of the topic of this article. Observations in situ have shown that hnRNA is present in the nucleus in the form of a beads on a string like structure of ribonucleoprotein fibrils with globular subunits spaced along their length. When the ribonucleoproteins are extracted under conditions which allow limited nucleolytic cleavage of the RNA, the hnRNA is released from the nuclei in the form of hnRNP particles. The exact number of different species of polypeptides present in the hnRNP particles is still controversial. In addition to minor proteins, the amount and the number of which vary with the type of hnRNP particle and extent of its purification, 70 to

90% of particle protein mass is represented by a few "core proteins". The apparent molecular weight of these proteins is dependent on conditions used during their electrophoresis, indicating some preserved secondary structure even in the presence of sodium dodecyl sulfate, such as can be caused by alpha-helical segments with a high percentage of hydrophobic residues. Their anomalous behavior during electrophoresis, and the presence of large numbers of charge isomers of these proteins, make it difficult to estimate not only their precise molecular weights but also the number of their species. In our experiments, we identified in rat liver three predominant molecular weight species of the "core" proteins (Mr = 33,000, 39,000 and 43,000 in the presence of sodium dodecyl sulfate and urea), each molecular weight species separating into a series of charge isomers. According to the nomenclature originally proposed by Bayer et al. (7), these three major molecular weight species would belong to the three groups of "core" proteins designated as A, B and C, in order of increasing molecular weights. The composition of all the "core" proteins is characterized by a high content of glycine residues and by the methylation of some of the arginine residues to N^G,N^G-dimethylarginine. As indicated by the name "core" proteins , it is assumed by most investigators that these proteins form the core of the hnRNP particles and that hnRNA is wrapped around this core. With these core hnRNP particles are associated additional minor proteins and ribonucleoproteins containing small nuclear RNA (snRNA).

METHYLATION OF THE PROTEINS OF THE HnRNP PARTICLES

As mentioned, one of the characteristic features of the "core" proteins of the hnRNP particles is the methylation of the guanidino group of some of their arginine residues (7, 8). The guanidino-group of arginine has been suggested as a possible site of specific interactions between proteins and nucleic acids (9). Besides the "core" proteins of the hnRNP particles, some proteins of ribosomes also contain the modified amino acid N^G,N^G-dimethylarginine (10). Therefore, we studied the incorporation of labeled methyl groups into the "core" proteins of the hnRNP particles and the changes in the methylation of these proteins after the inhibition of RNA synthesis by actinomycin D.

In two-dimensional electropherograms of the proteins of the hnRNP particles from rat liver (Fig. 1) the "core" proteins separated into three molecular weight species (Mr = 33,000, 39,000 and 43,000) which correspond to the A, B and C packing proteins of Beyer et al. (7) and to the Mr = 34,000, 36,000 and 41,000 major proteins of rat liver ribonucleoprotein particles described by Peters and Comings (11). The main difference between the described composition of "core" proteins of the hnRNP particles from HeLa cells and from rat liver is the predominant presence of only one molecular weight species of each of the three types (A, B and C) of the "core" proteins in the liver hnRNP particles. In the hnRNP particles from HeLa cells each of the three types of the "core" proteins is present as a doublet (13). Every

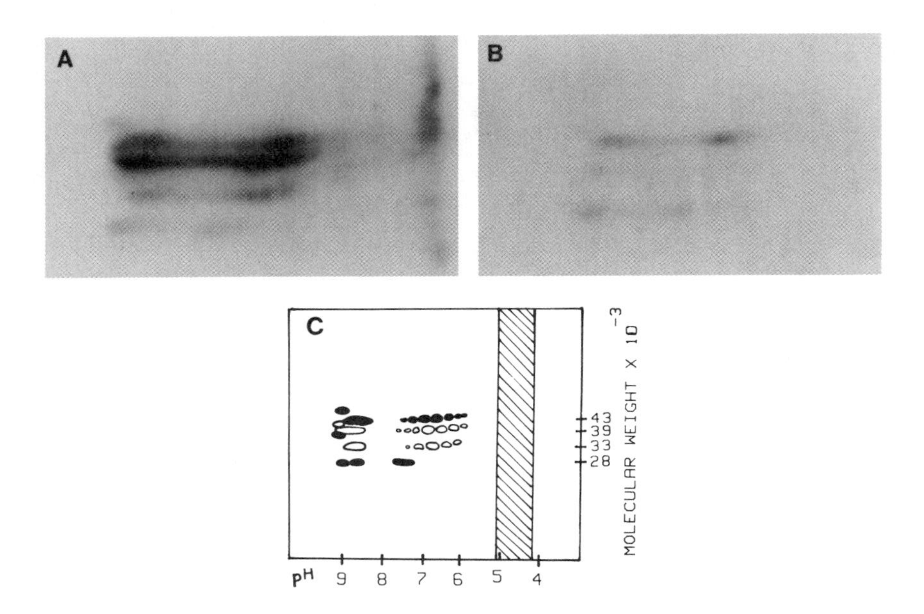

Fig. 1. Two-dimensional gel electrophoresis of the proteins of the hnRNP particles. The hnRNP particles were extracted from liver nuclei by 0.1M NaCl at pH 8.0 and further purified by centrifugation in a 15 to 30% sucrose density gradient (12). The proteins were separated by a non-equilibrium pH gradient electrophoresis in the first dimension and by sodium dodecyl sulfate-polyacrylamide gel electrophoresis in the second dimension. The hnRNP particles were isolated from liver of rats which had been injected intraperitoneally 12 hours earlier with 100uCi of [^{3}H-methyl]-S-adenosyl-L-methionine (spec. act. 83.5 Ci/mmol, New England Nuclear, Boston, MA) per 100g of body weight. (A) Electropherogram stained with Coomassie brilliant blue showing all the proteins of the hnRNP particles. (B) Autoradiogram showing proteins which have incorporated labeled methyl groups. (C) Diagrammatic representation of the electropherogram. The filled spots represent the proteins which have incorporated radiomethyl. The heterogeneous minor particle proteins separated in the crosshatched area.

molecular weight species of the "core" proteins separated into a series of charge isomers (Fig. 1). The minor acidic polypeptides, which can at least partly be removed from the hnRNP particles by repeated purification (12) are compressed into the acidic part of the gel as minor heterogeneous spots. The injection of [^{3}H-methyl]-S-adenosyl-L-methionine 12 hrs prior to the hnRNP isolation results in the incorporation of labeled methyl groups into all charge isomers of the 43,000 daltons protein. The C "core" proteins of the hnRNP particles are metabolically stable (14). S-adenosyl-L-methionine is a donor of methyl groups to proteins during their posttranslational methylation and does not participate in the formation of peptide bonds (15). Therefore it is likely that the observed incorporation of radiomethyl represents a turnover of methyl groups on the C "core" particle protein. In the group of the minor particle proteins all charge isomers of a 23,000 daltons protein and two basic 37,000 and 47,000 daltons proteins preferentially incorporate radiomethyl. To our surprise the incorporation of radiomethyl into all charge isomers of the other two major molecular weight species of the particle "core" proteins (Mr = 33,000 and 39,000) which also contain methylated orginine was minimal. The incorporation of radiomethyl into all charge isomers of the 43,000 daltons protein of the hnRNP particles indicates a much higher turnover of methyl groups on the C "core" protein than on the A and B "core" proteins.

The incorporation of radiomethyl into the total protein of the liver hnRNP particles decreased to 50% and 17% of control values 2 hrs and 6 hrs after the simultaneous injection of [^{3}H-methyl]-S-adenosyl-L-methionine and of actinomycin D (2 mg/kg of body weight) (16). This decrease was not observed in the methylation of histones, indicating that it does not represent a general suppression of protein methylation. After simultaneous injection of actinomycin D and [^{3}H]orotic acid the specific activity of RNA in the hnRNP particles of rat liver decreased to 85% and 50% 2 hrs and 6 hrs after the injection. The decrease observed after 6 hrs reflects the difference between a continuous increase in RNA specific activity in the hnRNP particles isolated from liver of control animals and the near constant specific activity of RNA in the hnRNP particles in liver of treated animals during the period from 2 hrs to 6 hrs after the injection of actinomycin D and [^{3}H]orotic acid. This constant specific activity of RNA in the hnRNP particles in the liver of animals treated with actinomycin D implies that, in addition to its effect on RNA synthesis, actinomycin D may directly block RNA processing. The double-stranded regions of hnRNA have been proposed to be involved in RNA processing (17). Intercalation of actinomycin D into these double-stranded stretches of hnRNA, in analogy to intercalation of actinomycin D into DNA (18), is structurally possible.

If rats were injected intraperitoneally with actinomycin D 12 hrs prior to sacrifice, the yield of hnRNP particles isolated from the liver nuclei decreased by 60 percent. The isolated particles were depleted of most minor proteins and of all charge isomers of the A and B "core" hnRNP particle proteins (19). The only abundant protein of the particles was the C "core" protein. Interestingly, all of the charge isomers of this protein were present in the particles. Only minimal incorporation of radiomethyl into particle proteins was observed in the liver of rats treated with actinomycin D 12 hrs and also injected with [^{3}H-methyl]-S-adenosyl-L-methionine 3 hrs before sacrifice (39 ± 28 dpm/mg particle proteins as compared with 1478 ± 293 dpm/mg particle proteins in controls). The processing of hnRNA is a dynamic process, and the binding of hnRNA to the core proteins must be only temporary. If the extent of the methylation of arginine residues of the "core" proteins modulates the RNA interaction, a turnover of the methyl groups would be expected. The involvement of the methylation of the proteins of hnRNP particles in RNA processing is in agreement with the suppression of the incorporation of radiomethyl into these proteins after the inhibition of RNA synthesis by actinomycin D.

C "CORE" PROTEIN OF THE HnRNP PARTICLES AND RNA-PROTEIN ASSOCIATION

The finding that the turnover of the methyl groups of all charge isomers of the C protein is much higher than that of other "core" proteins, together with other published results, point to a unique role of the C protein in the RNA-protein interactions in the hnRNP particles. The C proteins are phosphorylated (20, 21), and among the hnRNP proteins they have the highest affinity for RNA. Beyer et al. (7) found that C proteins are more resistant to dissociation by salt than other core proteins of the hnRNP particles. After an extensive digestion of the hnRNP particles of colon carcinoma cells by ribonuclease, Augenlicht et al. (22) found short nuclease-resistant pieces of hnRNA firmly associated with a peptide which probably corresponded to the C protein. In UV cross-linking experiments the C proteins are more efficiently cross-linked than other "core" proteins, both to poly(A)$^{+}$ and poly(A)$^{-}$ hnRNA (21, 23). The simultaneous rapid loss of RNA and C proteins after the exposure of intact HeLa cell hnRNP particles to ribonuclease (24) is likewise in accord with a close association of RNA and C proteins in the particle. These findings are consistent with a model in which a RNA-C protein complex occupies a peripheral position in the hnRNP particle.

It is possible that the individual charge isomers of the "core" C protein differ in their interaction with RNA. There are indications that the number of charge isomers of the individual "core" proteins in the hnRNP particles could be dependent on the physiological state of the cell. In our experiments (25) we observed only one charge isomer of the C protein in the hnRNP particles isolated from rat liver 18 hrs after partial hepatectomy, whereas by 30 hours after partial hepatectomy, all charge isomers of the C protein were found. Unfortunately, at present the reason for the presence of different charge isomers of the "core" proteins in the hnRNP particles is not known. The individual charge isomers have not yet been isolated and characterized. The formation of the charge isomers by glycosylation of the core proteins has been proposed (33) but not demonstrated.

After nearly complete inhibition of RNA synthesis by actinomycin D, only the C protein was found in the diminished number of hnRNP particles which could still be recovered from the cell. This result together with the results published in the literature imply that the C protein might be not only the sole "core" protein directly involved in the RNA-protein interaction, but also possibly the sole "core" protein required for the formation of the hnRNP particles. In brine shrimp three charge isomers of only one major protein (Mr = 40,000) are present in the hnRNP particles (26). But if only one core protein could facilitate the structural arrangements of hnRNA required for its processing and transport, what is the role of the other core proteins? In Drosophila cells hnRNA made after a heat shock is associated only with one predominant protein. Since the precursors of mRNA synthesized after a heat shock do not contain introns, Pederson proposed (27) that the "core" proteins found in hnRNP particles in cells not subjected to a heat shock are needed for the formation of specific RNA configurations required for processing. The transcripts of RNA present in the hnRNP particles of the brine shrimp were not analyzed for the presence of introns.

STRUCTURAL BASIS FOR RNA-PROTEIN INTERACTIONS IN HnRNP PARTICLES

It is of interest that the studies of the formation of hnRNP particles in the brine shrimp indicate that this formation is not dependent on any structural feature of the RNA (26). It is also possible to reconstitute RNP particles from single stranded homopolymeric RNA and from the proteins of hnRNP particles of HeLa cells (28). The reconstituted particles do contain all the "core" proteins present in native complexes. A general affinity of hnRNP particle "core" proteins for single-stranded nucleic acids might be expected, because antibodies against a pure subset of single-stranded DNA binding proteins react with the "core" proteins of the hnRNP particles, indicating extensive structural homology between these two groups of proteins (29). Notwithstanding this general affinity of particle "core" proteins for single-stranded polynucleotides, there are experimental data suggesting that the association of RNA with proteins in the hnRNP particles is specific. Bayer and coworkers (30) concluded from electron microscope studies of the morphology of nascent ribonucleoproteins in Drosophila embryos that the formation of hnRNP particles is nonrandom and depends on the nucleotide sequence of the RNA. A selective protection in the hnRNP particles from digestion by nucleases of these sequences of hnRNA which are homologous to mRNA (31, 32) is likewise indicative of sequence specific binding of RNA to proteins in the hnRNP particles.

Since it appears that hnRNA also interacts with the "core" proteins in the hnRNP particles in a sequence specific manner, which properties of the "core" proteins could facilitate such specific association? The experiment in which hnRNP particles were reconstituted *in vitro* from particle proteins and homopolymeric RNA (28) has shown that nonspecific binding forces between "core" proteins and RNA are adequate for the formation of the particle scaffold required for RNA processing.

Simultaneously, it also implies that the functional groups responsible for this nonspecific binding must be separated from the functional groups which are involved in specific associations between the hnRNA and the proteins in the hnRNP particles. In this reconstitution experiment the hnRNA in the isolated particles was degraded by reversibly activated nucleases and replaced by homopolymeric RNA. From the results published by Augenlicht et al. (22) it would be expected that the stretches of the original endogeneous RNA bound to the "core" proteins were protected from enzymatic degradation and remained associated with these proteins, thus blocking the functional groups involved in specific RNA-protein associations.

The "core" proteins have a unique amino acid composition containing a high level of glycine and the unusual modified arginine: N^G,N^G-dimethylarginine. Because of the very high content of the glycyl residues (22 to 26 mol %) Fuchs and coworkers proposed that the "core" hnRNP particle proteins be called glycine-rich proteins (33). A high content of glycine is indicative of high beta-pleated sheet content in these proteins. Circular dichroism studies by LeStourgeon and coworkers have shown that the "core" proteins may contain up to 30% of beta-pleated sheet regions (13). The parts of the polypeptide chains of the "core" proteins which are in the beta-pleated sheet configuration fit into the minor groove of double-stranded RNA and could interact specifically with RNA base pairs by hydrogen bonds. This type of association of RNA and "core" proteins in the hnRNP particle has been proposed (13). However, the selective affinity of the "core" proteins of the hnRNP particles for single-stranded nucleic acids (28) and the exclusion of double-stranded stretches of hnRNA from the hnRNP particles (34) are a contradiction to this otherwise attractive model. Rather, since for reasons not yet known double-stranded RNA will not interact with the particle "core" proteins, the exclusion of double-stranded stretches of hnRNA from such interactions could be one of the factors determining the specificity of RNA-protein association in the hnRNP particle.

Another factor which could specify RNA-protein association is the formation of hydrogen bonds between particular amino acids and bases of the RNA. Side chains of three amino acids could discriminate by hydrogen bonds between bases on single-stranded RNA. Unpaired guanine possesses two donor groups in a suitable position to bind by two hydrogen bonds with glutamate or aspartate, and the guanidino group of arginine can form two hydrogen bonds with unpaired guanine or cytosine (9). Despite their charge, the involvement of aspartate and glutamate in specific recognition of guanine would be possible, if by steric arrangement the repulsion between the phosphate groups on the RNA and the charges on the aspartate or glutamate side chains would be overcome. For example, this could occur by interaction of the ionized side chain of a neighboring lysine with the phosphate group. The guanidino group of arginine has five donor groups available for the formation of hydrogen bonds. But the guanidium cation will also preferentially interact with the negatively charged phosphates of the RNA and will not be available for hydrogen bonding. It has been proposed (9) that the positive charge on the side chain of arginine could be neutralized by a negative charge of a carboxylate anion (for steric reasons preferentially of glutamate). The two oxygen atoms of this anion could form hydrogen bonds with two NH groups of arginine directing the electrostatic interaction and thus facilitating the formation of two hydrogen bonds either with guanine or with cytosine. But the charge on the guanidino group could be removed also by methylation, likewise making the guanidino group available to engage in formation of hydrogen bonds. The presence of N^G, N^G-dimethylarginine in the "core" proteins, the decline of the methylation of the hnRNP particle proteins after the inhibition of RNA synthesis, and the higher turnover of methyl groups on the C "core" protein give credibility to the proposal that one of the mechanisms which control RNA-protein interaction in the hnRNP particles is the methylation of the arginine residues on the particle proteins. The involvement of N^G, N^G-dimethylarginine in RNA-protein association is likewise supported by the presence of this modified amino acid in the ribosomal proteins (10).

It is apparent that the RNA-protein association in the hnRNP particles is a result of complex and various interactions. Coulombic interactions between the positively charged side chains of amino acids and the negatively charged phosphates groups of nucleic acids are dominant and an order of magnitude larger than the energy associated with the hydrogen bonds (35). They determine the binding equilibrium between RNA and proteins and are crucially influenced by changes in the phosphorylation of the binding proteins. The distribution of positive and negative charges on the proteins will also determine the ability of proteins to interact with RNA by formation of specific hydrogen bonds. The removal of a charge from the guanidino group by methylation will make this group not only available for the formation of specific hydrogen bonds with guanine and cytosine in single-stranded RNA, but will also significantly modify the end result of all these complex interactions. Therefore, this methylation could be of fundamental importance for the changing specificity of RNA-protein association in the hnRNP particles during RNA processing.

ACKNOWLEDGEMENT

Our experimental work described in this article was supported by grant H-393 from the R.A. Welch Foundation.

REFERENCES

1. Wold, B.J., W.H. Klein, B.R. Hough-Evans, R.J. Britten, and E.H. Davidson: Sea urchin embryo mRNA sequences expressed in the nuclear RNA of adult tissues. Cell 14:941, 1978.

2. Esumi, H., Y. Takahashi, T. Sekiya, S. Sato, S. Nagase, and T. Sugimura: Presence of albumin mRNA precursors in nuclei of analbuminemic rat liver lacking cytoplasmic albumin mRNA. Proc Natl Acad Sci U.S.A. 79:734, 1982.

3. Holoubek, V.: Nuclear ribonucleoproteins containing heterogeneous RNA. In Chromosomal Nonhistone Proteins, Vol. 4, p. 21, Hnilica L.S., ed., CRC Press, Boca Raton, Florida, 1984.

4. Pederson, T. and N.G. Davis: Messenger RNA processing and nuclear structure: Isolation of nuclear ribonucleoprotein particles containing beta-globin messenger RNA precursors. J Cell Biol 87:47, 1980.
5. Jeanteur, Ph.: Enzymatic activities associated with hnRNP. In The Cell Nucleus, Vol. 9, p. 145, Busch, H., ed., Academic Press, New York, 1981.
6. Ramagli, L.S. and V. Holoubek: Decrease in the phosphorylation of the proteins associated with heterogeneous nuclear RNA after the application of 3'-methyl-4-dimethylaminoazobenzene. Chem-Biol Interactions 52:51, 1984.
7. Beyer, A.L., M.E. Christensen, B.W. Walker, and W.M. LeStourgeon: Identification and characterization of the packaging proteins of core 40S hnRNP particles. Cell 11:127, 1977.
8. Karn, J., G. Vidali, L.C. Boffa, and V.G. Allfrey: Characterization of the non-histone nuclear proteins associated with rapidly labeled heterogeneous nuclear RNA. J Biol Chem 252:7307, 1977.
9. Hélène, C. and G. Lancelot: Interactions between functional groups in protein-nucleic acid associations. Prog Biophys Molec Biol 39:1, 1982.
10. Chang, F.N., I.J. Navickas, C.N. Chang, and B.M. Dancis: Methylation of ribosomal proteins in HeLa cells. Arch Biochem Biophys 172:627, 1976.
11. Peters, K.E. and D.E. Comings: Two-dimensional gel electrophoresis of rat liver nuclear washes, nuclear matrix and hnRNA proteins. J Cell Biol 86:135, 1980.
12. Patel, N.T. and V. Holoubek: Dependence of the composition of the protein moiety of nuclear ribonucleoprotein particles on the extent of particle purification as studied by electrophoresis including a two-dimensional procedure. Biochim Biophys Acta 474:524, 1977.
13. LeStourgeon, W.M., A.L. Beyer, M.E. Christensen, B.W. Walker, S.M. Poupore, and L.P. Daniels: The packaging proteins of core hnRNP particles and the maintenance of proliferative cell states. Cold Spring Harbor Symp Quant Biol 42:885, 1977.
14. Martin, T., R. Jones, and P. Billings: HnRNP core proteins: Synthesis, turnover and intracellular distribution. Molec Biol Rep 5:37, 1979.
15. Liew, C.C. and D. Suria: Assessment of methylation in nonhistone chromosomal proteins. In Methods in Cell Biology, Vol. 19, p. 89, Stein, G., J. Stein, and L.J. Kleinsmith, eds., Academic Press, New York, 1978.
16. Upreti, R.K. and V. Holoubek: Methylation of proteins of the nuclear ribonucleoprotein particles in liver and kidney of rats injected with actinomycin D. Biochimie 64:435, 1982.
17. Naora, H.: Some aspects of double-stranded hairpin structures in heterogeneous nuclear RNA. Int Rev Cytol 56:255, 1979.
18. Chinsky, L. and P.Y. Turpin: Ultraviolet resonance Raman study of DNA and of its interaction with actinomycin D. Nuclei Acids Res 5:2969, 1978.

19. Upreti, R.K. and V. Holoubek: The effect of inhibition of RNA synthesis by actinomycin D on the population of basic polypeptides of the 30S nuclear ribonucleoprotein particles. Biochimie 64:247, 1982.
20. Holcomb, E.R. and D.L. Friedman: Phosphorylation of C-proteins of HeLa cell hnRNP particles. J Biol Chem 259:31, 1984.
21. Dreyfuss, G., Y.D. Choi, and S.A. Adam: Characterization of heterogeneous nuclear RNA-protein complexes in vivo with monoclonal antibodies. Mol Cell Biol 4:1104, 1984.
22. Augenlicht, L.H., M. McCormick, and M. Lipkin: Digestion of RNA of chromatin and nuclear ribonucleoprotein by staphylococcal nuclease. Biochemistry 15:3818, 1976.
23. van Eekelen, C.A.G. and W.J. van Venrooij: HnRNA and its attachment to a nuclear protein matrix. J Cell Biol 88:554, 1981.
24. Lothstein, L., H.P. Arenstorf, S-Y. Chung, B.W. Walker, J.C. Wooley, and W.M. LeStourgeon: General organization of protein in HeLa 40S nuclear ribonucleoprotein particles. J Cell Biol 100:1570, 1985.
25. Upreti, R.K. and V. Holoubek: The effect of partial hepatectomy and chronic alcohol administration on the population of basic polypeptides of rat liver nuclear ribonucleoprotein monoparticles. Pharmacol Res Commun 14:689, 1982.
26. Thomas, J.O., S.K. Glowacka, and W. Szer: Structure of complexes between a major protein of heterogeneous nuclear ribonucleoprotein particles and polyribonucleotides. J Biol Chem 171:439, 1983.
27. Pederson, T.: Nuclear RNA-protein interactions and messenger RNA processing. J Cell Biol 97:1321, 1983.
28. Wilk, H-E., G. Angeli, and K.P. Schäfer: In vitro reconstitution of 35S ribonucleoprotein complexes. Biochemistry 22:4592, 1983.
29. Valentini, O., G. Biamonti, M. Pandolfo, C. Morandi, and S. Riva: Mammalian single-stranded DNA binding proteins and heterogeneous nuclear RNA proteins have common antigenic determinants. Nucleic Acids Res 13:337, 1985.
30. Beyer, A.L., O.L. Miller, Jr., and S.L. McKnight: Ribonucleoprotein structure in nascent hnRNA is nonrandom and sequence-dependent. Cell 20:75, 1980.
31. Munroe, S.H. and T. Pederson: Messenger RNA sequences in nuclear ribonucleoprotein particles are complexed with proteins as shown by nuclease protection. J Mol Biol 147:437, 1981.
32. Ohlsson, R.I., C. van Eekelen, and L. Philipson: Non-random localization of ribonucleoprotein (RNP) structures within an adenovirus mRNA precursor. Nucleic Acids Res 10:3053, 1982.
33. Fuchs, J.-P., C. Judes, and M. Jacob: Characterization of glycine-rich proteins from ribonucleoproteins containing heterogeneous nuclear ribonucleic acid. Biochemistry 19:1087, 1980.
34. Calvet, J.P. and T. Pederson: Heterogeneous nuclear RNA double-stranded regions probed in living HeLa cells by crosslinking with psoralen derivative aminomethyltrioxsalen. Proc Natl Acad Sci U.S.A. 76:755, 1979.
35. Kumar, N.V. and G. Govil: Theoretical studies on protein-nucleic acid interactions II. Hydrogen bonding of amino acid side chains with bases and base pairs of nucleic acids. Biopolymers 23:1995, 1984.

13. cDNA Cloning and Structure-Function Relationships of a Mammalian Helix Destabilizing Protein: hnRNP Particle Core Protein A1

S. H. Wilson, F. Cobianchi, and H. R. Guy[a]

Laboratory of Biochemistry and [a]Laboratory of Mathematical Biology, National Cancer Institute, Bethesda, MD 20892

INTRODUCTION

Mammalian nucleic acid helix destabilizing proteins (HDPs) are a class of single-stranded nucleic acid binding proteins that are purified by virtue of their strong and selective binding to ssDNA-cellulose columns. Although purified HDPs are typified by the preparations from calf thymus and mouse myeloma (1,2), analogous proteins have been purified from Hela and a variety of other eukaryotic cell types. The calf and mouse proteins are purified as 24 to 27 KDa polypeptides that exist as monomers at salt concentrations in the physiological range (1,2). These proteins bind to ssDNA with much higher affinity than to dsDNA and are capable of depressing the melting temperature (Tm) of dsDNA (2-4). Binding to ssDNA is noncooperative; the binding site of each protein monomer spans 7 nucleotide residues, and a length of ssDNA can be almost fully covered by the protein, as protein-protein and protein-DNA interactions interfering with binding are weak (2). Binding affinity of the proteins to RNA and ssDNA is about the same, and the proteins can promote conformational changes in RNA secondary structure (3,4,5) equivalent to Tm depression of dsDNA. Hence, the proteins are properly considered as single-stranded nucleic acid binding proteinsor nucleic acid helix destabilizing proteins rather than as DNA binding proteins, exclusively (4-6).

Recent experiments involving antibody probing of Western blots of unfractionated extracts of mammalian cells have revealed that the native species of HDP is approximately 35,000-M_r (7,8). Since the purified protein is only 24,000 to 27,000-M_r, these results suggested proteolytic removal of a portion of the native HDP molecule during purification. Riva and coworkers further showed that HDP and the ~36,000-M_r 30S hnRNP particle protein, A1, are immunologically identical and that the typical 25,000-M_r HDP can be prepared from the purified A1 protein by <u>in vitro</u> proteolysis (8,9). Other slightly higher M_r HDPs (i.e. 38 to 40 KDa) had been purified earlier from 30S hnRNP particles of <u>Artemia salina</u> (10,11). Hence, HDPs are involved in mRNA metabolism as components of hnRNP particles, and their involvement in DNA metabolism as well is not excluded.

It has been widely reported that mammalian HDP stimulates the activity of purified DNA polymerase α, the replicative DNA polymerase of eukaryotic cells. This effect is probably through interaction with the template·primer rather than exclusively with the enzyme, as HDP levels stoichiometric with the template and vastly exceeding the enzyme are required (6,12,13). Yet, there is enzyme and protein specificity involved in the stimulation, since other types of DNA polymerase are not stimulated and other ssDNA binding proteins cannot substitute for the mammalian HDP (6). A direct involvement of HDP in DNA metabolism has not yet been demonstrated, however.

In this article, we discuss the isolation and characterization of a full-length cDNA for rat HDP. Sequencing of this cDNA (21) along with sequencing of the corresponding protein isolated from calf thymus (14,15) have now yielded new information about structure-function relationships of this protein and of its mRNA. Among other points, it is now apparent that the protein deduced from the cDNA is equivalent to the A1 protein of the ~30S heterogeneous nuclear ribonucleoprotein particle (23). The availability of this HDP cDNA will facilitate various additional studies of the role of HDP and the cloning of cDNAs for related proteins, such as other members of the 30S hnRNP particle core protein family (18).

RESULTS

Since HDPs of calf thymus and mouse myeloma are structurally related, we assumed that the primary structure of the calf protein could be the basis for cDNA cloning from a rodent library. The calf HDP is termed unwinding protein 1 (UP1) for historical reasons. Two sets of oligonucleotide probes corresponding to pentapeptide sequences in either the N-terminal or C-terminal portion of calf UP1 were used to probe a newborn rat brain cDNA library in λgt11. Six positive phage were identified and cloned by plaque purification. These clones next were examined for hybridization with the C-terminal probe. Only four were found to be positive, and one of the clones positive with both probes was chosen for detailed study. The cDNA insert in this clone, termed λHDP-182, was 300 to 400 bp longer than those of the other three clones positive with both probes.

Primary Structure of Rat HDP cDNA λHDP-182

The complete nucleotide sequence of the λHDP-182 cDNA is shown in Figure 1. The reading frame and translation start codon of the corresponding mRNA were assigned by comparison with the amino acid sequence of calf UP1 (14). The start codon is the first AUG from the 5' end corresponding to residues 29, 30 and 31. The triplet adjacent to the AUG encodes a serine residue, and it is noted that a blocked N-terminal serine residue is found in purified UP1 (14). The distance from the AUG start codon to the first in-phase termination codon is 988 bases. From this termination codon to the start of the poly A sequence at the 3' end there are 12 in-phase termination codons.

Analysis of the other two reading frames of the entire insert showed 38 and 24 termination codons, respectively, throughout the insert with a maximum open reading frame of 294 bases. The 3' end untranslated region of the HDP mRNA contains 718 bases, including an AAUAAA signal 21 bases before the poly A sequence.

The long open reading frame of the cDNA predicts a protein of 34,215 daltons. Perhaps surprisingly, amino acid residues 2 through 196 of this deduced protein are identical with the 195 amino acid sequence of purified calf UP1, indicating remarkable sequence conservation for this portion of the protein and verifying the identity of the cDNA. The 124 amino acid sequence of the deduced protein that is not present in purified calf UP1 (i.e. residues 197-320) has an unusual amino acid composition, in that it is 40% glycine, 15% serine, and 11% asparagine. The sequence contains 3 proline residues, only four acidic residues and is cysteine free.

The HDP cDNA Insert is Full-Length

Primer extension analysis of newborn rat brain polyA$^+$ RNA is shown in Figure 2. The primer was one strand of a 58 bp HhaI-HpaII fragment corresponding to residues 1 to 58 at the 5' end of the cDNA insert. Incubation conditions for AMV reverse transcriptase were adjusted to eliminate secondary structure in the template RNA. Only one band of elongated primer was observed. This corresponded to addition of about 35 bases to the primer, indicating that the mRNA contained only 35 residues at the 5' end that were not copied into the HDP-182 cDNA. The sequence of the first 31 nucleotides of the mRNA upstream from the start of the cDNA is shown in Figure 2, also. This portion of the mRNA does not contain an in-phase AUG codon, indicating that the AUG triplet 28 bases from the 5' end of the cDNA is the first in-phase start codon.

Northern blot analysis of newborn rat brain poly A$^+$ RNA was conducted using as probe a fragment of the λHDP-182 containing the first 1397 bp. After only 1 hour exposure of the autoradiogram a strong predominant signal was observed corresponding to approximately 2100 bases. A higher M_r species of ~4500 bases was noted and a smaller species of ~1600 bases was noted also. Similar results were obtained with total RNA, instead of poly A$^+$ RNA, and when the denaturation of RNA was performed in the presence of glyoxal, instead of formaldehyde and formamide (data not shown).

Based upon the length of the cDNA insert, the 5' sequence detected by primer extension analysis, and an assumed polyA tail of 200 residues, the overall length of the mRNA would be 1,927 bases. This number is consistent with the electrophoretic migration of the predominant mRNA species hybridizing with the probe. We conclude that the λHDP-182 cDNA insert is full-length with regard to the coding sequence and is nearly full-length overall.

Structure-Function Relationships of the HDP mRNA

The HDP mRNA has interesting structural features in both the 5' and 3' untranslated regions. The overall structure of the cDNA is summarized in Figure 3.

```
                10          20           30          40          50          60
  cgctgaacgctctcatcatcctaccgtc ATG TCT AAG TCA GAG TCC CCC AAG GAA CCG GAA CAG
                               Met Ser Lys Ser Glu Ser Pro Lys Glu Pro Glu Gln  12
    70              80              90             100             110             120
CTG CGG AAG CTC TTC ATT GGA GGG CTG AGC TTC GAA ACA ACC GAC GAG AGT CTG AGG AGC
Leu Arg Lys Leu Phe Ile Gly Gly Leu Ser Phe Glu Thr Thr Asp Glu Ser Leu Arg Ser  32
    130             140             150             160             170             180
CAT TTT GAG CAA TGG GGA ACA CTC ACC GAC TGT GTG GTA ATG AGA GAT CCA AAC ACC AAA
His Phe Glu Gln Trp Gly Thr Leu Thr Asp Cys Val Val Met Arg Asp Pro Asn Thr Lys  52
    190             200             210             220             230             240
AGA TCC AGA GGC TTT GGG TTT GTC ACA TAT GCC ACT GTG GAG GAA GTG GAT GCT GCC ATG
Arg Ser Arg Gly Phe Gly Phe Val Thr Tyr Ala Thr Val Glu Glu Val Asp Ala Ala Met  72
    250             260             270             280             290             300
AAT GCA AGA CCA CAC AAA GTG GAT GGA AGA GTT GTG GAA CCT AAG AGA GCT GTG TCA AGA
Asn Ala Arg Pro His Lys Val Asp Gly Arg Val Val Glu Pro Lys Arg Ala Val Ser Arg  92
    310             320             330             340             350             360
GAA GAT TCT CAG AGA CCA GGT GCC CAC TTA ACT GTG AAG AAG ATC TTT GTT GGC GGT ATT
Glu Asp Ser Gln Arg Pro Gly Ala His Leu Thr Val Lys Lys Ile Phe Val Gly Gly Ile 112
    370             380             390             400             410             420
AAA GAA GAC ACT GAA GAA CAT CAC CTA CGA GAT TAT TTT GAG CAG TAT GGG AAA ATT GAA
lys Glu Asp Thr Glu Glu His His Leu Arg Asp Tyr Phe Glu Gln Tyr Gly Lys Ile Glu 132
    430             440             450             460             470             480
GTG ATT GAA ATT ATG ACT GAC AGA GGC AGT GGA AAA AAG AGG GGA TTT GCG TTT GTC ACC
Val Ile Glu Ile Met Thr Asp Arg Gly Ser Gly Lys Lys Arg Gly Phe Ala Phe Val Thr 152
    490             500             510             520             530             540
TTT GAT GAC CAT GAC TCT GTG GAT AAG ATT GTT ATT CAG AAA TAC CAT ACT GTG AAT GGC
Phe Asp Asp His Asp Ser Val Asp Lys Ile Val Ile Gln Lys Tyr His Thr Val Asn Gly 172
    550             560             570             580             590             600
CAC AAC TGT GAA GTA AGA AAG GCT CTG TCG AAA CAA GAG ATG GCT AGT GCT TCA TCC AGC
His Asn Cys Glu Val Arg Lys Ala Leu Ser Lys Gln Glu Met Ala Ser Ala Ser Ser Ser 192
    610             620             630             640             650             660
CAG AGA GGT CGA AGT GGT TCC GGA AAC TTT GGT GGT GGT CGT GGA GGT GGT TTC GGT GGC
Gln Arg Gly Arg*Ser Gly Ser Gly Asn Phe Gly Gly Gly Arg Gly Gly Gly Phe Gly Gly 212
    670             680             690             700             710             720
AAT GAC AAT TTT GGT CGA GGA GGG AAC TTC AGT GGT CGT GGT GGC TTT GGT GGC AGC CGT
Asn Asp Asn Phe Gly Arg Gly Gly Asn Phe Ser Gly Arg Gly Gly Phe Gly Gly Ser Arg 232
    730             740             750             760             770             780
GGT GGT GGT GGA TAT GGT GGC AGT GGG GAT GGC TAT AAT GGA TTT GGC AAT GAT GGA AGC
Gly Gly Gly Gly Tyr Gly Gly Ser Gly Asp Gly Tyr Asn Gly Phe Gly Asn Asp Gly Ser 252
    790             800             810             820             830             840
AAT TTT GGA GGT GGT GGA AGC TAC AAT GAT TTT GGC AAT TAC AAC AAC CAG TCA TCA AAT
Asn Phe Gly Gly Gly Gly Ser Tyr Asn Asp Phe Gly Asn Tyr Asn Asn Gln Ser Ser Asn 272
    850             860             870             880             890             900
TTT GGA CCG ATG AAA GGA GGA AAC TTT GGA GGC AGG AGC TCT GGC CCT TAT GGT GGT GGA
Phe Gly Pro Met Lys Gly Gly Asn Phe Gly Gly Arg Ser Ser Gly Pro Tyr Gly Gly Gly 292
    910             920             930             940             950             960
GGC CAG TAC TTT GCT AAA CCA CGA AAC CAA GGT GGC TAT GGA GGT TCC AGC AGC AGC AGT
Gly Gln Tyr Phe Ala Lys Pro Arg Asn Gln Gly Gly Tyr Gly Gly Ser Ser Ser Ser Ser 312
    970             980             990        1000       1010       1020       1030
AGC TAT GGC AGT GGC AGG AGG TTC TAA ttacagccaggaaacaaagcttagcaggagagccagagaagtg
Ser Tyr Gly Ser Gly Arg Arg Phe ***
    1040       1050       1060       1070       1080       1090       1100       1110
acagggaagctacaggttacaacagatttgtgaactcagccaagcacagtggtggcagggcctagctgctacaaagaag
    1120       1130       1140       1150       1160       1170       1180       119
acatgttttagacaatactcatgtgtgtgggcaaaaactccaggactgtatctgtgactaattgtataacaggttattt
0       1200       1210       1220       1230       1240       1250       1260       12
tagtttctgttctgtggaaagtgtaaagcattccaacgaagggttttactgtagaccttttcacccatgctgttgatt
70       1280       1290       1300       1310       1320       1330       1340       1
gctaaatgtaaaagtctgatcatgacgctgaataaatgtgtcttttttttttttttttaatgtgctgtgtaaagttagt
350       1360       1370       1380       1390       1400       1410       1420
ctactctgaagccatcttggtaaacttccccaacagtgtgaagttagaattccttcagggtggtgccaaattccatttg
1430       1440       1450       1460       1470       1480       1490       1500
gaatttatttatggttgcttgggtggagaagccattgtcttcaaaaaccttggtgttgctaaactgccagttactgttg
 1510       1520       1530       1540       1550       1560       1570       1580
taacttaatgagtttcaccatttaaagggtcatccaagcaagatcacaatttggttataaaatggttgttggacaccta
  1590       1600       1610       1620       1630       1640       1650       1660
tgaagagaaaattgaataacggtctcagataaaataagagatgggaatgaagcttgtgtatcatccattatcatgtgta
   1670       1680       1690       1700
ctcaataaacgatttaattctcttgaatggaaaaaaaaaa
```

Figure 1. The nucleotide sequence of the λHDP-182 cDNA insert, and the deduced sequence of the 320 amino acid 34,215 dalton protein corresponding to the single long open reading frame. The coding strand is shown. The first amino acid residue in the deduced protein corresponds to the first AUG from the 5' end; the last amino acid residue is adjacent to the first in-phase termination codon. Residues 2 through 196 of the deduced protein are identical to the reported sequence of calf thymus UP1, the end of which is marked (*). In the 3' untranslated region a putative mRNA processing signal (AATAAA) and repeated sequences around a T sequence are noted. In the 5' untranslated region, a sequence complementary to a sequence in 18S rRNA is noted.

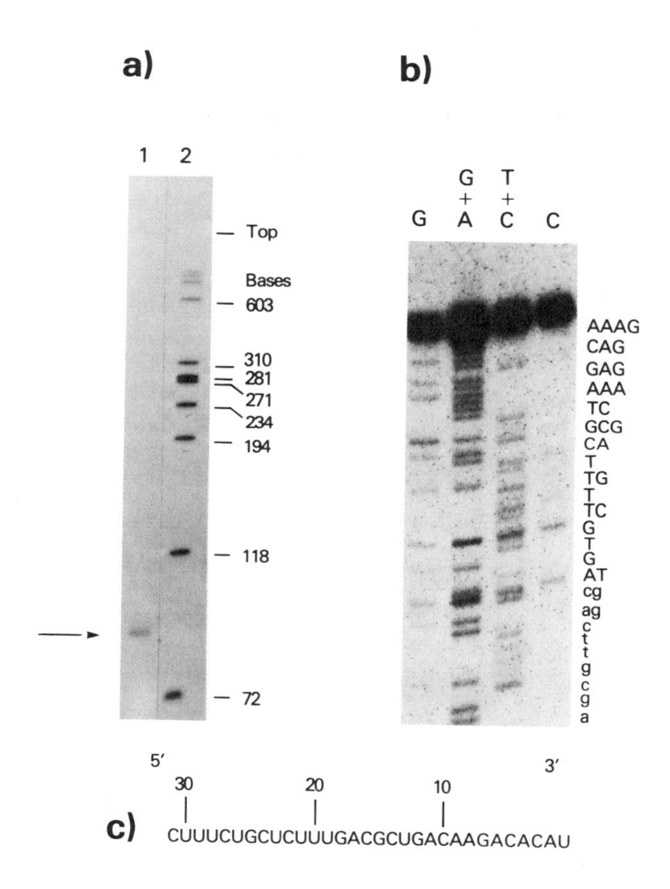

Figure 2. Autoradiogram showing results of primer extension analysis of newborn rat brain poly $A^{+}RNA$ (21). An HhaI-HpaII fragment of the λHDP-182 cDNA insert corresponding to residues 1 through 58 was gel purified, labeled at the 5' end with ^{32}P, and then annealed to RNA. Reverse transcriptase reaction products eventually were electrophoresed in a 8% polyacrylamide gel under denaturing conditions.
Panel a: Lanes 1 and 2, reaction products and molecular weight markers, respectively. The arrow marks the band due to elongated primer. Exposure was for 1 h. The primer migrated as a sharp band corresponding to 58 bases (not shown).
Panel b: Autoradiogram of a DNA sequencing gel showing a portion of the sequence complementary to the 5' end of HDP mRNA. Small and capital letters indicate the nucleotide sequence of the primer and extended product, respectively.
Panel c: Nucleotide sequence of the 5' end of the HDP mRNA not present in the HDP-182 cDNA. Numbering is from the 3' to 5' end.

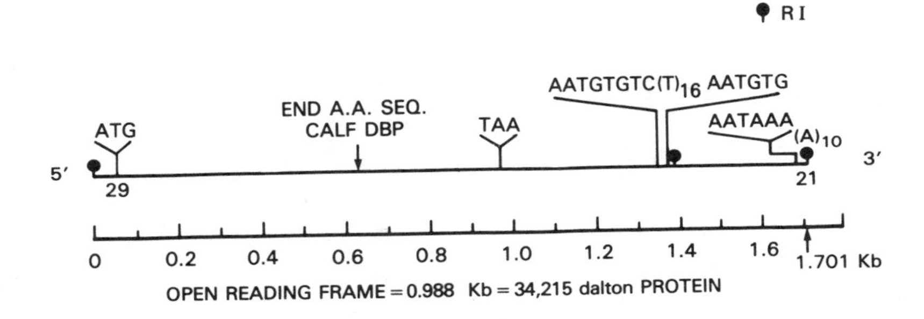

Figure 3. Diagram illustrating features of the primary structure of the HDP cDNA. The strand corresponding to the mRNA is shown. The long open reading frame discussed in the text is between the start (ATG) and stop (TAA) codons shown. The single internal EcoRI site is shown near the poly T sequence in the 3' untranslated region.

The 5' untranslated region contains an 11 residue sequence that is complementary to a sequence in the 3' end of 18S rRNA (Figure 4a). This sequence, which corresponds to residues 18 to 26 of the HDP cDNA, may be involved in control of translation. The 3' untranslated region (see Figure 3) contains a 16 residue poly T sequence that is flanked by oligonucleotide repeats; the primary structure of this region is illustrated in Figure 4b. Since the poly T sequence is flanked by direct repeats, the overall sequence in Figure 4b has similarity to a transposable element.

Computer search for rodent sequences homologous to the HDP mRNA revealed a number of partial, albeit significant, homologies. None of the homologous sequences identified formed a perfect match with the HDP cDNA, but all were significant at the level of 5 or more standard deviation units above mean. With the exception of two cases, all of the homologies corresponded to the 3' untranslated region of the HDP mRNA (Figure 5). The 100 residue region including and surrounding the T_{16} sequence shared homology with the cellular fos gene, a mobile element 5' of rRNA genes, and three other genes of no obvious interrelationship. Similarly, two other relatively discrete segments corresponding to the 3' untranslated region share homology with 4 other genes, and none of these have an obvious interrelationship.

Also shown in Figure 5 is that a 200 nucleotide segment of the coding region of the HDP mRNA shares partial homology with the mRNA for keratin intermediate filament subunits I and II. This segment of the cDNA falls near the end of the open reading frame and corresponds to the C-terminal portion of the deduced protein (see Table III).

Structure-Function Relationships of the Deduced Protein

Sequencing of a full-length cDNA for the mammalian nucleic acid HDP has yielded information on the structure of this protein not yet available from work with the purified protein, itself. The open reading frame of the cDNA, for example, indicated that the mRNA encodes a 34,215 dalton polypeptide of 320 amino acids. Computer-derived secondary structure plots of the 34,215 dalton HDP clearly indicate two major structural domains and a sharp transition between these domains occurring between amino acid residues 190 and 195 (Figure 6). The N-terminal domain of the 34,215 dalton protein appears to be tightly structured with prediction of 4 α-helix regions, 7 β-sheet regions, and alternating hydrophobic and hydrophilic regions. As noted above, the primary structure of this domain is identical between the rat cDNA-derived protein and the protein purified from calf thymus (UP1), indicating pressure for sequence conservation. The C-terminal domain of the 34,215 dalton protein has about 125 amino acids and is structurally flexible due to the high content of amino acid residues with strong β-turn and random coil prediction (16,17). This domain does not have regions of α-helix prediction and is slightly hydrophilic throughout (Figure 6). The high content of serine residues provides ample sites for phosphorylation; it is noted that a well known property of helix destabilizing protein isolated from hnRNP particles is that the protein is phosphorylated (10).

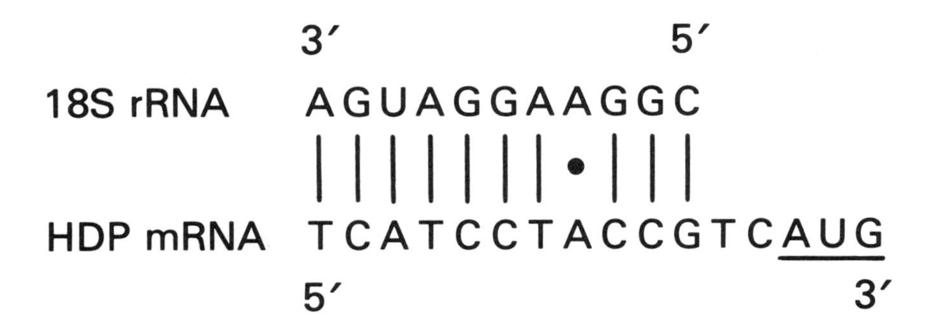

Figure 4: Diagrams illustrating sequences in the 3' and 5' untranslated regions of the HDP mRNA.
Panel a: Sequence homology between the 3' and of the 18S rRNA and a sequence in the 5' untranslated region of the HDP in RNA.
Panel b: Summary of direct and inverted repeats surrounding the poly T sequence in the HDP cDNA. Numbering corresponds to the HDP cDNA sequence in Figure 1.

Analysis of the primary structure of the C-terminal domain of the deduced protein revealed that it contains 16 repeating oligopeptide units (Table I) beginning with amino acid residue 202. These units consist of phenylalanine or tyrosine residues separated by 5 to 8 amino acid residues of likely random coil structure; seven of these repeating units contain an arginine residue.

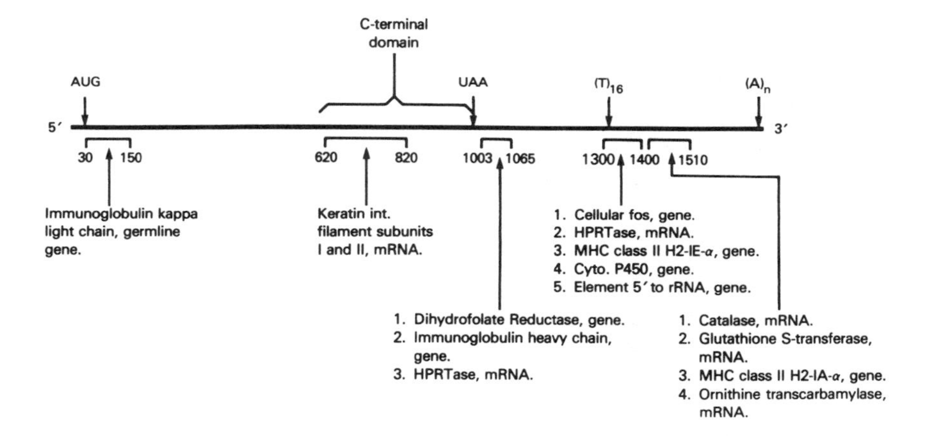

Figure 5. Scheme summarizing partial nucleotide sequence homologies between the HDP cDNA and sequences in the GeneBank Data Bank as of 8-1-85. The computer searches were as described in Table III. Numbering corresponds to the HDP cDNA sequence in Figure 1. All homologies illustrated were significant at the level of 5 or more standard deviation units above mean. The SRCHN program of D.J. Lipman and W.J. Wylbur was used with K-tuple size = 4, window size = 20 and gap penalty = 4.

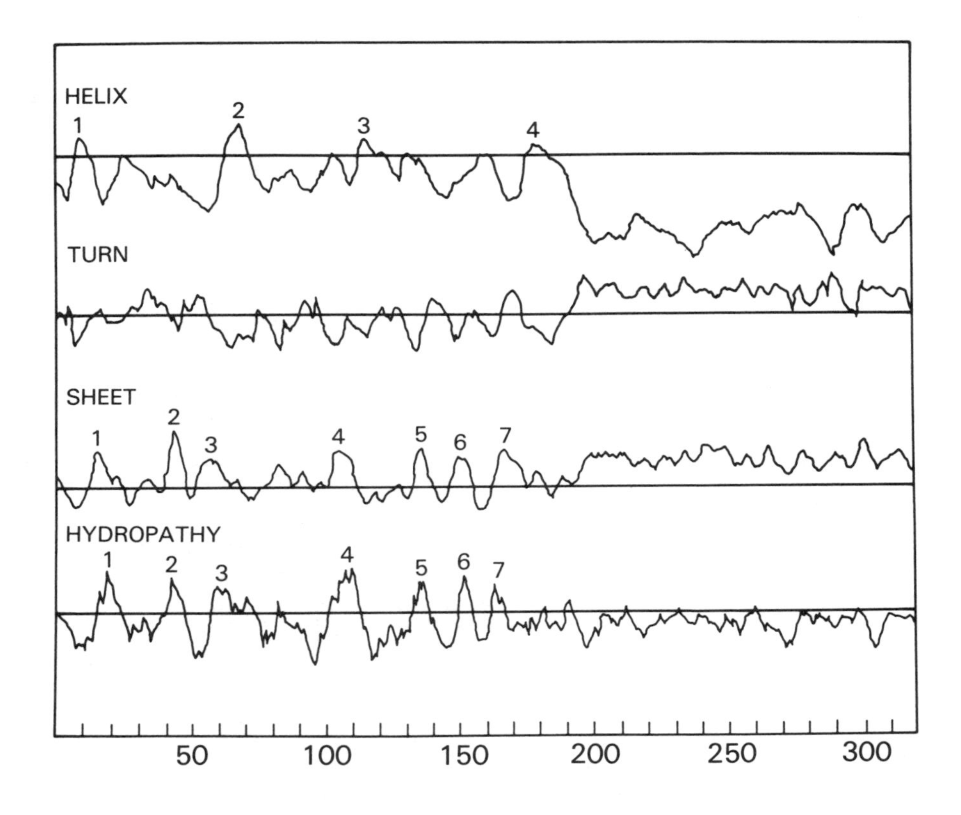

Figure 6. Computer derived secondary structure and hydropathy plots for rat helix destabilizing protein (21). Amino acid residue numbers are displayed relative to colinear secondary structure predictions and a hydropathy plot. For predictions of conformation, all values plotted above the line indicate a prediction for that conformational feature. All values above the line for the hydropathy plot indicate hydrophobic character, whereas values below the line indicate hydrophilic character of the polypeptide. Protein secondary structure prediction was based on information described by Garnier et al. (16).

This pattern of repeating units in the peptide sequence is suggestive of an evolutionary history of gene duplications with subsequent deletions and changes, as illustrated by the alignment in the right-hand column of Table I. The suggestion of gene duplications is evident also from inspection of the corresponding nucleotide sequence. The oligonucleotide CAATTTTGG is present twice, and the related sequences AAACTTTGG and ATTTTGG are present two times each as direct repeats.

The sequence ATTTTGG and a second closely related sequence are present once each as inverted repeats.

The functional significance of the unique structural domain in the HDP remains to be determined. It is interesting to note, however, that the largest subunit of both yeast and mouse RNA polymerase II contains a C-terminal domain with repeating heptapeptide units each with one aromatic residue (Table II). The primary structure of the consensus sequence of the repeats in the RNA polymerase subunits is much more represented among the various repeats than in the case of the consensus sequence for the HDP repeats. A common feature of the repeating structures with each protein is the periodicity of the aromatic residues. This and other structural features of the repeats provide interesting possibilities for binding to nucleic acids. A plausible conformation for the RNA polymerase repeating units, and the close interaction of this conformation with a single-stranded nucleic acid are illustrated in Figure 7. Each seven residue unit has two highly predicted (16,17) β turns leading to a helical configuration. The aromatic tyrosine residue in each unit is on the inside of the helix and would lie very close to the nucleic acid bases; thus, this structure and interaction with nucleic acid explains the periodicity of aromatic residues in the protein. This close fit would be disrupted by phosphorylation of serines, threonines, or tyrosines due to steric hinderance and electrostatic repulsion. The protein helix shown in Figure 7 has the same pitch and number of repeats per turn as A-DNA or RNA-DNA hybrids, although the protein could form helices with other pitches. The periodicity of aromatic residues and predictions of β turns in the C-terminal domain of HDP indicate that this protein may bind to nucleic acids in a similar manner. Unfortunately, the HDP repeating units are difficult to model because they are much less conserved than those of RNA polymerase and they contain many glycines, which should give the structure a great deal of conformational freedom. The possibility that these C-terminal domain repeating structures could bind specifically to certain types of DNA or RNA structures can be evaluated experimentally after appropriate subcloning in expression vectors and isolation of large amounts of the relevant proteins.

Finally, routine computer search of nucleic acid sequences related to the sequence for the 34,215 dalton HDP revealed only one partial match. A portion of the HDP cDNA (residues 604 to 792) corresponding to the C-terminal domain of HDP matches with a sequence in the cDNA for mouse epidermal keratin subunit I; excluding five apparent deletions or insertions, 92 of 128 residues are homologous (Table III).

DISCUSSION

Heterogeneous nuclear RNA (hnRNA) in eukaryotic cells is synthesized on chromatin and shortly thereafter associates with a discrete set of polypeptides to form an RNA-protein complex known as the heterogeneous nuclear ribonucleoprotein (hnRNP) fibril (for discussion see ref. 18). This structure seen in an electron micrograph resembles a linear fiber with small beads interspersed along the fiber from one end to the other. The small beads correspond to RNP particles, and under

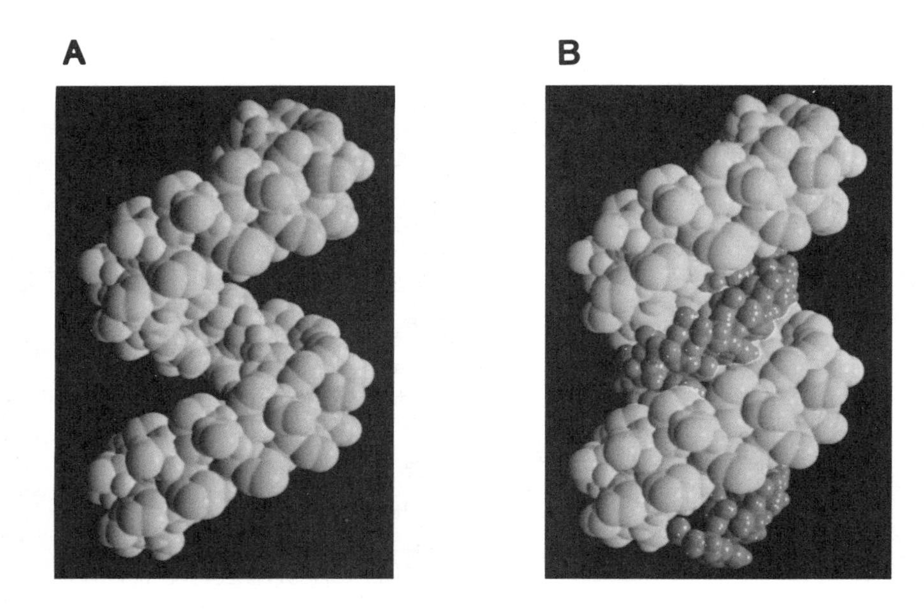

Figure 7. Computer-derived model of hypothetical structure of RNA polymerase II C-terminal domain and its interaction with a nucleic acid. (A) A protein composed of 17 repeats of the concensus sequence (Ser-Pro-Thr-Ser-Pro-Ser-Tyr) was given a conformation that forms a helix with 11 repeats per turn and 2.6 translation per repeat, which is the same as A-DNA or DNA-RNA hybrids. Tyr groups are oriented toward the center of the helix. (B) A single strand of A-DNA (yellow) is shown interacting with the RNA polymerase domain. The phosphate-sugar backbone is visable. The bases are covered by the protein and interact with the Tyr groups. It is possible that the protein conformation could be altered to interact with other nucleic acid structures such as that found for single stranded polycytidylic acid (26) or double-stranded A-DNA or DNA-RNA hybrids.

appropriate conditions these particles are released from the fibril and readily extracted from nuclei. The particles are characterized by their sedimentation rate in a sucrose gradient (~30S) and content of a set of 6 to 10 relatively abundant polypeptides. These so-called core polypeptides are 32,000- to 40,000-M_r, moderately basic, and have similar amino acid compositions including high glycine and low cysteine content and the unusual amino acid dimethylarginine. The proteins are immunologically cross-reactive and are thought to represent a family of polypeptides related through a certain extent of primary structure homology. These polypeptides all have blocked N-termini and contain a domain of ~12,000 daltons that is removed by mild proteolytic digestion (18). In the HeLa cell system, 7 major hnRNP core polypeptides are found and are termed A1, A2, B1, B2, C1, C2 and C3, respectively (23).

The relationship of the cDNA described here and the classical ~25,000-M_r HDP is clear from the precise match of the amino acid and nucleic acid sequences, and as noted above, a relationship between hnRNP particle proteins and the classical 25,000-M_r HDP has been indicated by immunological and by proteolytic digestion experiments. The individual hnRNP particle core protein actually giving rise to the 25,000-M_r HDP was unclear. Recently, Williams and Szer and their associates (23) determined the partial amino acid sequence for 30S hnRNP particle core protein A1 from HeLa cells and found that the sequence of this protein precisely matches the sequence of the HDP cDNA described here (23). Therefore, the λHDP-182 cDNA corresponds to the 30S hnRNP particle core protein A1. Since this protein appears to have functional nucleic acid binding capacity in both its N-terminal and C-terminal domains and can bind both mRNA and to DNA, the protein may be involved in regulation of transcription. A similar role has been proposed for the 40,000 dalton protein associated with the 5S RNA containing RNP particles of Xenopus laevis (24).

It will be interesting to investigate nucleic acid binding properties of the full-length 34,215 dalton HDP and contrast these with properties of the truncated proteins of 24 to 27 KDa. Nucleic acid binding properties such as Tm depression, cooperativity, binding site size and nucleotide specificity have not been determined for this protein and could be different from properties of the truncated proteins. It will be of particular interest to determine if the C-terminal domain alters or modulates binding properties, as in the case of the prototype ssDNA binding protein, gene 32 protein of T4 bacteriophage (22). As noted above, expression of the λHDP-182 cDNA using appropriate vector systems should facilitate such work by making available large amounts of the 34,215 dalton protein and of the N- and C-terminal domains.

REFERENCES

1. Herrick, G. and B. Alberts: Purification and physical characterization of nucleic acid helix-unwinding proteins from calf thymus. J Biol Chem 251:2124, 1976.
2. Planck, S.R. and S.H. Wilson: Studies on the structure of mouse helix-destabilizing protein-1. J Biol Chem 255:11547, 1980.
3. Herrick, G. and B. Alberts: Nucleic acid helix-coil transitions mediated by helix-unwinding proteins from calf thymus. J Biol Chem 251:2133, 1976.
4. Karpel, R.L. and A.C. Burchard: Physical studies of the interaction of a calf thymus helix-destabilizing protein with nucleic acids. Biochemistry 19:4674, 1980.
5. Karpel, R.L., W.S. Miller, and J.R. Fresco: Mechanistic studies of RNA denaturation by a helix destabilizing protein. Biochemistry 21:2102, 1982.
6. Omar, V., G. Biamonti, G. Mastromei and S. Riva: Structural and functional heterogeneity of single-stranded DNA-binding proteins from calf thymus. Biochim Biophys Acta 782:147, 1984.
7. Planck, S.R. and S.H. Wilson: Native species of helix destabilizing protein-1 in mouse myeloma identified by antibody probing of western blots. Biochem Biophys Res Comm 131:362, 1985.
8. Valentini, O., G. Biamonti, M. Pandolfo, C. Morandi, and S. Riva: Mammalian single-stranded DNA binding proteins and heterogeneous nuclear RNA proteins have common antigenic determinants. Nucleic Acids Res 13:337, 1985.
9. Pandolfo, M., O. Valentini, G. Biamonti, C. Morandi, and S. Riva: Single stranded DNA binding proteins derive from hnRNP proteins by proteolysis in mammalian cells. Nucleic Acids Res 13:6577, 1985.
10. DeHerdt, E., C. Thoen, L. Van Hove, E. Roggen, E. Piot, and H. Slegers: Identification and properties of the 38000-M_r poly (A)-binding protein of non-polysomal messenger ribonucleoproteins of cryptobiotic gastrulae of Artemia salina. Eur J Biochem 139:155, 1984.
11. Thomas, J.O., S.K. Glowacka, and W. Szer: Structure of complexes between a major protein of heterogeneous nuclear ribonucleoprotein particles and polyribonucleotides. J Mol Biol 171: 439, 1983.
12. Herrick, G., H. Delius, and B. Alberts: Single-stranded DNA structure and DNA polymerase activity in the presence of nucleic acid helix-unwinding proteins from calf thymus. J Biol Chem 251: 2142, 1976.
13. Detera, S.D., S.P. Becerra, J.A. Swack, and S.H. Wilson: Studies

on the mechanism of DNA polymerase α. J Biol Chem 256:6933, 1981.

14. Williams, K.R., K.L. Stone, M.B. LoPresti, B.M. Merrill, and S.R. Planck: Amino acid sequence of the UP1 calf thymus helix-destabilizing protein and its homology to an analogous protein from mouse myeloma. Proc Natl Acad Sci 82:5666, 1985.
15. Merrill, B.M., M.B. LoPresti, K.L. Stone, and K.R. Williams: HPLC purification of UP1 and UP2, two related single-stranded nucleic acid binding proteins from calf thymus. J Biol Chem 260, in press, 1985.
16. Garnier, J.D., D.J. Osguthorpe, and B. Robson: Analysis of the accuracy and implications of simple methods for predicting the secondary structure of globular proteins. J Mol Biol 120:97, 1978.
17. Chou, P.Y. and G.P. Fasnan: Conformational parameters for amino acids in helical, β-sheet, and random coil regions calculated from proteins. Biochemistry 13, 211, 1974.
18. Leser, G.P., J. Escara-Wilke, and T.E. Martin: Monoclonal antibodies to heterogeneous nuclear RNA-protein complexes. J Biol Chem 259:1827, 1984.
19. Corden, J.L., D.L. Cadena, J.M. Ahearn, Jr., and M.E. Dahmus: A novel structure at the carboxy terminus of the largest subunit of eucaryotic RNA polymerase II. Proc Natl Acad Sci USA, in press.
20. Allison, L.A., M. Moyle, M. Shales, and C.J. Ingles: Extensive homology among the largest subunits of eukaryotic and prokaryotic RNA polymerases. Cell 42:599, 1985.
21. Cobianchi, F., D.N. SenGupta, B.Z. Zmudzka, and S.H. Wilson: Structure of rodent helix destabilizing protein revealed by cDNA cloning. J Biol Chem 261:3536, 1986.
22. Williams, K.R. and W.H. Konigsberg: Gene Amplication and Analysis (Chirikjian, J.G. and Papas, T.S., eds.) Vol. 2, pp. 475-508, Elsevier-North Holland, 1981.
23. Kumar, A., K. Williams and W. Szer: Purificatoin and domain structure of core hnRNP proteins, A1 and A2, and their relationship to ssDNA binding proteins. J. Biol. Chem., in press, 1986.
24. Miller, J., A.D. McLachlan and A. Klug: Repetitive zinc-binding domains in the protein transcription factor IIIA from Xenopus oocytes. The EMBO Journal 4: 1609, 1985.
25. Steinert, P.M., R.H. Rice, D.R. Roop, B.L. Trus, and A.C. Steven: Complete amino acid sequence of a mouse epidermal keratin subunit and implications for the structure of intermediate filaments. Nature 302:794, 1983.
26. Arnott, S., R. Chandrasekaran, and A.G.W. Leslie: Structure of the single-stranded polyribonucleotide polycytidylic acid. J Mol Biol 106:735-748, 1976.

TABLE I

Two primary structure alignments of oligopeptides in the C-terminal domain of the 34,215 dalton HDP based upon spacing of phenylalanine or tyrosine residues.

Alignment[a] Illustrating Repeating Units	Alignment[a] Illustrating Conservation of Sequences
↓	↓ ↓ ↓
FGGGRGGG	GNFGGGRG
FGGNDN	GGFGG
FGRGGN	DNF--GRG
FSGRGA	GNF--SRG
FGGSRGGGG	GAFGGSRGG
YGGSGDG	GGYGGS-GDGYNGFGN
FGNDGSN	SNFGG--GGSYNDFGN
FGGGGS	SNFGPNKG
YNDFGN	GNFGG-RSS
YNNQSSN	GPYGGG--GQY--F
FGPMKGGN	GGYGGS-SSS
FGGRSSGP	SSYGSGRR--F
YGGGGQ	
FAKPRNQGG	
YGGSSSSS	
YGSGRR	
Consensus sequence:	
FGGGGS	GNFGG(S/G)RG

[a] Alignment of phenylalanine (F) or tyrosine (Y) residues is illustrated by arrows. The oligopeptides are listed in the order of their occurrance in the C-terminal domain starting with F 202 and extending to the C-terminus F.

TABLE II

Summary of C-Terminal Domain Oligopeptide Repeats In Nucleic Acid Proteins

Protein	Number and Length of Repeat Unit		Consensus Sequence and Number	Ref.
	#	length		
Mouse RNA Pol II largest subunit	52	7	TyrSerProThrSerProSer 22	Corden et al. 1985 (ref. 19)
Yeast RNA Pol II largest subunit	26	7	TyrSerProThrSerProSer 17	Allison et al. 1985 (ref. 20)
Rat Helix Destabilizing Protein	16	6 to 9	PheGlyGlyGlyGlySer 1	Cobianchi et al. 1985 (ref. 21)

TABLE III

Nucleotide sequence homology between cDNAs for rat HDP and mouse epidermal keratin

cDNA

Nucleotide Sequence Alignment[a]

Rat HDP (1706 bp)

Mouse Epidermal Keratin (1944 bp)

```
       600       610       620          630          640       650
GTTCCGGAAACTTTGGTGGTGGTCGTGGAGGTG---GTTTC---GGTGGCAATGACAATTTTGGTCGAGGAGG
          :::::: :: ::   :::::::    ::::    :::::::
GAGGAAGCAGCTTTGGGGGAGGCTATGGAGGTAGCAGTTTTGGGGGTGGCAG---------------------
       300       310       320       330       340

660       670       680       690       700       710       720       730
GAACTTCAGTGGTCGTGGTGGCTTTGGTGGCAGCCGTGGTGGTGGTGGATATGGTGGCAGTGGGGATGGCTAT
   :::: ::::: :  :   :  ::::::::::  ::::::          :::::: ::::::  ::::
---CTTCGGTGGTGGCAGCTTCGGTGGTGGCAGCTTTGGTGGC---------GGTGGCTGTGGGGGGGGCT--
        350       360       370       380                390       400

       740       750       760       770       780       790       800
AATGGATTTGGCAATGATGGAAGCAATTTTGGAGGTGGTGGAAGCTACAATGATTTTGGCAATTACAACA
                            :::: :::::::::  :      :::  ::::
----------------------------TTGGCGGTGGTGGATTCGGGGGGGATGGTGGCGGCCTTCTCT
                                  410       420       430       440
```

[a]Alignment was by the SRCHN program of D.J. Lipman and W.J. Wilbur, NIH. The GENBANK Data Bank as of 8/1/85 was evaluated using K-tuple size 4, window size 20 and Gap Penalty 4. The homology score for HDP and keratin (19) was 42 (22 standard deviations above mean score). The lengths of the cDNAs are shown in parentheses. Nucleotide sequences not shown did not exhibit significant homology. Numbering of residues was from the 5' end of the coding strand. The 1944 bp cDNA for keratin contained the entire coding region of the mRNA.

14. The Nuclear Matrix and Its Association with the RNA-Processing Machinery

R. Verheijen, F. Ramaekers, E. Mariman,[a] *H. Kuijpers, P. Vooijs, and W. J. van Venrooij*[a]

Department of Pathology, Radboud Hospital, University of Nijmegen, Geert Grooteplein Z24, 6525 GA Nijmegen, The Netherlands, and [a]Department of Biochemistry, University of Nijmegen, Geert Grooteplein N21, 6525 EZ Nijmegen, The Netherlands

INTRODUCTION

In recent years evidence has been obtained for the existence in eucaryotic cell nuclei of an intranuclear structure in addition to the nuclear envelope. This intranuclear structure is mostly refered to as "nuclear protein matrix" or "nuclear matrix".

In brief the procedure for isolating nuclear matrices from nuclei involves the removal of the nuclear membranes with detergents and solubilizaton of the chromatin with high salt buffers after treatment with DNase or DNase/RNase mixtures. The ultrastructure of the nuclear matrix reveals a peripheral layer composed of the nuclear lamina with its pore complexes, a nonhistone intranuclear network, and residual nucleoli (1).

The peripheral pore-complex lamina in higher eucaryotes is made up of mainly three polypeptides (2), the lamins A, B and C (60-74kD molecular weight). Literature dealing with the chromatin depleted nucleolar residue indicates that it contains a unique subset of proteins within nuclear matrix preparations (3,4). The protein composition and three-dimensional organization of the isolated nuclear matrix appears to be very complex (3,5). Accumulating data show that it represents a structure also existing in the intact cell (1,5-8). Corroborating evidence for this has been provided by electron microscopy on whole mount preparations. These studies also showed a distinct interaction between the cytoskeleton and the nuclear matrix (9).

Several studies indicate that the nuclear matrix is involved in various functional aspects of the nucleus such as DNA replication, transport and processing of RNA, steroid binding, protein phosphorylation and association with viral DNA synthesis and newly formed viral proteins (reviewed in 10-12). It has been reported that pre-mRNA is quantitatively attached to the nuclear matrix (6,13). Since also the intermediates of mRNA-processing are exclusively found in the matrix (14) it was suggested that the matrix is the location of RNA processing in vivo.

The protein composition of HeLa cell nuclear matrix preparations

The routine procedure that we use for the isolation of nuclear matrices does not essentially differ from methods described earlier (8,9) and has been reported in detail recently (see legend to Fig.1;15).

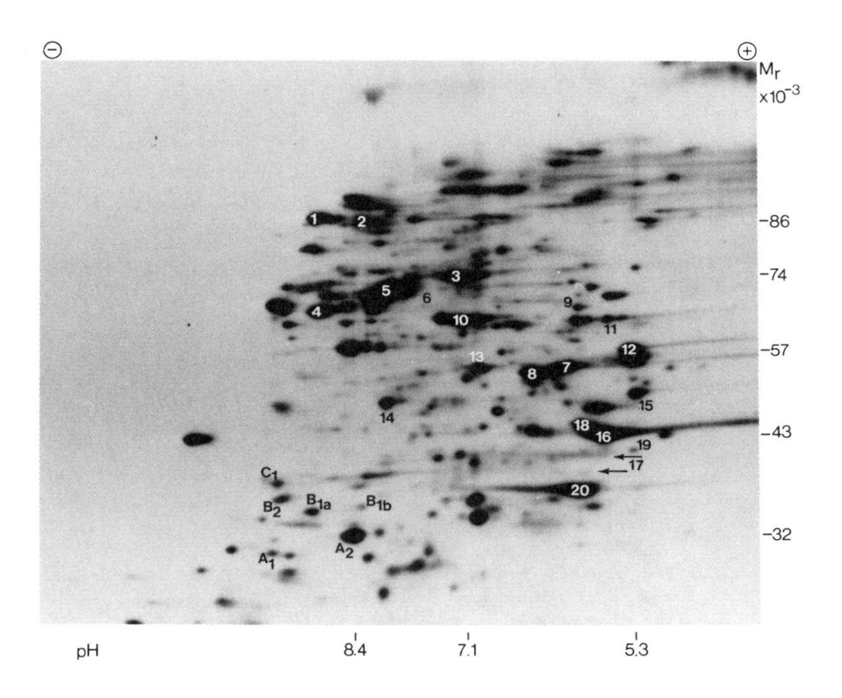

Figure 1. Two-dimensional gel electrophoresis of protein components present in the HeLa S3 nuclear matrix fraction. Cellular protein of HeLa cells was labeled with [^{35}S]-methionine. Nuclei were isolated by extracting the cells with 0.5% Triton X-100, followed by 0.5% Na-deoxycholate/1% Tween-40. Nuclear matrices were obtained by digestion of the nuclei with 800 µg/ml DNase I and 25 µg/ml RNase A for 15 min at 20°C, followed by solubilization of the digested chromatin with 0.4 M ammonium sulphate. The proteins were separated in the first dimension by non-equilibrium-pH-gradient electrophoresis (NEpHGE, + = acidic side, - = basic side), followed by SDS-gel electrophoresis in the second dimension.
Identification of proteins present in the nuclear matrix was performed by comigration with known proteins and by immunoblotting. As primary antibodies mouse monoclonal antibodies next to well defined human autoimmune sera were used. For details see (15).

Figure 1 shows a two-dimensional gel electrophoretic pattern of a nuclear matrix preparation from HeLa S3 cells. One of the major [^{35}S]-methionine labeled protein complexes present in the matrix fraction consists of a number of protein spots in the 65-72kD molecular weight region with isoelectric points ranging between 8 and 8.5 (spot nr. 5). The identity of this cluster, which consists of about 15-17 polypeptides is, as yet, unknown.
Several cytoskeletal proteins could be identified in these gels by comigration and immunoblotting studies. It was obvious that a 43kD polypeptide, comigrating in both dimensions with actin (spot nr. 16) occurs in the nuclear matrix fraction in a relatively high amount. Polypeptide spots comigrating with the intermediate filament proteins vimentin (spot nr. 12) and the four HeLa cytokeratins 7,8,18 and 19 (spot nrs. 7,8,18,19) were observed as well. Some of these could also be identified by immunoblotting using well defined antisera. Monoclonal antibodies to lamins A, B and C (15) identified the pore-complex lamina constituents in immunoblots (spot nrs. 3,9 and 10 respectively).
Pre-mRNA is present in the cell nucleus as fibrillar ribonucleoprotein (RNP) particles and granules (16). The major protein components of these RNP complexes are the core proteins of about 30-42kD molecular weight, represented by the A1, A2, B1a, B1b, B1c, B2, C1, C2 and C3 polypeptides (17). Most of these proteins could be identified by comigration with unlabeled core proteins of 35-40S pre-mRNP complexes. With this technique, however, we were unable to indicate the positions of the proteins C2 and C3. Using immunoblotting we could show their presence in the nuclear matrix fraction (15). These proteins were present over a relatively broad isoelectric point range in our blots (spots nr. 17), probably due to nucleic acid remainders still in tight interaction with these RNA binding proteins. In immunofluorescence studies, antibodies to such proteins stain the nuclear matrix, but not the nucleoli, rather diffusely (15).
Sera from patients suffering from connective tissue diseases often contain autoantibodies directed against nuclear proteins (18). Some of these autoantibodies have been shown to be directed against the internal fibrillar mass of the nuclear matrix or against nucleoli. Such antisera - carefully selected by immunofluorescence and one-dimensional immunoblotting - could identify three proteins with apparent molecular weights of 65kD, 70kD and 86kD (15) (spot nrs. 13, 4 and 2 respectively), specifically associated with the internal fibrillar mass. Furthermore five nucleolar proteins could be identified in this way (spot nrs. 1,11,14,15,20).

Association between the RNA-processing machinery and the nuclear matrix

It has been suggested that the nuclear matrix may function as a scaffold on which processing of pre-mRNA occurs (14). Both pre-mRNA as well as its cleavage products (14) and pre-mRNA-associated core proteins (Fig.1) can be found in the nuclear matrix. U1-snRNA, known to be involved in the splicing of pre-mRNA (19), and its associated proteins are also found in matrix preparations.
Western blots, prepared from various cellular fractions, were screened for the presence of U1-RNA-associated proteins using human autoantibodies (anti-(U1)RNP and anti-Sm). With these antibodies

proteins of 70kD, 33kD (the A-protein), 29kD (B´), 28kD (B) and 22kD (C) molecular weight could be detected in the nuclear fraction (Fig.2). Of these proteins the 70kD-, A-, and C-proteins are specifically associated with U1-RNA (20), while the B´- and B-protein are also present on other U-RNAs. When nuclei are treated with RNase and DNase the A-, B-, B´- and C-proteins are extracted to a large extent. However, the 70kD protein remains in the residual fraction. High salt extraction of the DNase and RNase treated nuclei only partly releases the 70kD protein (Fig. 2).

Using the monoclonal antibody 2.73 directed against this 70kD protein (15,21) on two-dimensional blots of nuclear matrix preparations, three discrete protein spots could be detected (Fig.3a). These three polypeptides, which have a very slow rate of [^{35}S]-methionine

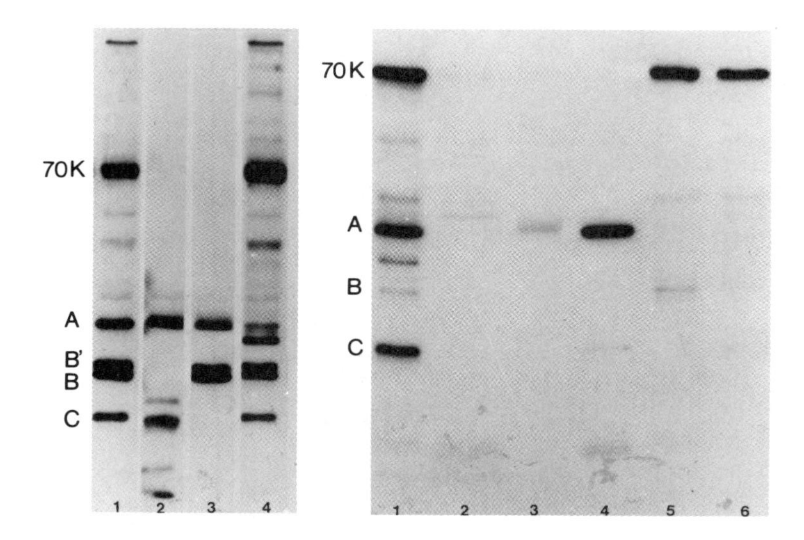

Figure 2. Detection of the 70kD U1-RNA associated polypeptide in cellular subfractions. One-dimenional Western blots were prepared from cellular fractions of HeLa cells. U-RNP proteins were detected by incubating such blots with human autoantibodies and [^{125}I]-protein A (20).

Left panel: Blot incubated with an anti-RNP/Sm serum.

Lane 1: untreated nuclei; lane 2: supernatant of RNase treated nuclei; lane 3: supernatant of subsequent DNase digested nuclei; lane 4: residual nuclear fraction.

Right panel: Blot incubated with a human anti-(U1)RNP serum. Lane 1: nuclei; lane 2: Triton X-100 soluble fraction of whole cells; lane 3: DOC/Tween soluble cellular fraction; lane 4: supernatant of DNase/RNase digested nuclei; lane 5: high salt extractable fraction of DNase/RNase treated nuclei; lane 6: nuclear matrix preparation.

incorporation and can be seen only in Coomassie-stained gels, migrate together within a short molecular weight- and pH range (spot nr. 6 in Fig.1). One- and two-dimensional immunoblots have demonstrated that the 70kD proteins recognized by the monoclonal antibody 2.73 and the human anti-(U1)RNP autoimmune sera are identical (data not shown). Immunofluorescence studies with the monoclonal antibody show a dot-like distribution of the 70kD antigen in the nuclear matrix of HeLa cells (Fig.3b; c). The finding that the 70kD protein apparently is the only U1-RNP-associated antigen which is not released from the nucleus or nuclear matrix with DNase and/or RNase digestion suggests to us that this 70kD protein directly interacts with components of the nuclear matrix. It may thus mediate the binding of U1-RNA to the nuclear matrix.

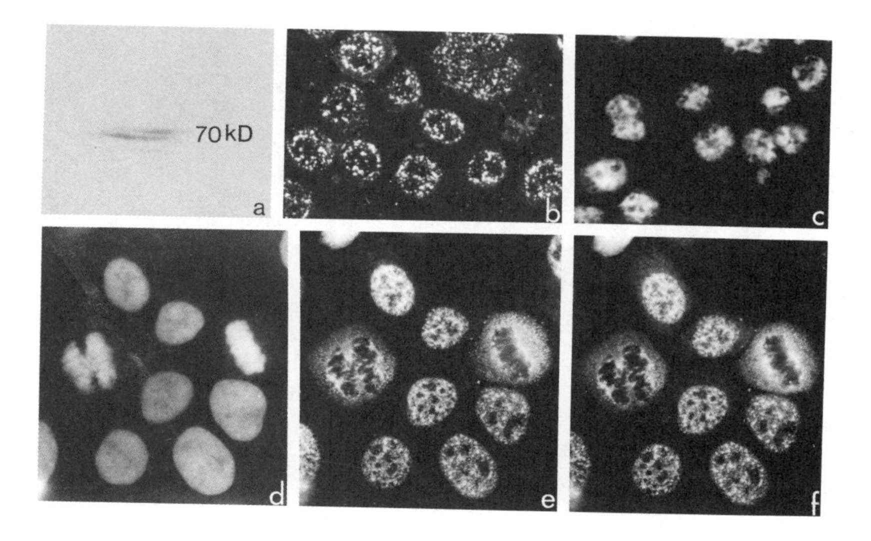

Figure 3. Immunoblotting and immunofluorescent patterns of the monoclonal antibody 2.73 directed against the U1-RNP specific 70kD polypeptide (21).
a) Reaction of the antibody on a two-dimensional Western blot of a HeLa nuclear matrix preparation.
b) Immunofluorescence pattern of antibody 2.73 on HeLa cells and c) HeLa cell nuclear matrix preparations.
d-f) Reactivity pattern of antibody 2.73 (e) and a human autoimmune serum (G15) recognizing specifically the 33kD U1-RNP component (f) on a monolayer of the human lung carcinoma cell line MR65 as demonstrated in a triple label immuno fluorescence assay. (d) DNA staining with Hoechst 33258.

In order to better understand the function and interactions of the various U1-RNP constituents and their association with the nuclear

matrix we have examined the subcellular localization of these components during both interphase and mitosis.
It is of special interest to study the behaviour of snRNPs during mitosis since hnRNA metabolism is repressed during this phase of the cell cycle and the integrity of the nucleus seems to be absent (22).
For the immunofluorescent localization of the 70kD U1-RNP polypeptide during mitosis we have used a cell line derived from a human pulmonary squamous cell carcinoma, MR65. These cells remain relatively flat during the mitotic cycle and contain large nuclei. The distribution of the 70kD component (as detected by the monoclonal antibody 2.73) was compared to the distribution of the nuclear DNA on the one hand and the U1-RNP associated A-protein (33kD) on the other hand. The latter polypeptide was specifically recognized by a human antiserum (G15) and has been shown not to interact directly with the nuclear matrix (see Fig.2). Figures 3d-f illustrate some of our findings. It is obvious from the double lable immunofluorescence staining patterns in Figs.3e and 3f that during interphase, metaphase and anaphase the distribution of the 70kD and the 33kD snRNP components are virtually identical. Furthermore, it is obvious that both components are restricted to the nucleus during interphase and are distributed throughout the cytoplasm during mitosis. During anaphase they seem to be located in the spindle columns. At all mitotic stages the two polypeptides are not associated with the condensed chromosomes. In interphase cells both antibodies stain the nucleus in a dot-like fashion with exclusion of the nucleoli. These data strongly suggest a close interaction of both components during several phases of the cell cycle.
Future studies will have to reveal whether or not the 70kD snRNP polypeptide can function as a mediator in the reorganization of the splicing machinery onto the nuclear matrix after the cell has gone through a mitotic cycle and the nucleus reforms at early telophase.

ACKNOWLEDGEMENTS

This work was supported by the Netherlands Cancer Foundation (Queen Wilhelmina Fund), the Netherlands Foundation for Chemical Research (S.O.N.), the Netherlands Organization for the Advancement of Pure Research (Z.W.O.), and the Netherlands League against Rheumatism.
The authors thank Dr. S. Hoch (La Jolla, Ca) for her gift of the hybrid cell line 2.73, and Dr. Gropp (Marburg) for kindly providing the cell line MR65.

REFERENCES

1. Kaufmann, S.H., D.S. Coffey, and J.H. Shaper: Considerations in the isolation of rat liver nuclear matrix, nuclear envelope, and pore-complex lamina. Exp Cell Res 132:105, 1981.
2. Franke, W.W., U. Scheer, G. Krohne, and E. Jarasch: The nuclear envelope and the architecture of the nuclear periphery. J Cell Biol 91:38S, 1981.
3. Peters, K.E., and D.E. Comings: Two-dimensional gel electrophoresis of rat liver nuclear washes, nuclear matrix and hnRNA proteins. J Cell Biol 86:135, 1980.
4. Comings, D.E., and K.E. Peters: Two-dimensional gel electrophoresis of nuclear particles. The cell nucleus (ed. H.

Busch), New York, Academic Press:89, 1981.
5. Staufenbiel, M., and W. Deppert: Preparation of nuclear matrices from cultured cells: subfractionation of nuclei in situ. J Cell Biol 98:1886, 1984.
6. van Eekelen, C.A.G., and W.J. van Venrooij: HnRNA and its attachment to a nuclear protein matrix. J Cell Biol 88:554, 1981.
7. Brasch, K.: Fine structure and localization of the nuclear matrix in situ. Exp Cell Res 140:161, 1982.
8. van Eekelen, C.A.G., M.H.L. Salden, W.J.A. Habets, L.B.A. van de Putte, and W.J. van Venrooij: On the existence of an internal nuclear protein structure in HeLa cells. Exp Cell Res 141:181, 1982.
9. Fey, E.G., K.M. Wan, and S. Penman: Epithelial cytoskeletal framework and nuclear matrix-intermediate filament scaffold; three-dimensional organization and protein composition. J Cell Biol 98:1973, 1984.
10. Shaper, J.H., D.M. Pardoll, S.H. Kaufmann, E.R. Barrack, B. Vogelstein, and D.S. Coffey: The relationship of the nuclear matrix to cellular structure and function. Adv Enz Reg 17:213, 1979.
11. Agutter, P.S., and J.C.W. Richardson: Nuclear non-chromatin proteinaceous structures; their role in the organization and function of the interphase nucleus. J Cell Sci 44:395, 1980.
12. Barrack, E.R., and D.S. Coffey: Hormone receptors and the nuclear matrix. Gene Regulation by Steroid Hormones II (ed. A. Roy, and J. Clark), New York, Springer-Verlag:239, 1983.
13. Long, B.H., C.-Y. Huang, and A.O. Pogo: Isolation and characterization of the nuclear matrix in Friend erythroleukemia cells: chromatin and hnRNA interactions with the nuclear matrix. Cell 18:1079, 1979.
14. Mariman, E.G.M., C.A.G. van Eekelen, R.J. Reinders, A.J.M. Berns, and W.J. van Venrooij: Adenoviral heterogeneous nuclear RNA is associated with the host nuclear matrix during splicing. J Mol Biol 154:103, 1982.
15. Verheijen, R., H. Kuijpers, P. Vooijs, W. van Venrooij, and F. Ramaekers: Protein composition of nuclear matrix preparations from HeLa cells; an immunochemical approach. J Cell Sci 1985, in press.
16. Holoubek, V.: Nuclear ribonucleoproteins containing heterogeneous RNA. Chromosomal nonhistone proteins - Biochemistry and Biology (ed. V. Hnilica), CRC Press 4, 1984.
17. Wilk, H.-E., H. Werr, D. Friedrich, H.H. Kiltz, and K.P. Schäfer: The core proteins of 35S heterogeneous nuclear ribonucleoprotein complexes; characterization of nine different species. Eur J Biochem 146:71, 1985.
18. Tan, E.M.: Autoantibodies to nuclear antigens (ANA): Their immunobiology and medicine. Adv Immunol 33:167, 1982.
19. Krämer, A., W. Keller, B. Appel, and R. Lührmann: The 5′ terminus of the RNA moiety of U1 small nuclear ribonucleoprotein particles is required for the splicing of messenger RNA precursors. Cell 38:299, 1984.
20. Habets, W., M. Hoet, P. Bringmann, R. Lührmann, and W. van Venrooij: Autoantibodies to ribonucleoprotein particles containing U2 small nuclear RNA. The EMBO Journal 4:1545, 1985.

21. Billings, P.B., R.W. Allen, F.C. Jensen, and S.O. Koch: Anti-RNP monoclonal antibodies derived from a mouse strain with lupus-like autoimmunity. J Immunology 128:1176, 1982.
22. Reuter, R., B. Appel, J. Rinke, and R. Lührmann: Localization and structure of snRNPs during mitosis. Exp Cell Res 1985, in press.

15. A Sequence-Specific, Methylated DNA-Binding Protein

Melanie Ehrlich and Richard Y.-H. Wang

Department of Biochemistry, Tulane Medical School, New Orleans, LA 70112

INTRODUCTION

Despite considerable evidence for a causal role of DNA methylation in vertebrate development (1-3), questions have been raised about whether this methylation at cytosine (C) residues has functional significance. We have shown that large tissue-specific differences in genomic 5-methylcytosine (m^5C) levels exist in vertebrates (4-7). That DNA methylation changes at so many different sites along the genome during differentiation might be due to the DNA recognition sequences for demethylation and *de novo* methylation (3) being quite short. This could lead to a kind of overshooting such that of $m^5C \rightarrow C$ or $C \rightarrow m^5C$ transitions occur at many more positions along the DNA than is necessary for vertebrate development.

To help define the roles of DNA methylation in vertebrate development, we looked for a protein which could bind preferentially to methylated DNA sequences. 5-Methylation at pyrimidine residues often (8-11), although not always (10,11), has profound effects upon recognition of DNA sequences by prokaryotic, sequence-specific DNA-binding proteins. Furthermore, a methyl group at the 5-position of m^5C and thymine (T) is relatively exposed in the major groove of the standard B-form DNA helix (12). Therefore, we thought it likely that some vertebrate DNA-binding proteins strongly differentiate between methylated and the analogous unmethylated DNA sequences. We have recently isolated from human placenta a protein which binds specifically to DNA sequences containing m^5C residues. This protein recognizes certain oligonucleotide sequences in double-stranded DNA and only when those sequences are methylated at some of the C residues. The existence of this previously unidentified type of protein indicates that indeed the 5-methyl group of certain m^5C residues of vertebrate DNA has important functional significance.

RESULTS

Recognition of m^5C-Rich DNA by Methylated DNA-Binding Protein (MDBP)

Methylated DNA-binding protein (MDBP) from human placenta was first identified by its preferential binding to human DNA enriched *in vitro* in m^5C versus the analogous DNA of a naturally low m^5C content (13). These ligands were prepared by extensive nick translation of human placental DNA with a radioactive mixture of dCTP, dATP, dGTP and dTTP to give m^5C-deficient (~0.5 mol% m^5C) DNA (4) or with m^5dCTP replacing dCTP to yield m^5C-rich (9 mol% m^5C) DNA. Binding of proteins to DNA was assessed by a nitorcellulose filter-binding assay. From an extract of placental nuclei, which was chromatographed on phosphocellulose, hydroxyapatite, and DEAE-cellulose, we obtained a DNA-binding activity having a ~3.5-fold preference for m^5C-enriched human DNA over the analogous m^5C-deficient DNA. Although still impure, this protein preparation does not have any detectable DNA methyltransferase activity (14). There are several other differences between MDBP and DNA methyltransferase, including that the latter has a strong preference for binding to single-stranded DNA and no increase in binding to double-stranded DNA upon its substitution with m^5C (14, 15). MDBP differs also from histones and HMG proteins in many properties including that those chromosomal proteins do not show preferential binding to m^5C-rich DNA sequences under similar assay conditions (L.-H. Huang *et al.*, unpublished results, 16). Also, the other chromosomal proteins are much more abundant than MDBP.

The binding specificity of MDBP was confirmed with many DNAs. The same preferential binding to methylated DNA was observed with nick-translated human DNA in double-label experiments using m^5C-rich ^3H-DNA and m^5C-deficient ^{32}P-DNA as when the radiolabels on the DNAs were reversed (13). Hemimethylated replicative form (RF) bacteriophage M13 recombinant DNA containing human β-globin sequences (17) was made by extension of an oligonucleotide primer on a viral strand template using m^5dCTP instead of dCTP. This RF (50% of C residues methylated) binds to MDBP three times better than does the nonmethylated RF (13). Even a hemimethylated RF with only 9% of its C residues methylated is bound 30% better by MDBP than is the nonmethylated RF. MDBP shows a strong preference for a naturally m^5C-rich DNA, phage XP12 DNA (34 mol% m^5C, <1 mol% C, ref. 18), over various m^5C-deficient DNAs in experiments with these unlabeled DNAs as competitors for binding with either radiolabeled XP12 or m^5C-enriched human DNA. Specificity for m^5C-rich DNA is lost if the competing DNA has been denatured (13).

Recognition of m^5CpG-Containing Sequences by MDBP

The strong binding of MDBP to DNA does not require complete substitution of C residues with m^5C residues. For example, *Micrococcus luteus* DNA (36 mol% C, 0.3 mol% m^5C) was enriched in m^5C by methylation with human DNA methyltransferase (14) so that it contained 6% of its bases as m^5C and most or all of this m^5C in CpG sequences. It competed for binding to radiolabeled XP12 DNA almost half as well as did an excess of nonlabeled XP12 DNA (Wang *et al.*, submitted for publication), even though the latter DNA contains 34% of its bases as m^5C. That retention by MDBP depends more on the concentration of m^5CpG sites in the DNA ligand than on the level of total m^5C residues was seen in comparisons of binding to MDBP by several DNAs extensively nick-translated in the presence of m^5dCTP and dCTP in various ratios (Table 1). After nick translation under comparable conditions, the concentration of m^5CpG dinucleotide sequences in human DNA was much less than in *M. luteus* DNA because of the ~4-fold underrepresentation of CpG sequences in mammalian DNA (3) and the higher A + T content of human DNA (58 *vs.* 28 mol%). The relative binding of these DNAs showed that the $m^5\overline{CpG}$ content correlated much better than the total m^5C content with the ability of the DNAs to be retained by MDBP (Table 1).

TABLE 1. Binding of MDBP to m^5C-Enriched DNAs with Different m^5CpG Contents.

m^5C-substituted ^3H-labeled DNA	% of bases as m^5C	% of dinucleotides as m^5CpG	Relative retention by MDBP[1]
Placenta	11	0.9	1.0
M. luteus	2.6	1.0	1.0
M. luteus	4.0	1.5	1.3
M. luteus	6.5	2.5	1.8
M. luteus	13	5.1	5.8
XP12	34	12	11
λ	11	2.9	2.5

[1] Percentage of the given ^3H-labeled DNA bound ÷ percentage of an m^5C-subsituted, ^{32}P-labeled human DNA bound by MDBP in a competition experiment (13) using approximately equimolar amounts of both DNAs.

The above results suggest that m^5CpG, the predominant site of m^5C in vertebrate DNA (3), is part of the recognition sequence for binding MDBP. That m^5CpG sites are necessary but not sufficient for specific binding by MDBP was shown in experiments with *M. luteus* DNA nick-translated with the four major deoxynucleoside triphosphates (and no m^5dCTP) and then exhaustively methylated with both *HpaII* methylase and *HhaI* methylase. Although this *in vitro* methylated DNA with a fragment size distribution of 1-9 kilobase pairs (kb) contained ~3% of its bases as m^5C residues at CpG sites (CCGG and GCGC), it bound MDBP no better than did analogously prepared mock-methylated DNA (X.-Y. Zhang *et al.*, unpublished results). In contrast, human DNA enriched in m^5C by nick translation with m^5dCTP so that it had ~1% of its residues as m^5C in the sequence m^5CpG was bound 3- to 4-fold better by the same preparation of MDBP than was the corresponding human DNA sample depleted in m^5C residues by nick translation with dCTP (13).

Sequence-Specific Binding of Methylated DNA by MDBP

We first examined whether certain *MboI* fragments (13) of naturally m^5C-rich XP12 DNA were preferentially retained by MDBP. After radiolabeling with [^{32}P]dATP at their 3'ends and deproteinization, the DNA fragments were incubated with MDBP. Those binding to MDBP were obtained by filtration through a nitrocellulose membrane and subsequent elution with 0.2% sodium dodecyl sulfate and then analyzed by polyacrylamide gel electrophoresis and autoradiography (Wang *et al.*, submitted for publication). As seen in Fig. 1, there was preferential retention of certain groups of DNA fragments by MDBP. Because virtually all of the C residues are 5-methylated in XP12 DNA, the selective retention of a subset of restriction fragments implies that MDBP has a considerable degree of sequence specificity.

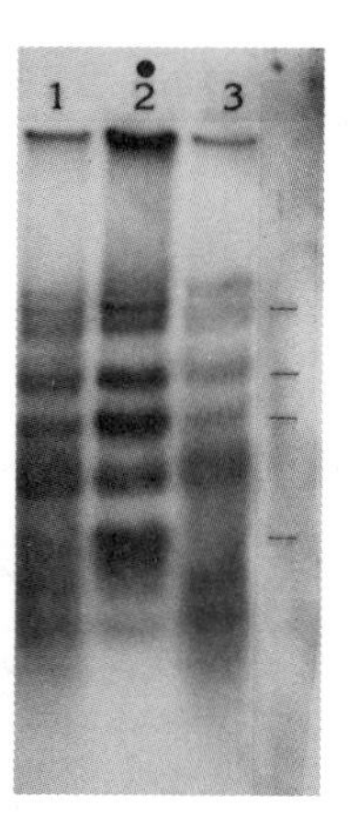

Figure 1. Analysis of *MboI*-digested XP12 DNA fragments bound by MDBP. Electrophoretic and autoradiographic analysis of DNA fragments retained by a nitrocellulose filter after incubation with non-specific DNA-binding proteins (lane 1) or with MDBP (lane 2). Lane 3, input DNA not previously incubated with DNA-binding proteins. In this and subsequent figures, methods were similar and the lanes with a dot on top contained DNA fragments that were m^5C-rich and incubated with MDBP. The sizes of the enriched DNA fragments indicated by the dashes were ∿900, 520, 400 and 250 base pairs (bp).

We next used much smaller DNAs of known sequence, phage M13mp8 RF and plasmid pBR322 DNA, for detecting MDBP-binding sites. From a *HinfI* digest of the hemimethylated, ^{32}P-labeled RF of M13mp8 prepared as described above, only 4 out of 25 fragments were preferentially retained by MDBP (Fig. 2). Much less total retention of fragments and no selectivity was seen when the analogous unmethylated fragments were incubated with MDBP. In the case of pBR322 DNA, *in vitro* enrichment for m^5C was obtained in a reaction catalyzed by human DNA methyltransferase (14). A mock-methylated DNA sample was prepared under the same conditions except that S-adenosylmethionine was omitted. After methylation to give 5% of the bases as m^5C, restriction with *Hin*fI and radiolabeling at the 3'-ends, the DNA fragments were incubated with MDBP and then analyzed for binding to MDBP (Fig. 3). One methylated fragment (298-bp) showed a strong preferential retention by MDBP (lane 3). For such selective binding it was necessary that the fragment be methylated; this fragment was not enriched in the mock-methylated DNA digest incubated with MDBP (Fig. 3).

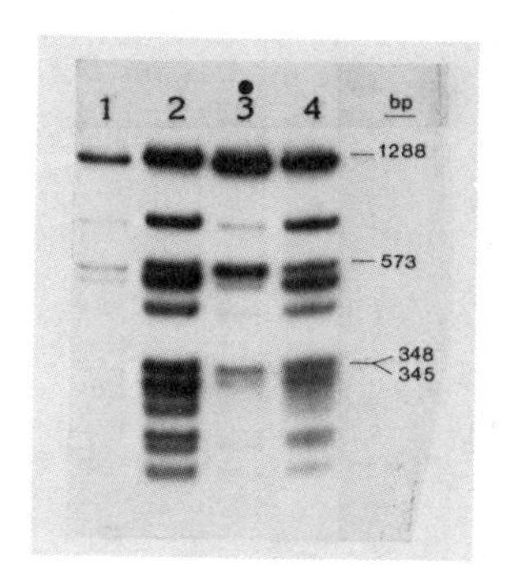

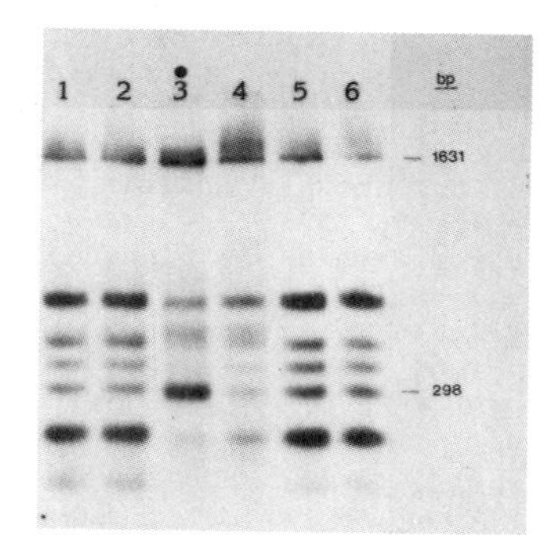

Figure 2. Analysis of *Hin*fI-digested M13mp8 RF fragments retained by MDBP. Lanes 1 & 2, unmethylated DNA; 3 & 4, hemimethylated DNA (primer-extended with m^5dCTP and no dCTP). Lanes 1 & 3, incubation with MDBP; 2 & 4, incubation with non-specific DNA-binding proteins.

Figure 3. Analysis of *Hin*fI-digested pBR322 fragments retained by MDBP. Lanes 1-3, methylase-methylated DNA; 4-6, mock-methylated DNA. Lanes 1 & 6, input DNA (no DNA-binding proteins); 2 & 5, DNA incubated with non-specific DNA-binding proteins or, 3 & 4, with MDBP.

To further analyze the MDBP-binding sites on the 298-bp fragment from the *Hin*fI-digested pBR322, this fragment was purified, methylated *in vitro*, and radiolabeled at its 3'-ends. Digestion with *Alu*I released a 214-bp DNA subfragment that was assayed for MDBP-binding in the standard nitrocellulose filter-binding assay (13). In equivalent incubations, ~60 times more of this methylated fragment than of the mock-methylated fragment was bound by MDBP, thereby confirming that a strong binding site for MDBP is in this region of pBR322. To define the location of such a site, this subfragment was used for DNase I footprinting (19). MDBP protected a 22-bp region containing three m^5CpG sites and this protection was observed only in the methylated DNA fragment (Wang *et al.*, submitted for publication). Sequencing the entire 214-bp fragment in its *in vitro* methylated and its unmethylated forms (20) indicated that methylation was occurring only at CpG sites. It is not sufficient for a sequence to be rich in m^5CpG sites to bind MDBP selectively as shown MDBP's poor binding to methylated cloned human α-globin DNA sequences which, after methylation, have an unusually high density of m^5CpG sites at their 5'-end (21). Therefore,

MDBP specifically recognizes certain m^5CpG-containing sequences. The frequency of MDBP-binding fragments from the examined small, m^5C-enriched prokaryotic DNAs and the fact that a considerable fraction of m^5C-enriched human DNA can be preferentially bound by MDBP precludes MDBP having a large oligonucleotide recognition sequence. Our data suggest that MDBP-specific binding sites comprise a family of related sequences with some binding MDBP more strongly than others.

CONCLUSION

We have demonstrated that MDBP from human placental nuclei is a sequence-specific, DNA-binding protein which can bind selectively to its corresponding DNA sequence only when that sequence is methylated. To elucidate the role of MDBP in modulating the effects of DNA methylation on cellular functions, it will be important to determine the frequency, distribution and state of methylation of MDBP recognition sites in vertebrate DNA. For those genes that have an MDBP recognition site appropriately positioned in a regulatory region, MDBP might mediate the control of transcription of certain genes by DNA methylation. Alternatively, since the small oligonucleotide sequences recognized by MDBP should be fairly evenly distributed in the genome, this protein might regulate DNA replication, rearrangements, or determine local DNA conformation changes. Given the presence of m^5C in all vertebrate genomes (3,6), the dramatic changes in DNA methylation patterns occurring during vertebrate development (1-7, 22-24), the correlation of some of these changes with gene expression (1-3), the modulation of gene expression by certain experimental alterations in DNA methylation (1-3), and, as described here, the existence of _at least_ one type of MDBP in mammals, it is concluded that DNA methylation must play a pivotal role in vertebrate development.

ACKNOWLEDGEMENTS

We thank Drs. L.-H. Huang and X.-Y. Zhang for use of their data and X.-Y. Zhang for critical reading of the manuscript. This research was supported in part by USPHS grant CA-19942.

REFERENCES

1. Jaenisch, R. and D. Jahner: Methylation, expression and chromosomal position of genes in mammals. _Biochim Biophys Acta_ 782:1, 1984.

2. Doerfler, W.: DNA methylation and gene activity. _Ann Rev Biochem_ 52:93, 1983.

3. Ehrlich, M., and R.Y.-H. Wang: 5-Methylcytosine in eukaryotic DNA. _Science_ 212:1350, 1981.

4. Ehrlich, M., M.A. Gama-Sosa, L.H. Huang, R.A. Midgett, K.C. Kuo, R.A. McCune, and C.W. Gehrke: Amount and distributuion of 5-methylcytosine in human DNA from different types of tissues or cells. _Nucleic Acids Res_ 10:2709, 1982.

5. Gama-Sosa, M.A., R.Y.-H. Wang, K.C. Kuo, C.W. Gehrke, and M. Ehrlich: The 5-methylcytosine content of highly repeated sequences in human DNA. _Nucleic Acids Res_ 11:3087, 1983.

6. Gama-Sosa, M.A., R.M. Midgett, V.A. Slagel, S. Githens, K.C.

Kuo, C.W. Gehrke, and M. Ehrlich: Tissue-specific differences in DNA methylation in various mammals. Biochim Biophys Acta 740:212, 1983.
7. Gama-Sosa, M.A., V.A. Slagel, R.W. Trewyn, R. Oxenhandler, K.C. Kuo, C.W. Gehrke, and M. Ehrlich: The 5-methylcytosine content of DNA from human tumors. Nucleic Acids Res 11:6883, 1983.
8. Goeddel, D.V., D.G. Yansura, and M.H. Caruthers: Studies on gene control regions. VI. The 5-methyl of thymine, a lac repressor recognition site. Nucleic Acids Res 4:3039, 1977.
9. Hofer, B., and H. Koster: On the influence of thymidine analogues on the activity of phage fd promoters in vitro. Nucleic Acids Res 8:6143, 1980.
10. Huang, L.H., C.M. Farnet, K.C. Ehrlich, and M. Ehrlich: Digestion of highly modified bacteriophage DNA by restriction endonucleases. Nucleic Acids Res 10:1579, 1982.
11. McClelland, M.: The effect of site-specific methylation on restriction endonuclease cleavage. Nucleic Acids Res 11:r169, 1983.
12. Seeman, N.C., Rosenberg, J.M., and A. Rich: Sequence-specific recognition of double helical nucleic acids by proteins. Proc Natl Acad Sci USA 73:804, 1976.
13. Huang, L.H., R.Y.-H. Wang, M.A. Gama-Sosa, S. Shenoy, and M. Ehrlich: A protein from human placental nuclei binds preferentially to 5-methylcytosine-rich DNA. Nature 308:293, 1984.
14. Wang, R.Y.-H., L.H. Huang, and M. Ehrlich: Human placental DNA methyltransferase: DNA substrate and DNA binding specificity. Nucleic Acids Res 12:3473, 1984.
15. Taylor, S.M., and P.A. Jones: Mechanism of action of eukaryotic DNA methyltransferase: Use of 5-azacytosine-containing DNA. J Mol Biol 162:679, 1982.
16. Felsenfeld, G., J. Nickol, M. Behe, J. McGhee, and D. Jackson: Methylation and chromatin structure. Cold Spring Harbor Symp. Quant. Biol 47:577, 1983.
17. Ley, J.T., N.P. Anagnou, and A.W. Nienhuis: RNA processing errors in patients with β-thalassemia. Proc Natl Acad Sci USA 79: 4775, 1982.
18. Ehrlich, M., K. Ehrlich, and J.A. Mayo: Unusual properties of the DNA from Xanthomonas phage XP12 in which 5-methylcytosine completely replaces cytosine. Biochim Biophys Acta 395:109, 1975.
19. Galas, D., and A. Schmitz: DNase footprinting: a simple method for the detection of protein-DNA binding specificity. Nucleic Acids Res 5:3157, 1978.
20. Maxam, A.M., and W. Gilbert: Sequencing end-labeled DNA with base-specific chemical cleavages. Meth Enzymol 65:499, 1980.
21. Michelson, A.M., and S.H. Orkin: Boundaries of Gene Conversion within the Duplicated Human α-Globin Genes. J Biol Chem 258: 15245, 1983.
22. Ponzetto-Zimmerman, C., and D.J. Wolgemuth: Methylation of satellite sequences in mouse spermatogenic and somatic DNAs. Nucleic Acids Res 12:2807, 1984.
23. Chapman, V., L. Forrester, J. Sanford, N. Hastie, and J. Roussant: Cell lineage-specific undermethylation of mouse repetitive DNA. Nature 307:284, 1984.
24. Zhang, X.-Y., R.Y.-H. Wang, and M. Ehrlich: Human DNA sequences exhibiting gamete-specific hypomethylation. Nucleic Acids Res 13:4837, 1985.

16. A Review of Protein Factors and Nucleoprotein Complexes Involved in Specific Transcription by Eukaryotic RNA Polymerase II

Randolph J. Hellwig[ab] *and Salil K. Niyogi*[a]

[a]University of Tennessee–Oak Ridge Graduate School of Biomedical Sciences and Biology Division, Oak Ridge National Laboratory, Oak Ridge, TN 37831; [b]Laboratory of Cell Biology, Rockefeller University, New York, NY 10021

INTRODUCTION

The elucidation of the molecular mechanisms of accurate transcription initiation by eukaryotic RNA polymerase II requires knowledge of the properties of the enzyme, the structure and transcription of chromatin, the DNA sequences of transcription control signals, and the protein (and other) factors that mediate promoter recognition through their interaction with the DNA template and/or the RNA polymerase enzyme. This review will focus primarily on *in vitro* studies utilizing current cell-free extracts and the soluble fractions (factors?) thereof, to examine the protein-DNA interactions and nucleoprotein complexes that are involved in accurate transcription initiation on eukaryotic class II genes by RNA polymerase II.

Development of Soluble Cell-Free Transcription Extracts

Progress in the study of eukaryotic gene expression has been greatly aided by the development and extensive use of soluble cell-free systems that mediate the accurate initiation of specific RNA synthesis from a variety of class II gene promoters (reviewed in 1,2). Satisfactory extracts have been prepared by enrichment of cytoplasm with nuclear components presumably liberated at low salt concentrations (3), by lysis of whole cultured cells followed by ultracentrifugation to remove the residual chromatin (4), and more recently by high salt lysis and extraction of "transcription components" from purified nuclei (5). In combination with recombinant DNA technology, these extracts have already been quite useful for elucidating the role of specific DNA sequences in gene transcription. In addition, specific transcription by RNA polymerase II in the soluble extracts seems susceptible to known down-regulatory proteins, such as SV40 T antigen (6) and an adenovirus EII gene-encoded DNA binding protein (7), when added in the presence of appropriate homologous templates. Chick embryos, chick embryo fibroblasts, and Rous sarcoma virus-transformed chick embryo fibroblasts were shown to contain a factor that preferentially blocked the accumulation of RNA polymerase II-directed transcripts in a cell-free system (8). Also, specific synthesis directed by RNA polymerase II was inhibited in extracts prepared from poliovirus-infected cells that appeared to lack a factor(s) required for specific transcription (9). These results illustrate the versatility and suitability of these soluble extracts for analyzing gene transcription and identifying factors involved in specific transcription.

Studies with Fractionated Extracts

Recently, advances in the identification of the component factors for promoter recognition and transcription initiation have come through the separation of soluble transcription extracts into fractions, subfractions and even putative factors by column chromatographic methods (see ref. 10 for a review). One of the first such studies was that described by Matsui *et al.* (11). In this protocol, a cytoplasmic-type extract from human KB cells was first separated on phosphocellulose into four fractions (*a*,*b*,*c* and *d*) of which three (*a*,*c* and *d*) were required to reconstitute specific RNA synthesis with RNA polymerase II from the Ad2 major late promoter. Fraction *c* was separated on DEAE-cellulose into two fractions (*e* and *f*) of which only *e* contained important transcription factors. Fraction *e* was then separated into four additional fractions (*g*,*h*,*i* and *j*) by chromatography on DNA-cellulose. Fractions *h* and *j* along with *a* and *d* were the only ones necessary to reconstitute specific transcription *in vitro*. The transcription fractions ("factors") were designated TFIIA (*a*), TFIIB (*h*), TFIIC (*j*) and TFIID (*d*) (11).

TFIIA contained an activity that stimulated the low level of specific transcription seen in the presence of template DNA, RNA polymerase II, TFIIB/C (unfractionated), and TFIID. A fraction that similarly stimulated specific transcription was prepared by Davison *et al.* (12) using a somewhat different procedure, and is thought to contain the same activity as that in TFIIA. The "TFIIA-like" stimulatory fraction was then purified through Heparin-Ultrogel, DEAE-cellulose, Cibacron-Blue agarose, and chromatographed twice through AcA34-Ultrogel to yield predominantly a factor designated AcAIIP43 (13). AcAIIP43 was characterized as a 43 kd protein that appears to function as a general, not a gene-specific, transcription factor. In addition, it possessed DNase I-inhibitory activity, but not RNase-inhibitory activity, and had properties similar to those of actin (13). This was most interesting in view of recent results indicating that nuclear actin may be involved in transcriptional processes *in vivo* (14). However, although these observations are exciting, Sawadogo and Roeder (15) have now found that TFIIA stimulates transcription by virtue of an RNase-inhibitory activity, and there is no absolute requirement for TFIIA in the reconstitution of specific transcription with other more purified fractions. Further work is needed to show whether or not nuclear actin has a true role in transcription.

Fraction TFIIB of Matsui *et al.* (11) was originally described as containing a stimulatory activity in addition to an actual factor involved in the process of promoter selection or specific initiation.

This fraction has been further separated into "factors" TFIIB and TFIIE (15), both of which are essential for the reconstitution of specific initiation and remain to be further characterized.

TFIIC does not appear to have a direct role in the process of promoter selection or specific initiation (11). The activity in TFIIC has been purified to homogeneity and was shown to be identical to the enzyme poly(ADP-ribose)polymerase (16). It appears that TFIIC functions only indirectly by suppressing non-specific initiation at nicks on the DNA template (16), a well known propensity for RNA polymerase II (reviewed in ref. 17). TFIIC had no effect on specific transcription of a circular template (15).

Finally, TFIID appeared to direct specific transcription (11) and contained an essential transcription factor, as was demonstrated by its absolute requirement in a reconstituted system containing TFIIB, TFIIE, RNA polymerase II and template DNA (15). It will be of great interest to determine if this fraction contains an activity which directs the RNA polymerase II enzyme to the promoter site.

Tsai *et al.* (18) described a fractionation protocol in which chromatography removed the detrimental RNase activity and allowed for the reconstitution of a specific *in vitro* transcription system from HeLa cells and chick oviduct tissue. Their fractionation scheme involved the separation of a HeLa whole cell extract into three fractions (DE50, DE175 and DE500) by column chromatography through DEAE-Sephadex. The flow-through fraction (DE50) was further separated on phosphocellulose to yield three fractions (P100, P350 and P1000). The numbers indicate the millimolar salt concentrations at which the fractions were eluted. Of these fractions, four (DE175, DE500, P100 and P1000) were required to reconstitute a specific transcription system with a cloned ovalbumin DNA template. Fraction DE175 contained at least the RNA polymerase II, DE500 acted to suppress background synthesis of low molecular weight RNA, P100 served to enhance the level of specific RNA synthesis, and P1000 was essential for specific transcription. Similar results were obtained upon fractionation of a chick oviduct extract, with P350 containing the RNase activity. It was also found that the HeLa and chick oviduct fractions were completely interchangeable. Essentially all of the active fractions contained DNA binding proteins. The P1000 fraction displayed the highest DNA binding activity. It was subsequently shown by a DNA binding assay incorporating exonuclease footprinting methods that this fraction contained an activity that bound to the chick ovalbumin gene CAAT box-containing distal promoter (19). Additional work should resolve whether or not this binding activity is an important transcription component. More recently, the same group has utilized affinity column chromatography involving antibodies to purified RNA polymerase II in order to search for polymerase-specific factors in soluble extracts (20).

Another fractionation scheme for the HeLa whole cell extract was reported by Samuels *et al.* (21). The soluble extract was again separated first on phosphocellulose into four fractions ([A], [B], [C] and [D]). Only [A], [C] and [D] were required to reconstitute specific RNA synthesis from the Ad2 major late promoter. [A] was further fractionated on DEAE-Sephacel to give three fractions ([AA],[AB] and [AC]), of which only [AB] contained factor(s) important for transcription. Fractionation of [C] on single-stranded DNA-cellulose resulted in four fractions ([CA],[CB],[CC] and [CD]), with [CB] essential and [CC] and [CD] useful for reconstituting specific transcription. Finally, fraction [D] was fractionated on DEAE-Sephacel to give two fractions ([DA] and [DB]), of which only [DB] was needed. Specific transcription could therefore be reconstituted with [AB], [CB], and [DB] in combination with exogeneous purified RNA polymerase II and the DNA template. The requirement for additional RNA polymerase II was somewhat surprising since the original soluble extract normally contains sufficient enzyme. In fact, we have similarly fractionated the soluble extract and found no requirement for exogeneous RNA polymerase II (22; Hellwig and Niyogi, unpublished observations).

Fraction [AB] (described above) contained "promoter-specificity" factor(s) in addition to the endogenous HeLa RNase inhibitor activity. The transcriptionally important DNA-binding activity was presumed to correspond to TFIIA of Matsui et al. (11), while the RNase inhibitor activity could be replaced with highly purified human placental RNase inhibitor with no loss in specific transcription. Fraction [CB] is presumably similar to TFIIB of Matsui et al. (11) and it stimulated specific transcription. It also appears to be composed of at least two factors, which have not been adequately separated as yet. Fractions [CC] and [CD] suppressed nonspecific background synthesis although they were not essential for specific transcription. It appeared that [CD] contained poly(ADP-ribose)polymerase and therefore would correspond to TFIIC of Matsui et al. (11). Fraction [DB] contained a stimulatory activity that was not necessarily essential for specific transcription and it appears to be similar to TFIID of Matsui et al. (11). It was subsequently shown by Fire et al. (23) that fractions [AB] and [DB] interact with the template DNA in a promoter-specific fashion to generate a stable pre-initiation complex. These complexes become "activated" upon the addition of RNA polymerase II and fraction [CB], and require only the four ribonucleoside triphosphates to rapidly initiate specific RNA synthesis.

Yet another protocol has been developed by Dynan and Tjian (24,25) to fractionate a HeLa whole cell extract. In this case, heparin-agarose was used to separate the soluble extract into a flow-through fraction that contained a stimulatory activity, while the salt step eluate was fractionated through DEAE-Sepharose to yield transcription factors in the flow-through and the endogenous RNA polymerase II in the salt step. The DEAE-Sepharose flow-through was further resolved on Sephacryl S-300 and phosphocellulose to give two fractions designated factors Sp1 and Sp2. Therefore, four fractions — the stimulatory fraction, the RNA polymerase II fraction, factors Sp1 and Sp2 — were required to reconstitute specific transcription from the SV40 early and late promoters.

Inclusion of the stimulatory fraction (described above) was not necessary when further purified fractions were used. Interestingly, Sp1 was shown to be an essential component only for transcription from both SV40 promoters (24) and from the BK virus early promoter (25). It was shown that Sp1 binds to 5'-flanking DNA sequences upstream from several viral and cellular promoters, and it appears that multiple copies of the sequence GGGCGG are important for binding (26,27). This factor is not required for the in vitro transcription from the Ad2 major late promoter and is in fact, inhibitory (24). Factor Sp2, however, appears to be a general transcription factor in the sense that it is required to reconstitute specific RNA synthesis from all of the promoters examined (24). Apparently, Sp2 functions to confer promoter recognition ability on the partially purified RNA polymerase II-containing fraction. It is very interesting that along with the expected general transcription factor(s), HeLa cells would contain a very specific factor (Sp1). It is probable that this type of factor is involved in some aspect of the regulation of expression not only from SV40 (which transforms HeLa cells) promoters but from certain cellular promoters (26). Indeed, recent results show not only that monkey DNA contains Sp1 binding regions but that these sites, which are similar to SV40 promoters, can actively direct transcriptional initiation (28).

In a fractionation protocol described by Parker and Topol (29), a Drosophila nuclear extract was first separated on DEAE-cellulose into three fractions (A,B and C) of which A and B could reconstitute specific transcription on cloned Drosophila histone H3, H4, and actin genes. Fractionation of fraction B through DEAE-Sephadex A25 and phosphocellulose yielded two fractions (B and D) with only B being essential for transcription. An alternate separation of DEAE-cellulose fraction B through Biorex 70 and DNA-cellulose finally yielded a "B" fraction similar to that above. Therefore, reconstitution of specific transcription required the purified fractions A and B, purified RNA polymerase II and template DNA. It appeared that fraction A, although absolutely required for reconstitution of transcription with purified components, contained a general transcription factor and has not been characterized further. Fraction B, another essential fraction, contained DNA-binding activity after extensive fractionation. This DNA-binding activity was shown by DNase I footprint analysis to bind specifically to a 65-base pair region of DNA in all of the promoters examined, and this binding protected the TATA box, the initiation site, and part of the RNA leader. It was inferred from these results that the binding activity and the transcription factor are probably one and the same, and are involved in promoter recognition by RNA polymerase II. Fraction C contains partially purified RNA polymerase II, but it can be replaced by purified enzyme. Finally, fraction D contained a stimulatory activity that was not needed when more concentrated preparations of fraction B were used in the reactions. The reason for this is not known. No direct correlations could be made between these fractions (factors) and those from the other laboratories described above.

Subsequently, Parker and Topol (30) described the very interesting results obtained from a similar fractionation of a soluble extract from heat-shocked Drosophila. The first of these was the identification of a factor, termed heat-shock transcription factor (HSTF), that was required for specific transcription of a heat-shock gene in the presence of fraction A and purified RNA polymerase II. Second, HSTF was found in both shocked and nonshocked cells, and appeared to have more transcriptional activity when prepared from heat-shocked cells. This was not totally unexpected since activation of heat-shock genes *in vivo* does not appear to require *de novo* protein synthesis (31). Third, HSTF and the fraction B activity described above (29) could be chromatographically separated. Fraction B was not needed for specific heat-shock gene transcription, and "B" activity appeared to be reduced when it was prepared from heat-shocked cells. Lastly, it was shown that HSTF bound specifically to a 55-base pair region of DNA 5' to the heat-shock gene TATA sequence. From these observations it was suggested that both the fraction B activity and the HSTF may bind to the heat-shock gene DNA in order to regulate the level of gene expression.

Fractionation of a nuclear extract from *Drosophila* tissue culture cells revealed the presence of multiple components involved in the accurate transcription of both distal and proximal promoters of the alcohol dehydrogenase (Adh) gene (32). Promoter selectivity was maintained after fractionation of the extract. A combination of transcription and DNA binding experiments, coupled with promoter mutagenesis, led to the conclusion that selectivity depends on the presence of novel transcription factors. It appears that multiple sequence-specific DNA binding proteins interact differentially with the proximal and distal promoters of Adh to activate transcription.

In summary of this section, we observe that putative transcription factors (fractions) have been prepared by column chromatographic procedures. Of these, several have been found to bind DNA specifically, either proximal to TATA sequences or to sequences further upstream from the initiation site, and thus appear to be involved in the processes of promoter recognition/selection and gene regulation. Studies of this type will surely be continued and expanded to further isolate and characterize transcription components that bind to DNA as well as those that interact with RNA polymerase II.

Studies of Transcription Complexes

Another experimental approach that could prove to be important for both the identification of transcription factors and the study of initiation mechanics and kinetics is the formation and analysis of "transcription complexes." For our purposes, this term will be defined as those nucleoprotein complexes that are composed of DNA, RNA polymerase II, and any of the transcription factors that are involved in one or more of the processes of promoter selection, recognition of DNA sequences, the initiation of accurate RNA synthesis, and RNA chain elongation.

The observation of Sinha *et al.* (33) that incubation of SV40 DNA with three different HeLa soluble transcription extracts, under transcription conditions minus the ribonucleoside triphosphates, resulted in the formation of histone-deficient, yet chromatin-like nucleoprotein complexes suggested that such complexes and the proteins (factors) contained therein might be important in the transcription process. This contention was further supported by results showing that promoter-containing nucleoprotein complex templates were transcribed more efficiently than the corresponding "naked" DNA (33). Further studies demonstrated that purified nucleoprotein complexes containing the adenovirus major late promoter were able to serve as template for accurate transcription in vitro (22). Use of such nucleoprotein complexes eliminated the need for bulk DNA (poly[d(I-C)]) in transcription reactions performed at template concentrations too low to yield a detectable signal from "naked" DNA (22). It also led to more linear responses to transcription extract, DNA and cation concentrations, unlike the stringent (all or none) responses displayed with the "naked" DNA template (Hellwig and Niyogi, unpublished observations). In addition, it was shown (22) that the nucleoprotein complexes could serve as template for specific transcription in vitro by an RNA polymerase II-containing chromatographic fraction similar to fraction [C] of Samuels et al. (21). Specific RNA synthesis was also found in reactions that contained only nucleoprotein template and fraction [D] (22). Since fractions [A], [C] and [D] had been required for transcription of "naked" DNA in vitro (21), the results described above appear to be good evidence for the presence of essential transcription factors in the chromatin-like nucleoprotein complexes. The association of RNA polymerase II-containing transcription complexes with high molecular weight nucleoprotein structures was also recently reported by Culotta et al. (34).

Although most studies to date have found that the majority of the transcriptionally important fractions contain DNA binding activity, it appears that fraction [D] of Samuels et al. (21) may contain an activity which associates either with the preinitiation complex or more probably with the RNA polymerase II (22; Hellwig, Ayer, and Niyogi, unpublished observations). It is not yet known if this activity is similar to the RNA polymerase II-binding stimulatory factor S-II described by Horikoshi et al. (35).

Additional studies have examined the nature of the transcription complexes formed upon incubation of template DNA in the presence of soluble transcription extracts. Bunick et al. (36) have found that the hydrolysis of rATP is essential for the formation of an initial ternary transcription complex. Such complexes, when formed in the presence of a HeLa cell extract, template DNA, and ribonucleoside triphosphates, were classified as initiation complexes, were stable during agarose gel electrophoresis, and were capable of elongating the initiated RNA chain in the presence of 0.25% Sarkosyl (37). This approach appears to be better suited for studying elongation parameters than it is for the actual steps of complex formation and promoter recognition.

A somewhat different Sarkosyl-sensitivity assay was used by Tolunay et al. (38) to also demonstrate that initiation complexes could be formed by incubation of template DNA containing the adenovirus 2 major late promoter with a HeLa cell-free extract and rATP. These complexes were rapidly converted to the expected Sarkosyl-resistant elongation complexes upon addition of the missing riboucleoside triphosphates. The initiation complexes were stable and could be purified on glycerol gradients. It was concluded that the template DNA essentially catalyzed the preinitiation complex assembly and the complexes were fully competent for initiation and elongation. Subsequently, Safer et al. (39) used gel filtration to isolate stable preinitiation, initiation and elongation complexes. A preinitiation complex, formed in the absence of exogenous nucleotides, was converted into an initiation complex upon the addition of rATP or dATP. The initiation complex, when incubated with all four ribonucleoside triphosphates, was converted into an elongation complex capable of producing the correct run-off transcript. Based on their results, they postulated that transcriptional activity was controlled in part by regulating the association of transcription factors at each initiation event.

It was shown by Triadou et al. (40), using electrophoretic analysis to identify specific transcription-initiation complexes, that tissue-specific factors are involved in the formation of transcriptionally competent complexes. Only extracts from β-globin gene expressing tissue were capable of producing specific RNA synthesis from a cloned β-globin gene.

Studies with somewhat more direct application to the deduction of mechanisms for promoter recognition and the initiation of transcription have utilized fractionated soluble extracts. Davison et al. (12) found that the transcription fraction from heparin-Ultrogel contained factor(s) that could form specific promoter DNA-containing pre-initiation complexes. Interestingly, these complexes were formed in the absence of both RNA polymerase II and ribonucleoside triphosphates, were stable for 20 min under their reaction conditions, and appeared to be "primed" for the rapid initiation of specific RNA synthesis.

Utilizing a preincubation-pulse-chase protocol and different soluble extract fractions, Fire et al. (23) found that "transcription complexes" could be formed in both the absence and presence of RNA polymerase II. In the absence of enzyme, fractions [AB] and [DB] interacted with the promoter-containing template DNA to form a complex. In turn, RNA polymerase II could bind to this DNA-protein complex and after association with fraction [CB], the nucleoprotein complex was turned into a stable "activated complex." All of these steps were ribonucleoside triphosphate-independent, and specific transcription was rapidly initiated upon the addition of ribonucleoside triphosphates to the "activated complexes." A similar model for the overall transcription process has been suggested by Hawley and Roeder (41) as a result of their work with human RNA polymerase II, transcription factor-containing fractions, and a modified Sarkosyl-sensitivity assay. In both cases, no obligatory protein-protein interactions were found, and it was concluded again that DNA-protein interactions were solely responsible for control of the entire transcription process. However, neither study was designed to rule out DNA-mediated protein-protein interactions or any type of concerted protein-protein-DNA interactions. In addition, these studies were performed with fractions, not individual purified factors, and therefore the involvement of multiprotein-complex "factors" cannot be ruled out.

CONCLUSIONS

A great deal of progress in elucidating the mechanisms of gene transcription and regulation has been made as a result of the development of accurate soluble transcription extracts and the analysis of the transcription factors contained therein. It appears that much of the transcription process is controlled at the level of sequence-specific

DNA-protein binding interactions. It is encouraging to note that similar specific transcription complexes have now been found in vivo (42). There is still much to be learned in regard to promoter selection and recognition, but it appears that investigators have found the systems whose manipulations will give us the answers we desire.

ACKNOWLEDGEMENTS

Sponsored by the Office of Health and Environmental Research, U.S. Department of Energy, under contract DE-AC05-84OR21400 with Martin Marietta Energy Systems, Inc. R.J. Hellwig was a predoctoral investigator supported by National Institutes of Health Grant GM7431 and by an Oak Ridge Associated Universities Laboratory Graduate Research Fellowship. Currently, he is a postdoctoral investigator supported by a grant from the Department of Energy to Dr. Anthony R. Cashmore at Rockefeller University.

REFERENCES

1. Heintz, N. and R.G. Roeder: Transcription of eukaryotic genes in soluble cell-free systems. In Genetic Engineering-Principles and Methods vol. 4:57, 1982. J.K. Setlow and A. Hollaender (Eds.), Plenum Press, New York.
2. Manley, J.L., A. Fire, M. Samuels, and P.A. Sharp: In vitro transcription: whole cell extract. Methods Enzymol 101:568, 1983.
3. Weil, P.A., D.S. Luse, J. Segall, and R.G. Roeder: Selective and accurate initiation of transcription at the Ad2 major late promoter in a soluble system dependent on purified RNA polymerase II and DNA. Cell 18:469, 1979.
4. Manley, J.L., A. Fire, A. Cano, P.A. Sharp, and M.L. Gefter: DNA-dependent transcription of adenovirus genes in a soluble whole-cell extract. Proc Natl Acad Sci USA 77:3855, 1980.
5. Dignam, J.D., R.M. Lebovitz, and R.G. Roeder: Accurate transcription initiation by RNA polymerase II in a soluble extract from isolated mammalian nuclei. Nucl Acids Res 11:1475, 1983.
6. Hansen, U., D.G. Tenen, D.M. Livingston, and P.A. Sharp: T antigen repression of SV40 early transcription from two promoters. Cell 27:603, 1981.
7. Handa, H., R.E. Kingston, and P.A. Sharp: Inhibition of adenovirus early region IV transcription in vitro by a purified viral DNA binding protein. Cell 302:545, 1983.
8. Crawford, N., A. Fire, M. Samuels, P.A. Sharp, and D. Baltimore: Inhibition of transcription factor activity by poliovirus. Cell 27: 555, 1981.
9. Tyagi, J.S., G.T. Merlino, B. de Crombrugghe, and I. Pastan: Chicken embryo extracts contain a factor that preferentially blocks the accumulation of RNA polymerase II transcripts in a cell-free system. J Biol Chem 257:13001, 1982.
10. Dignam, J.D., P.L. Martin, B.S. Shastry, and R.G. Roeder: Eukaryotic gene transcription with purified components. Methods Enzymol 101:582, 1983.
11. Matsui, T., J. Segall, P.A. Weil, and R.G. Roeder: Multiple factors required for accurate initiation of transcription by purified RNA polymerase II. J Biol Chem 255:11992, 1980.

12. Davison, B.L., J.-M. Egly, E.R. Mulvihill, and P. Chambon: Formation of stable preinitiation complexes between eukaryotic class B transcription factors and promoter sequences. Nature 301:680, 1983.
13. Egly, J.M., N.G. Miyamoto, V. Moncollin, and P. Chambon: Is actin a transcription initiation factor for RNA polymerase B? The EMBO J 3:2363, 1984.
14. Scheer, U., H. Hinssen, W.W. Franke, and B.M. Jockusch: Microinjection of actin-binding proteins and actin antibodies demonstrates involvement of nuclear actin in transcription of lampbrush chromosomes. Cell 39:111, 1984.
15. Sawadogo, M., and R.G. Roeder: Factors involved in specific transcription by human RNA polymerase II: Analysis by a rapid and quantitative in vitro assay. Proc Natl Acad Sci USA 82:4394, 1985.
16. Slattery, E., J.D. Dignam, T. Matsui, and R.G. Roeder: Purification and analysis of a factor which suppresses nick-induced transcription by RNA polymerase II and its identity with poly(ADP-ribose)-polymerase. J Biol Chem 258:5955, 1983.
17. Chambon, P.: Eukaryotic nuclear RNA polymerases. Ann Rev Biochem 44:613, 1975.
18. Tsai, S.Y., M.-J. Tsai, L.E. Kops, P.P. Minghetti, and B.W. O'Malley: Transcription factors from oviduct and HeLa cells are similar. J Biol Chem 256:13055, 1981.
19. Elbrecht, A., S.Y. Tsai, M.-J. Tsai, and B.W. O'Malley: Identification by exonuclease footprinting of a distal promoter-binding protein from HeLa cell extracts. DNA 4:233, 1985.
20. Tsai, S.Y., P. Dicker, P. Fang, M.-J. Tsai, and B. O'Malley: Generation of monoclonal antibodies to RNA polymerase II for the identification of transcriptional factors. J Biol Chem 259:11587, 1984.
21. Samuels, M., A. Fire, and P.A. Sharp: Separation and characterization of factors mediating accurate transcription by RNA polymerase II. J Biol Chem 257:14419, 1982.
22. Hellwig, R.J., S.N. Sinha, and S.K. Niyogi: Specific transcription of preformed nucleoprotein complexes containing the adenovirus major late promoter with a chromatographic fraction containing RNA polymerase II. Proc Natl Acad Sci USA 82:6769, 1985.
23. Fire, A., M. Samuels, and P.A. Sharp: Interactions between RNA polymerase II, factors, and template leading to accurate transcription. J Biol Chem 259:2509, 1984.
24. Dynan, W.S., and R. Tjian: Isolation of transcription factors that discriminate between different promoters recognized by RNA polymerase II. Cell 32:669, 1983.
25. Dynan, W.S., and R. Tjian: Recognition of upstream sequences in the SV40 promoter requires a promoter-specific transcription factor. In Gene Expression, p. 53, 1983. D.H. Hamer and M.J. Rosenberg (Eds.), Alan R. Liss, Inc., New York.
26. Gidoni, D., W.S. Dynan, and R. Tjian: Multiple specific contacts between a mammalian transcription factor and its cognate promoters. Nature 312:409, 1984.
27. Dynan, W.S., and R. Tjian: The promoter-specific transcription factor Spl binds to upstream sequences in the SV40 early promoter. Cell 35:79, 1983.
28. Dynan, W.S., J.D. Saffer, W.S. Lee, and R. Tjian: Transcription factor Spl recognizes promoter sequences from the monkey genome

that are similar to the simian virus 40 promoter. Proc Natl Acad Sci USA 82:4915, 1985.
29. Parker, C.S., and J. Topol: A Drosophila RNA polymerase II transcription factor contains a promoter-region-specific DNA-binding activity. Cell 36:357, 1984.
30. Parker, C.S., and J. Topol: A Drosophila RNA polymerase II transcription factor binds to the regulatory site of an HSP 70 gene. Cell 37:273, 1984.
31. Ashburner, M., and J.J. Bonner: The induction of gene activity in Drosophila by heat shock. Cell 17:241, 1979.
32. Heberlein, U., B. England, and R. Tjian: Characterization of Drosophila transcription factors that activate the tandem promoters of the alcohol dehydrogenase gene. Cell 41:965, 1985.
33. Sinha, S.N., R.J. Hellwig, D.P. Allison, and S.K. Niyogi: Conversion of simian virus 40 DNA to ordered nucleoprotein structures by extracts that direct accurate initiation by eukaryotic RNA polymerase II. Nucl Acids Res 10:5533, 1982.
34. Culotta, V.C., R.J. Wides, and B. Sollner-Webb: Eucaryotic transcription complexes are specifically associated in large sedimentable structures: Rapid isolation of polymerase I, II, and III transcription factors. Mol Cell Biol 5:1582, 1985.
35. Horikoshi, M., K. Sekimizu, and S. Natori: Analysis of the stimulatory factor of RNA polymerase II in the initiation and elongation complex. J Biol Chem 259:608, 1984.
36. Bunick, D., R. Zandomeni, S. Ackerman, and R. Weinmann: Mechanism of RNA polymerase II-specific initiation of transcription in vitro: ATP requirement and uncapped runoff transcripts. Cell 29:877, 1982.
37. Ackerman, S., D. Bunick, R. Zandomeni, and R. Weinmann: RNA polymerase II ternary transription complexes generated in vitro. Nucl Acids Res 11:6041, 1983.
38. Tolunay, H.E., L. Yang, W.F. Anderson, and B. Safer: Isolation of an active transcription initiation complex from HeLa cell-free extract. Proc Natl Acad Sci USA 81:5916, 1984.
39. Safer, B., L. Yang, H.E. Tolunay, and W.F. Anderson: Isolation of stable preinitiation, initiation, and elongation complexes from RNA polymerase II-directed transcription. Proc Natl Acad Sci USA 82:2632, 1985.
40. Triadou, P., J.-C. Lelong, F. Gros, and M. Crepin: Tissue-specific formation of transcription-initiation complexes at the 5' end of the mouse major globin gene. Eur J Biochem 135:163, 1983.
41. Hawley, D.K., and R.G. Roeder: Separation and partial characterzation of three functional steps in transcription initiation by human RNA polymerase II. J Biol Chem 260:8163, 1985.
42. Mattaj, I.W., S. Lienhard, J. Jiricny, and E.M. De Robertis: An enhancer-like sequence within the Xenopus U2 gene promoter facilitates the formation of stable transcription complexes. Nature 316:163, 1985.

17. Heavy Metals as Probes for Nonhistone Protein-DNA Interactions

Ryszard Olinski, Zainy M. Banjar, Warren N. Schmidt, Robert C. Briggs, and Lubomir S. Hnilica

Departments of Biochemistry and Pathology, A. B. Hancock, Jr., Memorial Laboratory of the Vanderbilt University Cancer Center, and Center in Molecular Toxicology, Vanderbilt University School of Medicine, Nashville, TN 37232

Assuming that the term "nonhistone protein" applies to any nuclear protein which is not a histone, the number of individual members of this class in all the plants or animals must be truly enormous, even if most species share at least some of these macromolecules. Many of these proteins associate temporarily or permanently with DNA, performing various enzymatic, structural and gene regulatory functions. It is especially this last propensity which attracts increasing attention of many scientists. Yet, as compared with the gene regulatory proteins in prokaryotes, little is known about protein-mediated gene regulation in higher, eukaryotic organisms. However, this scientific area is rapidly evolving with considerable progress in several gene regulatory systems (e.g. the polymerase III transcription factor IIIa, the heat shock protein expression, etc.).

There are at least two major approaches to the studies of DNA-binding proteins. One detects the DNA-protein complexes either through their resistance to nuclease digestion (a system first developed to study repressor proteins in bacteria), through sequence specific interactions of proteins with cloned DNA attached to a solid support, or, through their unique immunological behavior. The second approach attempts to "freeze" the DNA-protein associations by forming covalently linked bonds, usually with the aid of a chemical DNA-protein crosslinking agent or radiation. Because of some experience in our laboratory with covalent DNA-protein crosslinking, we will focus on this second approach.

Initial evidence indicating the presence of cell-specific DNA-protein complexes in higher animals came from the experiments with dehistonized chromatin. When used as immunogen, dehistonized chromatin elicited antisera reacting, by complement fixation, in tissue specific fashion either in tissues undergoing differentiation (1) or carcinogenesis (2). Dissociation and reconstitution experiments showed that these antisera recognized complexes of DNA with chromosomal nonhistone proteins (3-6). Crosslinking of chromatin by bifunctional alkylating agents or radiation rendered these complexes non-dissociable (7,8).

Since cis-diaminedichloroplatinum II (cis-DDP) as well as its trans isomer are known DNA-protein crosslinking agents (9-12), we have initiated studies on mechanisms by which these crosslinks are formed in vivo. The increasing use of cis-DDP in cancer chemotherapy provided us with additional incentives. While the cis-DDP exhibits a potent antitumor activity, its trans isomer is much less toxic and essentially inactive (13). Both these platinum coordination complexes form interstrand DNA crosslinks (11). However, because of its stereospecificity, only the cis-DDP can crosslink two adjacent guanine residues on the same DNA strand (15) (intrastrand crosslinking). In addition to DNA crosslinks, both isomers can crosslink nuclear proteins to DNA either in isolated nuclei or in living cells (9-12). The biological significance of these DNA-protein crosslinks is not known. Because most of the experimental evidence for DNA-protein crosslinking has been obtained by the method of alkaline elution of nicked DNA retained on filters, the identity of the crosslinked proteins is also unknown.

The development of immunoblotting methods of proteins, first separated electrophoretically and then transferred to nitrocellulose sheets, permits positive identification of individual antigens with appropriate antisera or monoclonal antibodies (16,17). Although this experimental approach is not suitable for studies on non-covalently bound DNA-protein complexes which dissociate during electrophoresis, it is applicable to the analysis of covalently DNA-bound proteins. In a relatively simple experimental outline, live cells or isolated nuclei can be incubated with the crosslinker (coordination complexes of platinum or other metals, bifunctional alkylating agents, hexavalent chromium salts, etc.) or irradiated (UV, γ or x-rays). Following solubilization in buffered sodium dodecyl sulfate, the cell lysate is centrifuged at high speed for a prolonged time period to pellet the DNA. Additional solubilization in sodium dodecyl sulfate containing 5 M urea, followed by ultracentrifugation removes essentially all remaining proteins which are not covalently bound to the DNA. The final pellet is then sonicated and digested with DNase I. The released proteins are concentrated, separated by polyacrylamide electrophoresis, and either stained directly or, for more positive identification, transferred to nitrocellulose sheets and incubated with appropriate antiserum or monoclonal antibody (18). The main limitation of the immunoblot analysis is possible denaturation of the antigen(s) by sodium dodecyl sulfate necessary for the separation of covalently bound DNA-protein complexes and their subsequent polyacrylamide gel electrophoresis. However, according to the literature, very few antigens are denatured by this treatment (19). Indeed, in several laboratories, including ours, strong immune response was obtained to proteins eluted from sodium dodecyl sulfate-containing gels (20,21).

Using the technique outlined above, we analyzed HeLa-S_3 proteins crosslinked by relatively high (0.2-2.0 mM) concentrations of cis- or trans-DDP. Coomassie stained gels of the nuclease digested DNA pellets showed that many proteins became crosslinked to the DNA, especially at higher metal concentrations (12). Immunoblots of these proteins reacted with rabbit antisera to 0.35 M NaCl extract or residue of isolated HeLa-S_3 nuclei revealed a number of crosslinked antigens, some unique for either the NaCl soluble or residual fraction, others detectable in both (12). There was a noticeable absence of histones among the crosslinked protein species and essentially identical crosslinking patterns were obtained for live cells or isolated nuclei (12). Interestingly, greater number of crosslinked antigens reacted with antiserum to the 0.35 M NaCl nuclear extract as compared with that to the nuclear residue, thus indicating that there must be numerous proteins relatively weakly associated with DNA. The crosslinking distance of cis-DDP is approximately 4 Å.

A similar approach can be used in more dymanic experiments since relatively short time is needed to "fix" the proteins in their positions relative to the DNA. In collaboration with Drs. Janet and Gary Stein at the University of Florida School of Medicine, we have analyzed the DNA-protein crosslinking pattern during the HeLa-S_3 cell cycle (22). Cells released from double thymidine block were incubated at hourly intervals with 1 mM cis-DDP and the distribution of the crosslinked proteins was determined by immunoblotting with antisera to either 0.35 M NaCl extract or residue of HeLa-S_3 nuclei from the mid-S period of the cell cycle. Again, the 0.35 M NaCl antiserum recognized many more antigens than that to the 0.35 M NaCl nuclear residue. The results in Fig. 1A (antibody to 0.35 M NaCl nuclear extract) show that while most of the antigenic proteins did not change qualitatively during the cell cycle, several exhibited marked quantitative cell cycle dependence in their DNA crosslinking. Most notably, antigens at approx. M_r 34 kD and 120 kD began to crosslink only 8 hrs after the release of cells from G_1/S while another antigen, approx. M_r 44 kD was crosslinked only during the first 8 hrs after the release (see arrows in Fig. 1A). Although very prominent in immunoblots, the approx. M_r 34 kD antigen was barely detectable in parallel gels stained with Coomassie Brilliant Blue. Control samples, not incubated with cis-DDP, taken at the same time intervals as the experimental points did not show any detectable DNA-protein crosslinks

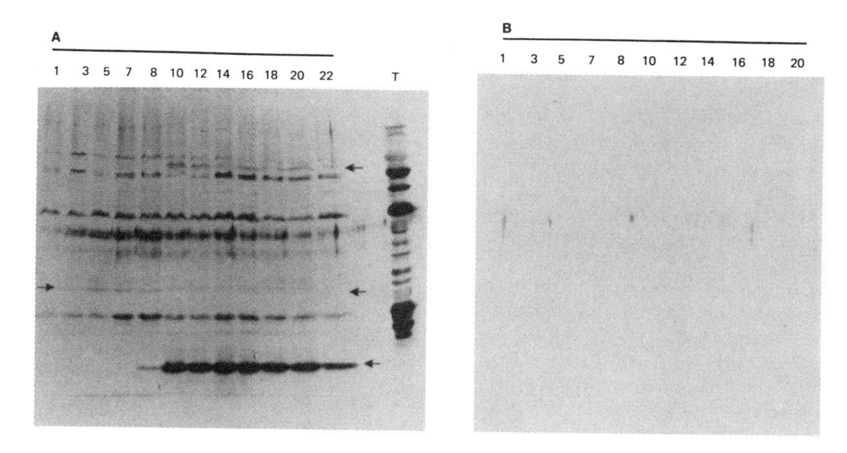

Figure 1. Immunochemical detection of protein antigens crosslinked to the DNA during the cell cycle. (A) The electrophoretogram was transferred to a nitrocellulose sheet which was incubated with antiserum to 0.35 M NaCl extract of isolated HeLa nuclei and immunochemically stained with the peroxidase-antiperoxidase method. Each experimental lane represents protein equivalent to 60 µg pelleted DNA. The experiments were repeated four times with essentially identical results. One of these experiments used hydroxyurea to synchronize the cells. The approx. M_r 34 kD, 44 kD, and 120 kD antigens are indicated by arrows. (B) Controls of the immunochemical detection of proteins crosslinked to the DNA during HeLa cell cycle. The cells were incubated in the absence of cis-DDP and treated as in (A). Each lane represents protein equivalent to 60 µg pelleted DNA. (Reprinted with permission from reference 22.)

(Fig. 1B). Analysis of the antigens reactive with antiserum to the 0.35 M NaCl nuclear residue revealed that most of the antigens reacting in Fig. 1A could not be detected. However, two prominent antigens, the approx. 34 kD and another, approx. 48 kD, started to crosslink at 8 hrs after the G_1/S release. Presently, we are raising monoclonal antibodies to the approx. 34 kD protein to attempt its further characterization.

Although instructive, the HeLa-S_3 experiments did not provide us with information about the identity of the crosslinked proteins and their crosslinking at therapeutic cis-DDP (7 mg/kg) or isotoxic trans-DDP (40 mg/kg) concentrations. To obtain answers to these questions, we utilized some of the various antisera available in our laboratory to nuclear components of Novikoff hepatoma. In these experiments, rats bearing the ascites form of Novikoff hepatoma were injected (six days after the tumor transplant) with either 7 mg/kg cis-DDP or 40 mg/kg trans-DDP. Ascites fluid (3 ml) was removed from the animals by aspiration at 1, 8, 24, 48 and 72 hrs after injection of the drug and processed in a fashion similar to that described for the HeLa-S_3 experiments. Antisera to dehistonized chromatin, to nuclear matrix and to Novikoff hepatoma cytoskeletal preparation were employed to detect the crosslinked antigens.

When administered in therapeutically significant dose (7 mg/kg), cis-DDP crosslinked only four major protein bands. Essentially identical crosslinking patterns were obtained in animals injected with an isotoxic (40 mg/kg) dose of trans-DDP. As can be seen in Fig. 2A-C, the same four crosslinked antigens reacted with either of the three antisera and were identified as the principal components of Novikoff hepatoma intermediate filaments. These three cytokeratins (M_r 39, 49, and 56 kD) and an additional band, reactive with all the three antisera at approx. M_r 68 kD, became crosslinked already 1 hr after the administration of the drug. Our preliminary experiments indicate that although the trans-DDP (in isotoxic dose) crosslinked the cytokeratin proteins more extensively, these crosslinks were removed rapidly (in 24 hrs) as compared with those mediated by the cis-DDP isomer (72 hrs).

Some of our earlier experiments indicated that other metals, when incubated with Novikoff hepatoma cells, also produced DNA protein crosslinks. Using the immunoblotting analysis we have shown that some, but not all members of protein families comprising nuclear matrix, lamina-envelope complex, or dehistonized chromatin became also crosslinked, in addition to the principal cytokeratin fractions (23,24). The most efficient crosslinking was observed for hexavalent chromium compounds. However, various forms of Cu, Hg, Cd, Ni, Pb and Al did also crosslink some of the nuclear proteins to DNA. Each metal exhibited a crosslinking pattern of its own, different from the other metals (24).

It is noteworthy that when the platinum coordination complexes were administered in a fashion approximating the treatment of patients, principal intermediate filament proteins, the p39, p49 and p56 cytokeratins, were among the most prominently crosslinked cellular components. Since intermediate filaments are notable for their cell-specific distribution, changing with differentiation and carcinogenesis (25-27), this finding may be biologically important. Because of their in vivo DNA crosslinking patterns, the intermediate filaments must penetrate the cell nucleus in close proximity to DNA (cis-DDP is a short distance crosslinker), thus providing a continuum from desmosomal junctions on the cell membrane to its nucleus and DNA.

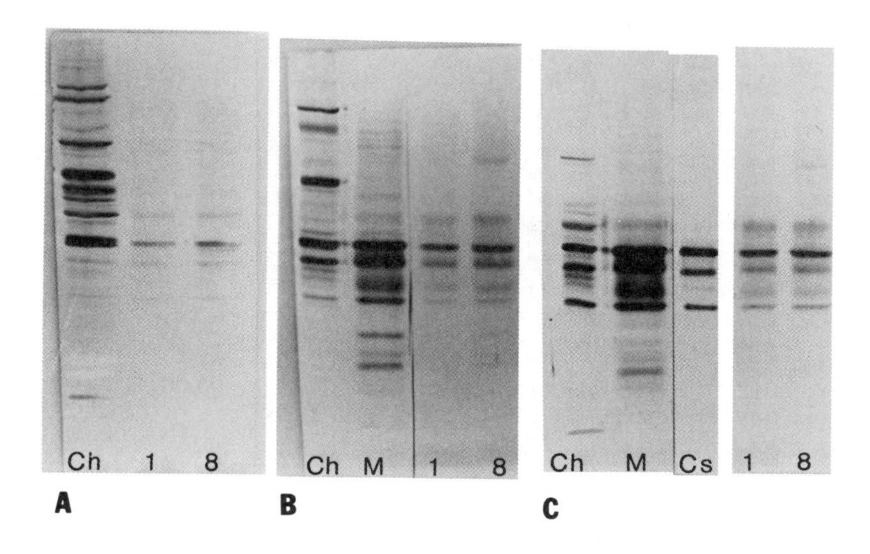

Figure 2. Immunochemical detection of antigens crosslinked to the DNA in Novikoff hepatoma cells after the administration of 40 mg/kg of trans-DDP. Electrophoretically separated proteins were transferred to nitrocellulose sheets and reacted with antisera to: A - dehistonized Novikoff hepatoma chromatin; B - Novikoff hepatoma nuclear matrix; C - Novikoff hepatoma cytoskeletal preparation. Ch = chromatin (10 μg DNA), M = matrix (5 μg protein), Cs = cytoskeletal protein preparation (1 μg protein). 1 and 8 = proteins associated with the DNA pellet 1 hr or 8 hrs, respectively, after the administration of the drug.

Based on their propensity to form DNA-protein crosslinks, heavy metals, especially platinum coordination complexes and salts of hexavalent chromium can become useful in studies of DNA-protein interactions. They appear to enter the cell nucleus with relative ease (hexavalent chromium must be first converted, by the cell, to its trivalent form which is reactive in crosslinking but does not penetrate the cell membrane) and in concentrations up to 2 mM do not form significant amounts of protein-protein crosslinks (23,24). The crosslinked DNA-protein complexes are stable and can be isolated by a variety of methods to recover either the crosslinked protein or DNA.

ACKNOWLEDGMENTS

The authors wish to acknowledge the excellent editorial assistance of Ms. Doris Harris in preparation of this manuscript. Supported by grants from the NCI (CA-26412 and CA-36459), NIEHS (ES-00267) and the Polish Academy of Sciences. ZMB was supported by a grant from King Abdulaziz University, Jeddah, Saudi Arabia.

REFERENCES

1. Chytil, F., and T.C. Spelsberg: Tissue differences in antigenic properties of nonhistone protein-DNA complexes. Nature New Biol 233:215, 1971.
2. Wakabayashi, K., and L.S. Hnilica: The immunospecificity of nonhistone protein complexes with DNA. Nature New Biol 242:153, 1973.
3. Wakabayashi, K., S. Wang, and L.S. Hnilica: Immunospecificity of nonhistone proteins in chromatin. Biochemistry 13:1027, 1974.
4. Chiu, J.F., M. Hunt, and L.S. Hnilica: Tissue-specific DNA-protein complexes during azo dye hepatocarcinogenesis. Cancer Res 35:913, 1975.
5. Wang, S., J.F. Chiu, L. Klyszejko-Stefanowicz, H. Fujitani, and L.S. Hnilica: Tissue specific chromosomal non-histone protein interactions with DNA. J Biol Chem 251:1471, 1976.
6. Campbell, A.M., R.C. Briggs, R.E. Bird, and L.S. Hnilica: Cell specific antiserum to chromosome scaffold proteins. Nucleic Acids Res 6:205, 1979.
7. Olinski, R., R.C. Briggs, L.S. Hnilica, J. Stein, and G. Stein: Gamma-radiation-induced crosslinking of cell-specific chromosomal nonhistone protein-DNA complexes in HeLa chromatin. Radiation Res 86:102, 1981.
8. Olinski, R., R.C. Briggs, L.S. Hnilica, J. Stein, and G. Stein: Cross-linking of chromosomal non-histone proteins to DNA by UV radiation and some antitumor drugs. Chem-Biol Interactions 34:173, 1981.
9. Filipski, J., K.W. Kohn, and W.M. Bonner: Differential crosslinking of histones and non-histones in nuclei by cis-Pt(II). FEBS Lett 152:105, 1983.
10. Lippard, S.J., and J.D. Hoeschele: Binding of cis- and trans-dichlorodiammineplatinum(II) to the nucleosome core. Proc Natl Acad Sci USA 76:6091, 1979.
11. Zwelling, L.A., T. Anderson, and K.W. Kohn: DNA-protein and DNA interstrand cross-linking by cis- and trans-platinum(II) diamminedichloride in L 1210 mouse leukemia cells and relation to cytotoxicity. Cancer Res 39:365, 1979.
12. Banjar, Z.M., L.S. Hnilica, R.C. Briggs, J. Stein, and G. Stein: Cis- and trans-diamminedichloroplatinum(II)-mediated cross-linking of chromosomal non-histone proteins to DNA in HeLa cells. Biochemistry 23:1921, 1984.
13. Hill, J.M., and R.J. Speer: Organo-platinum complexes as antitumor agents (review). Anticancer Res 2:173, 1982.
14. Fichtinger-Schepman, A.M.J., J.L. van der Veer, J.H.J. den Hartog, P.H.M. Lohman, and J. Reedijk: Adducts of the antitumor drug cis-diamminedichloroplatinum(II) with DNA: formation, identification and quantitation. Biochemistry 24:707, 1985.
15. Sherman, S.E., D. Gibson, A.H.J. Wang, and S.J. Lippard: X-Ray structure of the major adduct of the anticancer drug Cisplatin with DNA: cis-[$Pt(NH_3)_2$ {α (pGpG) }]. Science (Washington) 230:412, 1985.
16. Towbin, H., T. Staehelin, and J. Gordon: Electrophoretic transfer of proteins from polyacrylamide gels to nitrocellulose sheets: procedure and some applications. Proc Natl Acad Sci USA 76:1350, 1979.
17. Glass, W.F., R.C. Briggs, and L.S. Hnilica: Identification of tissue-specific nuclear antigens transferred to nitrocellulose from polyacrylamide gels. Science (Washington) 211:70, 1981.
18. Ward, W.S., W.N. Schmidt, C.A. Schmidt, and L.S. Hnilica: Association of cytokeratin p39 with DNA in intact Novikoff hepatoma cells. Proc Natl Acad Sci USA 81:419, 1984.
19. Stumph, W.E., S.C.R. Elgin, and L. Hood: Antibodies to proteins dissolved in sodium dodecyl sulfate. J Immunol 113:1752, 1974.

20. Silver, L.M., and S.C.R. Elgin: Immunological analysis of protein distributions in Drosophila polytene chromosomes. The Cell Nucleus 5:215, 1978.

21. Glass, W.F., J.A. Briggs, M.J. Meredith, R.C. Briggs, and L.S. Hnilica: Enzymatic modification of Novikoff hepatoma lamins A and C. J Biol Chem 260:1895, 1985.

22. Banjar, Z.M., L.S. Hnilica, R.C. Briggs, E. Dominques, J.L. Stein, and G.S. Stein: Cis-diamminedichloroplatinum-mediated crosslinking of nuclear proteins to DNA is cell cycle specific. Arch Biochem Biophys 237:202, 1985.

23. Wedrychowski, A., W.S. Ward, W.N. Schmidt, and L.S. Hnilica: Chromium-induced crosslinking of nuclear proteins and DNA. J Biol Chem 260:7150, 1985.

24. Wedrychowski, A., W.N. Schmidt, and L.S. Hnilica: The in vivo crosslinking of proteins and DNA by heavy metals. J Biol Chem, in press.

25. Babbiani, G., T. Kapanci, P. Barrazone, and W.W. Franke: Immunochemical identification of intermediate filaments in human neoplastic cells: a diagnostic aid for the surgical pathologist. Am J Pathol 104:206, 1981.

26. Lazarides, E.: Intermediate filaments, a chemically heterogeneous developmentally regulated class of proteins. Annu Rev Biochem 51:219, 1982.

27. Osborn, M., M. Altmannsberger, E. Debus, and K. Weber: Conventional and monoclonal antibodies to intermediate filament proteins in human tumor diagnosis, in Cancer Cells. The Transformed Phenotype, A.J. Levine, G.F. Vande Woude, W.C. Topp, and J.D. Watson, eds., Cold Spring Harbor Laboratory, Cold Spring Harbor, NY, pp. 191-200, 1984.

18. Topoisomerase II DNA-Complexes: Novel Targets of Antineoplastic Drug Action

Leonard A. Zwelling

Section of Pharmacology, Department of Chemotherapy Research, University of Texas System Cancer Center M. D. Anderson Hospital and Tumor Institute at Houston, Box 52, 1515 Holcombe Boulevard, Houston, TX 77030

INTRODUCTION

The major problem facing investigators in cancer therapy today is the elucidation of exploitable differences between normal and malignant cells; genetic, immunologic, biochemical, and pharmacologic. That certain chemotherapeutic agents can successfully cure patients with disseminated malignancies suggests that differences exist between normal and malignant cells that active antineoplastic agents can somehow discern. An understanding of the biochemical basis for the differences revealed by the drugs may provide insight into the nature of malignancy itself as well as provide pathways toward better therapies.

Although the list of effective antineoplastic drugs has increased over the past decades, the development of these compounds has been, for the most part, empiric. The actual critical intracellular targets of the majority of the active, clinically useful chemotherapeutic agents remain to be defined.

Recently, an intracellular DNA-binding enzyme, whose biologic function is still unclear, has become an object of intense scrutiny in regard to its potential role as a critical target of several classes of antineoplastic drugs. This enzyme is DNA topoisomerase II. Although many of its biochemical functions are clearly elucidated in chemical systems, which of these functions are acting within living cells, how these actions play essential roles in cell function, how anticancer drugs interfere with these functions, and whether this interference is, in fact, the manner by which these drugs selectively kill malignant cells remain unanswered questions. In this review we will summarize the work which has linked topoisomerase II to the action of certain classes of antineoplastic drugs.

DNA INTERCALATING AGENTS

Among the most active agents used in the treatment of human cancer are DNA intercalating agents. DNA intercalators include adriamycin and 4'-(9-acridinylamino)-methanesulfon-*m*-anisidide (*m*-AMSA) (Figure 1). These drugs have a planar configuration which allows them to interdigitate between adjacent DNA base pairs and untwist the DNA helix. This untwisting is not necessarily accompanied by an interruption in the integrity of the phosphodiester backbone of the helix. Isolated DNA exposed to intercalators is not cleaved. However, the DNA of cells exposed to these drugs is cleaved. It was the study of this cleavage of cellular DNA that led to the identification of topoisomerase II as a target of these drugs.

PROTEIN-ASSOCIATED DNA CLEAVAGE

The cleavage of the DNA of cells exposed to adriamycin was, at one point, believed to be a free radical mediated reaction (1-4). The intercalating capacity of the drug was thought to localize the free radical generating quinone moiety (Figure 1) to the DNA target. Ross *et al*.(5-6) made a critical discovery that tended to favor an alternative model for the production of DNA cleavage in cells exposed to intercalating compounds. These authors found that treatment of mammalian cells with ellipticine, an intercalator with virtually no free radical generating capacity, resulted in the production of DNA cleavage which was identical to that produced by adriamycin (quantified by the filter elution method of Kohn) (7). Thus, the free radical generating capacity of intercalating agents could not readily explain their capacity to produce DNA cleavage in cells.

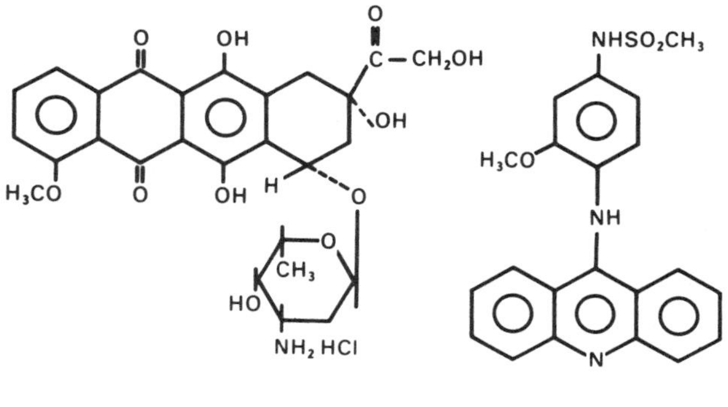

ADRIAMYCIN

m-AMSA

Figure 1. The structures of adriamycin and m-AMSA.

A second important observation by Ross et al. was that intercalator-induced DNA cleavage was accompanied by a strong, probably covalent, attachment of protein to DNA. The magnitude of this "DNA-protein crosslinking" was equal to that of the DNA cleavage. Further, the cleavage was undetectable in this assay without proteolytic digestion of the cell lysate prior to cleavage quantification. The cleavage was thus termed "protein-associated" and "protein-concealed". These results tended to favor an etiologic and physical proximity between intercalator-induced DNA-protein crosslinking and intercalator-induced DNA cleavage. As the two drugs, adriamycin and ellipticine, share the capacity to intercalate, it seemed reasonable to suppose that it was the intercalation per se, that somehow led to protein-associated DNA cleavage.

Intercalation untwists DNA. This action would distort cellular DNA, producing torsional stress. An activity that could correct that torsion by transiently binding to and breaking DNA, thus allowing the DNA to swivel and return to its pretreatment, supercoiled state could be a defense against the DNA-damaging effects of intercalation. Enzymes called topoisomerases were known to have such properties and the products of their reactions with DNA were altered in the presence of intercalators (8). Could intercalators somehow elicit a topoisomerase-mediated response at the cellular level?

EVIDENCE FOR AN ENZYMATIC ORGIN OF INTERCALATOR-INDUCED DNA CLEAVAGE

Experiments with a new intercalator, *m*-AMSA, began to reveal the initial evidence that an enzyme might be involved in the production of protein-associated DNA cleavage. The magnitude of the DNA cleavage produced by *m*-AMSA in mouse leukemia L1210 cells was greater than ten-fold that produced by lethal doses of adriamycin (9). The frequency of cleavage reached a maximum level beyond which the addition of more drug produced no further breakage. This saturation in cleavage was not due to a saturation of *m*-AMSA uptake which was linear over the dose range within which DNA cleavage saturation occurred (10).

Following the removal of L1210 cells from *m*-AMSA, protein-associated DNA cleavage rapidly disappeared. Both the rate of cleavage formation and the rate of cleavage disappearance could be slowed at reduced temperature or brought to zero at ice temperature (Figure 2). The DNA cleavage produced by *m*-AMSA, and by the other intercalators (9,11,12), was both single- and double-stranded.

Results of assays of DNA cleavage using alkaline sucrose sedimentation (without proteolytic digestion of cell lysates) rather than filter elution produced quantitatively comparable results to those obtained using filter elution (9). Thus, it appeared that the protein that concealed the cleavage in filter elution assays without proteinase did not do so in sedimentation assays without proteinase, suggesting that the DNA-bound protein did not bridge the cleavage site.

The production of *m*-AMSA-induced, protein-associated DNA cleavage in L1210 cells was saturable, reversible, and temperature-sensitive. The cleavage was both single- and double-stranded, and the cleavage sites appeared not to be bridged by the bound proteins.

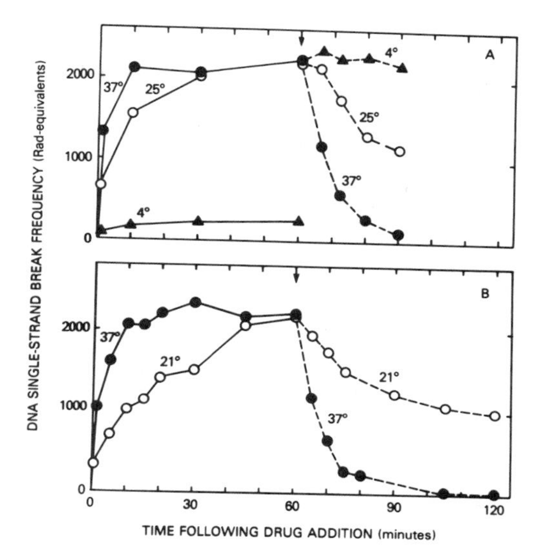

Figure 2. The temperature sensitivity of the rates of formation and disappearance of DNA cleavage in murine leukemia cells treated with 1 µM m-AMSA. Cells were treated at the indicated temperatures and DNA cleavage quantified at various times up to 60 min. At this point (arrow) cells were removed from the drug by centrifugation and those cells treated at 37° were resuspended in drug-free medium at various temperatures. A and B are separate experiments (reprinted from ref. 9 with permission).

DNA CLEAVAGE ACTIVITY WAS LOCALIZED TO THE CELL NUCLEUS

By isolating cell nuclei (13-15), the m-AMSA-induced DNA cleaving activity could be demonstrated to reside in a component of the cell nucleus extractable in 0.35 M salt (16). Returning this salt extract to the salt-depleted nuclei restored the susceptibility of those nuclei to intercalator-induced DNA cleavage. Additionally, the dependence of the cleavage on magnesium, the identification of a pH optium of approximately 6.4 for cleavage, and the potentiating effects of ATP and nonhydrolyzable ATP analogues on intercalator-induced cleavage could all be demonstrated in isolated nuclei (15). The protein-associated, intercalator-induced cleavage activity appeared to be embodied in a protein component of the cell nucleus extractable by salt. The cleavage reaction had the characteristics of an enzyme-mediated reaction such as temperature, salt, magnesium and pH optima.

CHROMATIN CONFORMATION IS CHANGED IN CELLS TREATED WITH m-AMSA

The results of alkaline elution and alkaline sedimentation assays are governed by the behavior of denatured, linear DNA strands. As it was speculated that *m*-AMSA was somehow producing alterations in DNA three-dimensional structure, we wished to examine the behavior of cellular DNA using a less disruptive assay in which results might be governed by the resultant three-dimensional structure of cellular DNA. The nucleoid sedimentation assay fulfilled this requirement (17).

Nucleoids are protein-DNA complexes that sediment in neutral 1.9 M NaCl-sucrose gradients following cell lysis atop the gradient (18,19). Most cellular proteins, including nucleosomal histones, remain atop the gradient. Some chromatin structural components remain with the DNA, retaining its compact structure. This compact DNA-protein complex sediments rapidly. Small amounts of DNA cleavage decompact the sedimenting nucleoid and retard its sedimentation.

Nucleoids from m-AMSA-treated cells were expected to sediment much more slowly than nucleoids from untreated cells due to the large amount of DNA cleavage detected in alkaline elution and alkaline sedimentation assays. Surprisingly, the nucleoids from *m*-AMSA-treated cells sedimented as rapidly as those from untreated cells. However, limited proteinase digestion of the lysed cells prior to sedimentation unmasked some of the breaks (as indicated by a slowing of nucleoid sedimentation). Thus, the breaks existed, but were protein-concealed. The absence of detectable alteration in DNA compaction following *m*-AMSA treatment indicated that the proteins prevented swiveling at the DNA cleavage sites within the sedimenting nucleoid. Thus, the morphology at the break site must be as follows: at neutral pH the protein associated with the break site bridges the break site and, at least within the sedimenting nucleoid, prevents DNA from swiveling about the break site. At alkaline pH, one end, but not both ends, of the protein-bridged DNA site dissociates from the protein, implying noncovalent linkage. The other binding site withstands alkaline pH and is probably covalent in nature (20).

The intercalating dye ethidium can be incorporated into nucleoid gradients at various concentrations. Such titration gradients can be used to monitor the structure of sedimenting nucleoids. Increasing concentrations of ethidium decrease DNA twist (Tw: angle between adjacent DNA base pairs), thus decompacting and slowing the sedimentation of the nucleoids. With increasing ethidium concentrations the nucleoid will recompact and sediment rapidly. Thus, ethidium titration curves are biphasic (Figure 3). The ethidium titration curve described by the nucleoids from *m*-AMSA-treated cells was not identical to that described by the nucleoids from untreated cells. Without ethidium, the treated and untreated nucleoids sedimented identically (21). However, the minimum sedimentation distance (complete nucleoid decompaction) was attained at a lower ethidium concentration in nucleoids from *m*-AMSA-treated cells. Thus, it took a smaller reduction in twist (by ethidium) to fully relax the sedimenting nucleoids from *m*-AMSA treated cells.

The distance a nucleoid sediments is a measure of the supercoiled density or Writhe (Wr) of its DNA. Wr is usually negative in mammalian cells (negative supercoiling). Twist (Tw) and linking number (Lk) (the number of times one DNA strand winds about the other) are related through the equation:

$$Wr = Lk - Tw \qquad \text{(Equation 1)}$$

At the minimum nucleoid sedimentation point, the nucleoid is relaxed, Wr=0 and Lk=Tw. In as much as no swiveling occurs in sedimenting nucleoids (*vide supra*) from *m*-AMSA-treated cells and because swiveling is required to change Lk, Lk is invariant in nucleoids regardless of the ethidium concentration within the gradient. Thus, if it takes less ethidium to reduce twist to the point where Wr = 0, Lk in *m*-AMSA-treated cells is higher than in untreated cells. So although swiveling (changing Lk) did not occur within the nucleoids from *m*-AMSA-treated cells, the altered Lk detected in these nucleoids indicates that Lk had changed within the cell prior to cell lysis and nucleoid sedimentation. Lk changes in living cells are the result of topoisomerase action. Topoisomerases are enzymes that, by definition, alter the Lk of DNA (22).

Apparently *m*-AMSA blocks an enzyme that reduces Lk in cells. Such an enzyme must be a DNA gyrase-like topoisomerase II. The ability of novobiocin, a type II topoisomerase inhibitor, to slow the restoration of Lk to its pretreatment level following *m*-AMSA treatment and removal (21) substantiates the involvement of type II topoisomerase in *m*-AMSA's actions within cells.

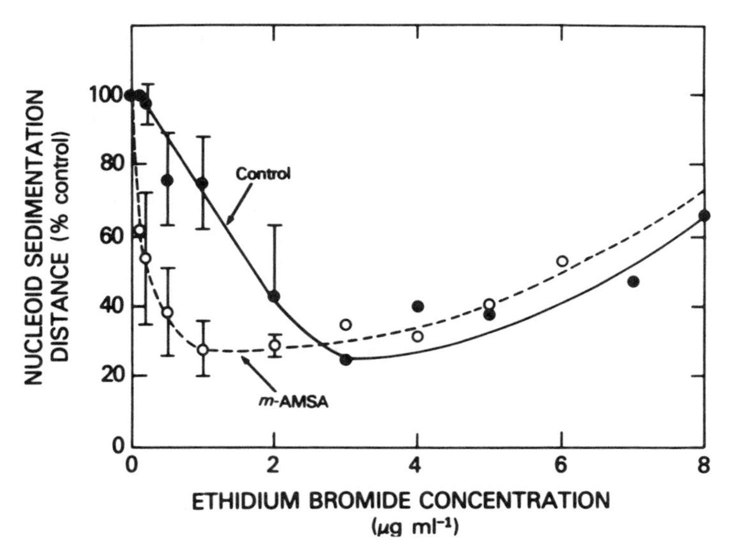

Figure 3. Ethidium bromide titration curves of nucleoids from untreated or m-AMSA-treated (2 μM x 30 min) cells. Error bars are standard deviations of at least three independent determinations (reprinted with permission, from ref. 21).

m-AMSA-STIMULATED DNA-PROTEIN CROSSLINKING ACTIVITY COPURIFIES WITH TYPE II TOPOISOMERASE FROM L1210 CELLS

The production of protein-associated DNA cleavage by m-AMSA in mammalian cells appeared consistent with the drug's action on a type II DNA topoisomerase. Proof would require enzyme purification. Rather than purify topoisomerase II, we sought to purify "m-AMSA-stimulated DNA-protein crosslinking activity". In this way we could focus on the activity of interest without introducing a bias as to what the resultant protein would be. Filipski *et al.* (16) demonstrated that this activity was localized to a 0.35 M NaCl extract of L1210 cell nuclei; we began our purification there. A simple filter binding assay was devised using ^{3}H-SV-40 DNA as a target and chemical conditions identical to those employed to quantify DNA-protein crosslinking within cellular DNA (filter elution) (7).

Gel filtration, DNA cellulose chromatography, and glycerol gradient sedimentation were employed. The resultant m-AMSA-stimulated DNA-protein crosslinking activity copurified with topoisomerase II (Figure 4). Double-stranded DNA cleavage was also produced by this enzyme and enhanced by m-AMSA. The enzyme is covalently bound to the 5' side of the cleavage site as previously demonstrated in whole cells (23). During our performance of this work, Dr. Leroy Liu of the Johns Hopkins School of Medicine published findings with topoisomerase II (24) and m-AMSA (25) virtually identical to our own. The findings of our work with cells, nuclei and isolated biochemical systems are summarized in Table 1.

m-AMSA stabilizes a normal intermediate in the DNA breaking-rejoining cycle of topoisomerase II (Figure 5) in a fashion analogous to the effects of nalidixic acid on bacterial DNA gyrase (26). Treatment of this stabilized complex with detergents or alkali results in covalent DNA-protein binding and DNA cleavage. The drugs simply increase the quantity of this normally-formed complex (E*DNA) and thus enhance the magnitude of DNA-protein crosslinking and DNA cleavage by increasing the pool of "cleavable complexes".

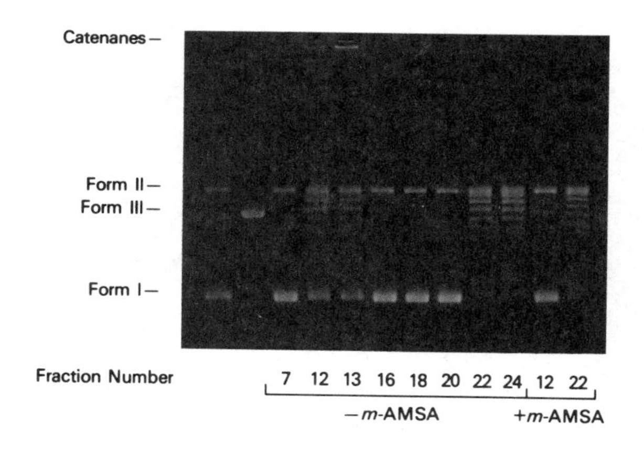

Figure 4. The more rapidly sedimenting of two topoisomerases from L1210 cell nuclei contained catenating and m-AMSA-stimulated DNA cleaving activities. Topoisomerase activity was found in glycerol gradient fractions 12-13 and 22-24, but only the former had DNA cleaving activity in the presence of 10 μM m-AMSA. Markers for Forms I, II, and III are in the two left lanes.

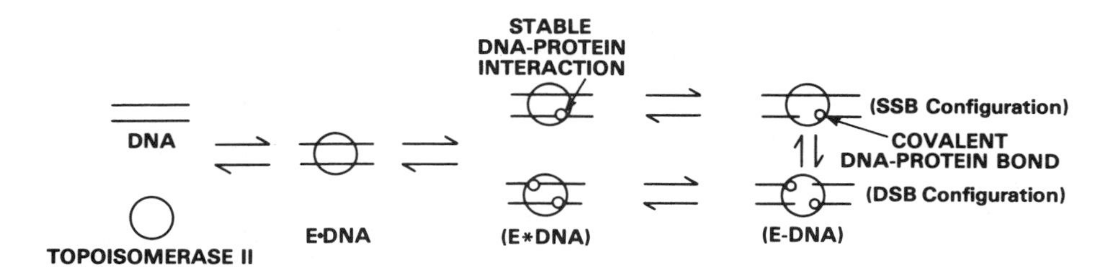

Figure 5. The configurations of topoisomerase(E)-DNA interaction. E·DNA = noncovalent interaction; E*DNA, the "cleavable complex" stabilized by m-AMSA; E-DNA, the cleaved complex derived from the cleavable complex following SDS treatment.

DNA INTERCALATION IS NOT A REQUIREMENT OF A DRUG WHICH CAN STABILIZE TOPOISOMERASE II-DNA COMPLEXES

The epipodophyllotoxins VP-16 and VM-26 are extremely active antineoplastic drugs that do not intercalate into DNA. These compounds, however, do stabilize topoisomerase II-DNA complexes in a fashion precisely like that of m-AMSA (27,28). These data suggest that these drugs, as well as the intercalating agents, probably exert their DNA effects through interactions with topoisomerase itself or the DNA-topoisomerase complex rather than through DNA binding as the primary event.

THE POSSIBLE SIGNIFICANCE OF THE ACTION OF ANTINEOPLASTIC DRUGS ON TOPOISOMERASE II

Effective chemotherapeutic agents appear to be able to distinguish between normal and malignant cells *in vivo*. At some level this is likely to reflect important biochemical differences between phenotypically distinct cell types. Evidence has now accumulated that identifies topoisomerase II as a target of several active antineoplastic drugs. Can we use this information to gain insight into the actual functions of topoisomerase II in mamamlian cells or to understand whether interference with these functions can explain the preferential cytotoxicity of the drugs toward malignant cells?

A number of pharmacologic and hormonal agents have been demonstrated to enhance the formation of m-AMSA-induced cleavable complexes in mammalian cells when the cells were exposed to these agents prior to m-AMSA treatment. All these agents altered the growth and DNA synthesis rates of the cell populations. The agents include 1-β-D-arabinofuranosylcytosine (ara-C) (29), hydroxyurea (HU) (29), 5-azacytidine (5-aza-CR) (30), α-difluoro-methylornithine (DFMO) (31) and 17-β-estradiol (E_2) (32). 5-aza-CR and E_2 not only perturb growth, but they also produce specific gene transcription. Alterations in DNA conformation also accompany exposure of cells to these agents. Thus, three kinds of cellular functions may be associated with topoisomerase II action, at least as indicated by enhanced m-AMSA-induced DNA cleavability: DNA replication, gene transcription, and chromosome remodeling, as for example, would be required for post-synthetic chromosome segregation prior to mitosis.

In a recent study of quiescent and proliferating normal human fibroblasts and malignant human glioblastoma cells, we found that quiescent cells of both phenotypes exhibited a low sensitivity to the DNA cleaving and cytotoxic actions of m-AMSA. Proliferating cells of both phenotypes exhibited a greater m-AMSA sensitivity than did their quiescent counterparts. However, proliferating malignant cells were more sensitive to the DNA cleaving actions of m-AMSA than were proliferating normal fibroblasts. This greater cleavability of the DNA of malignant cells in response to m-AMSA was not reflected in a greater cytotoxic sensitivity when compared with proliferating fibroblasts. However, if the majority of the normal cells within a cancer patient behave as did the quiescent fibroblasts, and the rapidly growing malignant cells resemble proliferating cells, the observed therapeutic index may have its biochemical basis in the actions of m-AMSA on topoisomerase II. The normal cells most susceptible to destruction by intercalators and epipodophyllotoxins are the rapidly proliferating bone marrow elements. This tends to indicate that the growth rate or proliferative capacity, not the cellular phenotype, may be the critical determinant of the cytotoxic potency of these agents.

By examining the development of enhanced m-AMSA DNA cleavability over time, following the conversion of cells from quiescence to proliferation, we found that the onset of DNA synthesis preceded the onset of enhanced cleavability by several hours. Although enhanced cleavability reached its peak in mid- to late S-phase, cleavability did not decrease in G_2. Thus, the enhanced topoisomerase II-mediated DNA cleavability was acquired by newly proliferating cells in S-phase, and maintained in G_2. This led us to speculate that topoisomerase II may play a role in premitotic chromosome segregation. Studies of DNA replication of intracellular viruses (33), yeast (34), and bacteria (35) all support the idea that topoisomerase II may be critical for chromosome segregation. The chromosomes of mammalian cells treated with topoisomerase II interactive compounds have also been demonstrated to exhibit "chromosome stickiness", which may represent unsegregated chromosomes (36).

Drugs that disrupt chromosome segregation would be likely to result in cell death and would be most devastating to rapidly dividing cells. This description is consistent with the spectrum of clinical activity displayed by the compounds known to interact with topoisomerase II.

Although the involvement of topoisomerase II with gene transcription is not as clearly documented as is the enzyme's role in chromosome segregation, recent studies of cleavage sites produced by topoisomerase show good agreement with nuclease hypersensitivity sites at the 5' or 3' ends of genes (37). Future developments in this field are likely to focus on obtaining a clearer biochemical picture of the precise mechanism of topoisomerase II's function as well as how drugs interfere with it. At the cellular level, studies with malignant cells resistant to these drugs may aid our cellular biological, pharmacological, and biochemical understanding of topoisomerase II-mediated processes and their physiologic role in normal and malignant cell function.

ACKNOWLEDGEMENTS

We thank Rhonda Briscoe for typing this manuscript.

REFERENCES

1. Bachur, N.R., M.Y. Gee and R. D. Friedman: Nuclear catalyzed antibiotic free radical formation. Cancer Res. 42:1078, 1982.

2. Bachur, N.R., S.L. Gordon and M.Y. Gee: A general mechanism for microsomal activation of quinone anticancer agents to free radicals. Cancer Res. 38:1745, 1978.

3. Bachur, N.R., S.L. Gordon, M.Y. Gee, and K. Kon: NADPH cytochrome P-450 reductase activation of quinone anticancer agents to free radicals. Proc. Natl. Acad. Sci USA 76:954, 1979.

4. Berlin, V. and W. A. Haseltine: Reduction of adriamycin to a semiquinone free radical by NADPH cytochrome P-450 reductase produces DNA cleavage in a reaction mediated by molecular oxygen. J. Biol. Chem. 256:4747, 1981.

5. Ross, W.E., D.L. Glaubiger and K.W. Kohn: Protein-associated DNA breaks in cells treated with adriamycin or ellipticine. Biochim. Biophys. Acta. 519:23, 1978.

6. Ross, W.E., D. Glaubiger and K.W. Kohn: Qualitative and quantitative aspects of intercalator-induced DNA strand breaks. Biochim. Biophys. Acta. 562:41, 1979.

7. Kohn, K.W., R.A.G. Ewig, L.C. Erickson and L.A. Zwelling: Measurement of strand breaks and cross-links by alkaline elution in DNA Repair: A Laboratory Manual of Research Procedures, Friedberg, E.C. and P.C. Hanawalt, Eds., Marcel Dekker, New York, 1981, 379.

8. Keller, W.: Determination of the number of superhelical turns in simian virus 40 DNA by gel electrophoresis. Proc. Natl. Acad. Sci. USA 72:4876, 1975.

9. Zwelling, L.A., S. Michaels, L.C. Erickson, R.S. Ungerleider, M. Nichols and K.W. Kohn: Protein-associated deoxyribonucleic acid strand breaks in L1210 cells treated with the deoxyribonucleic acid intercalating agents 4'-(9-acridinylamino)methanesulfon-m-anisidide and adriamycin. Biochemistry 20:6553, 1981.

10. Zwelling, L.A., D. Kerrigan, S. Michaels and K.W. Kohn: Cooperative sequestration of m-AMSA in L1210 cells. Biochem. Pharmacol. 31:3269, 1982.

11. Ross, W.E. and M.O. Bradley: DNA double-strand breaks in mammalian cells after exposure to intercalating agents. Biochim. Biophys. Acta 654:129, 1981.

12. Zwelling, L.A., Y. Pommier, D. Kerrigan and M.R. Mattern: Intercalator-induced protein-associated DNA strand breaks in mammalian cells in Developments in Cancer Chemotherapy, Glazer, R.I., ed., CRC Press, Boca Raton, FL, 1984, 181.

13. Filipski, J. and K.W. Kohn: Ellipticine-induced protein-associated DNA breaks in isolated L1210 nuclei. Biochem. Biophys. Acta 698:280, 1982.

14. Pommier, Y., D. Kerrigan, R. Schwartz, and L. Zwelling: The formation and resealing of intercalator-induced DNA strand breaks in isolated L1210 cell nuclei. Biochem. Biophys. Res. Commun. 107:576, 1982.

15. Pommier, Y., R.E. Schwartz, K.W. Kohn and L.A. Zwelling: Formation and rejoining of deoxyribonucleic acid double-strand breaks induced in isolated cell nuclei by antineoplastic intercalating agents. Biochemistry 23:3194, 1984.

16. Filipski, J., J. Yin and K.W. Kohn: Reconstitution of intercalator-induced DNA scission by an active component from nuclear extracts. Biochim. Biophys. Acta 741:116, 1983.

17. Cook, P.R., and I.A. Brazell: Supercoils in human DNA. J. Cell Sci. 19:261, 1975.

18. Mattern, M.R. and R. B. Painter: Dependence of mammalian DNA replication on DNA supercoiling. I. Effects of ethidium bromide on DNA synthesis in permeable Chinese hamster ovary cells. Biochim. Biophys. Acta 563:293, 1979.

19. Mattern, M.R., L.A. Zwelling, D. Kerrigan and K.W. Kohn: The reconstitution of higher-order DNA structure after x-irradiation of mammalian cells. Biochem. Biophys. Res. Commun. 112:1077, 1983.

20. Pommier, Y., M.R. Mattern, R.E. Schwartz and L.A. Zwelling: Absence of swiveling of sites of intercalator-induced protein-associated deoxyribonucleic acid strand breaks in mammalian cell nucleoids. Biochemistry 23:2922, 1984.

21. Pommier, Y., M.R. Mattern, R.E. Schwartz, L.A. Zwelling, and K.W. Kohn: Changes in deoxyribonucleic acid linking number due to treatment of mammalian cells with the intercalating agent 4'-(9-acridinyl-amino)methanesulfon-m-anisidide. Biochemistry 23:2927,1984.

22. Gellert, M.: DNA topoisomerases. Ann Rev Biochem. 50:879, 1981.

23. Marshall, B., R.K. Ralph, and R. Hancock: Blocked 5'-termini in the fragments of chromosomal DNA produced in cells exposed to the antitumor drug 4'-[(9-acridinyl)-amino]methanesulphon-m-aniside (mAMSA). Nucleic Acids Research 11:4251, 1983.

24. Liu, L.F., T.C. Rowe, L. Yung, K.M. Tewey and G.L. Chen: Cleavage of DNA by mammalian DNA topoisomerase II. J. Biol. Chem. 258:15365, 1983.

25. Nelson, E.M., K.M. Tewey and L.F. Liu: Mechanism of antitumor drug action: Poisoning of mammalian DNA topoisomerase II on DNA by 4'-(9-acridinylamino)-methanesulfon-m-anisidide. Proc. Natl. Acad. Sci. USA 81:1361, 1984.

26. Higgins, N.P. and N.R. Cozzarelli: The binding of gyrase to DNA analysis by retention by nitrocellulose filters. Nucleic Acids Research 10:6833, 1982.

27. Chen, G. L., Yang, T.C. Rowe, B.D. Halligan, K.M. Tewey and L.F. Liu: Nonintercalative antitumor drugs interfere with the breakage reunion reaction of mammalian DNA topoisomerase II. J. Biol. Chem. 259:13560, 1984.

28. Ross, W., T. Rowe, B. Glisson, J. Yalowich and L. Liu: Role of topoisomerase II in mediating epipodophyllotoxin-induced DNA cleavage. Cancer Research 44:5857, 1984.

29. Minford, J., D. Kerrigan, M. Nichols, S. Shackney and L.A. Zwelling: Enhancement of the DNA breakage and cytotoxic effects of intercalating agents by treatment with sublethal doses of 1-β-D-arabinosylcytosine or hydroxyurea in L1210 cells. Cancer Research 44:5583, 1984.

30. Zwelling, L.A., J. Minford, M. Nichols, R.I. Glazer and S. Shackney: Enhancement of intercalator-induced deoxyribonucleic acid scission and cytotoxicity in murine leukemia cells treated with 5-aza cytidine Biochem. Pharmacol. 33:3903, 1984.

31. Zwelling, L.A., D. Kerrigan, and L. J. Marton: Effect of difluoromethylornithine, an inhibitor of polyamine biosynthesis, on the topoisomerase II-mediated DNA scission produced by 4'-(9-acridinylamino)methane-sulfon-m-anisidide in L1210 murine leukemia cells. Cancer Research 45:1122, 1985.

32. Zwelling, L.A., D. Kerrigan and M.E. Lippman: Protein-associated intercalator-induced DNA scission is enhanced by estrogen stimulation in human breast cancer cells. Proc. Natl. Acad. Sci. USA 80:6182, 1983.

33. Sundin, O. and A. Varshavsky: Arrest of segregation leads to accumulation of highly intertwined catenated dimers: dissection of the final stages of SV40 DNA replication. Cell 25:659, 1981.

34. DiNardo, S., K. Voelkel, and R. Sternglanz: DNA topoisomerase II mutant of Saccharomyces cerevisiae: Topoisomerase II is required for segregation of daughter molecules of the termination of DNA replication. Proc. Natl. Acad. Sci. USA 81:2616, 1984.

35. Steck, T.R. and K. Drlica: Bacterial chromosome segregation: evidence for DNA gyrase involvement in decatenation. Cell 36:1081, 1984.

36. Rosenberg, L.J. and W. N. Hittelman: The direct and indirect clastogenic activity of anthracenedione. Cancer Research 43:3270, 1983.

37. Udvardy, A., P. Schedl, M. Sander and T.S. Hsieh: Novel partitioning of DNA cleavage sites for Drosophila topoisomerase II. Cell 40:933, 1985.

TABLE 1. Characteristics of DNA Cleavage Produced by *m*-AMSA in L1210 Cells, in Their Isolated Nuclei, or in an Isolated Chemical System Containing SV-40 DNA and L1210 Cell Topoisomerase II.

	Whole Cells	Nuclei	Chemical System
Protein-concealment	+	+	+
Reversibility	+	+	+
Temperature Sensivity	+	+	+
Saturability	+	+	+
pH Dependence		+	+
Mg^{+2} Dependence		+	+
NaCl Dependence		+	+
ATP Dependence		–	–

19. Regulation of Gene Expression by Thyroid Hormone Nuclear Receptors

Herbert H. Samuels, Ana Aranda,[a] Juan Casanova, Richard P. Copp, Zebulun D. Horowitz, Laura Janocko, Angel Pascual,[a] Hadjira Sahnoun, Frederick Stanley, Bruce M. Raaka, and Barry M. Yaffe

Division of Molecular Endocrinology, Rose F. Tishman Laboratories for Geriatric Endocrinology, Department of Medicine, New York University Medical Center, New York, NY 10016, and [a]Departmento de Endocrinología Experimental, Instituto de Investigaciones Biomédicas, Del C.S.I.C., Facultad de Medicina de la U.A.M., Arzobispo Morcillo 4., 28029 Madrid, Spain

INTRODUCTION

The thyroid hormones have marked effects on the growth, development and metabolism of essentially all tissues of higher organisms (1,2). The action of L-thyroxine (L-T4) and L-triiodothyronine (L-T3) has been examined both in intact animals and in cultured cells (1). Studies over the past decade indicate that the thyroid hormones act to regulate the expression of specific genes by stimulating the accumulation of mRNA's which encode for specific proteins. The regulation of several genes has been studied in detail (3-16). These include; stimulation of growth hormone synthesis in the anterior pituitary *in vivo* (3); regulation of growth hormone gene expression in cultured rat pituitary cells (4-12); and stimulation of malic enzyme mRNA (13-15) as well as several other genes which encode for hepatic proteins whose function is currently unknown (16).

Abundant evidence indicates that most if not all of the significant cellular responses regulated by the thyroid hormones in mammalian cells is mediated by a cellular receptor localized to the cell nucleus (1). Evidence to support this notion has been derived from studies in intact animals (17) and cell culture (18). Several different clonal strains of growth hormone producing rat pituitary cells (GH_1, GH_3, and GC) have been shown to be highly effective cell culture models to study thyroid hormone action (4-12). In these cells physiological concentrations of L-T3 and L-T4 stimulate growth hormone synthesis and growth hormone mRNA accumulation (6-10), and the kinetics of stimulation is similar to that in the anterior pituitary after hormone injection (3,8). These cell lines contain thyroid hormone nuclear receptors which have affinity and hydrodynamic properties similar to receptors in various tissues *in vivo* (19,20). Therefore, cultured

GH_1, GH_3 and GC cells appear to be highly useful cell culture models to study the molecular basis of thyroid hormone action. In this chapter we review the physical and biologic properties of thyroid hormone receptors and describe studies which have explored the detailed mechanisms by which the thyroid hormone-receptor complex influences growth hormone gene expression.

THYROID HORMONE RECEPTORS

General Properties

Table 1 lists the properties of thyroid hormone nuclear receptors in GH_1 cells. Similar iodothyronine affinity and hydrodynamic properties have also been observed for receptor derived from rat liver nuclei (19-22). The affinity of the nuclear receptor for L-T3, L-T4, and other iodothyronine analogues in intact cells parallel the iodothyronine biologic potency in cultured cells (23) and in intact animals (24). Scatchard analysis demonstrates linear plots with no evidence of positive or negative cooperativity (7,21,22). GH_1 and GC cell nuclei contain approximately 15,000-20,000 copies of receptor (7,25) while the anterior pituitary and rat liver have approximately 8,000-10,000 receptors per cell nucleus (26). Other tissues have lower levels of receptor abundance (26). Unlike steroid hormone receptors no cytosolic forms of the receptor have been identified and in the absence of ligand the thyroid hormone receptor is only identified in the nuclear fraction (27,28).

TABLE 1. Properties of Thyroid Hormone Receptors in GH_1 Cells

1. Single class of high-affinity binding sites with no evidence for cooperativity (18,22,23,27).
2. Iodothyronine affinity parallels biologic potency (23).
3. 15,000 receptors per cell nucleus (22,23,25).
3. No cytoplasmic counterpart (22,28).
4. DNA binding protein (20).
5. Receptor is extracted from nuclei by 0.4 M KCl.
   (a) sedimentation coefficient - 3.8 S (20).
   (b) Stokes radius - 3.3 nm (20).
   (c) estimated molecular weight - 55,000 (20).
6. Receptor has a half-life of 4 to 5 hours and a synthetic rate of about 2,000 molecules/hour/cell (25,28).
7. Nuclear receptor levels are reduced by thyroid hormone incubation (23,25).

The receptor is a DNA binding protein (20,29) which is primarily associated with linker DNA regions although a subset of receptor (approximately 5%) appears to be enriched in mononucleosome particles which have characteristics reported for transcriptionally competent chromatin (20,30). Micrococcal nuclease digestion excises the receptor along with a protected linker DNA fragment of approximately 30-40 base pairs (bp) and this species has a sedimentation coefficient of 6.5 S, a Stokes radius of 6.0 nm, and a particle density of 1.42 g/cm^3 (20). Extraction of nuclei with 0.4 M KCl solubilizes the receptor in a stable form which has a sedimentation coefficient of 3.8 S, a Stokes radius of 3.3 nm, and a particle density of 1.36 g/cm^3 (20).

Receptors extracted with high salt show identical affinities for iodothyronines as receptor in isolated nuclear preparations (27) suggesting that the interaction of receptor with DNA does not alter the affinity for hormone. Using the Stokes radius, sedimentation coefficient, and particle density of the salt extracted receptor we have estimated its molecular weight (M_r) to be about 55,000 (20). Photoaffinity labeling studies using N-2-diazo-3,3,3-trifluoro-propionyl-L-triiodothyronine in intact cells (28,31) indicates that receptor exists as an abundant 47,000 M_r species (75% of total receptor) as well as a 57,000 M_r doublet form (approximately 25% of receptor). We have presented evidence to support the notion that the 57,000 M_r form may be a precursor of the 47,000 M_r species which is more abundant because of its longer half-life (28).

<u>Regulation of Nuclear Receptor Levels by L-T3 and Estimation of Receptor Half-Life and Synthetic Rates</u>

Nuclear associated receptor is in a dynamic steady state and in the absence of hormone has a half-life of approximately 4-5 hours and a synthetic rate of 1,500-2,000 molecules per cell per hour (25). In GH_1 and GC cells, L-T3 or other iodothyronines elicit a reduction in nuclear receptor levels which is directly related to the extent of occupancy of receptor by hormone (23,25). To assess the mechanism of L-T3 regulation of receptor levels, we developed a dense amino acid labeling technique to quantitate receptor half-life and synthetic rates and the influence of hormone on these processes (25). To estimate receptor half-life and the rate of receptor synthesis, cells are incubated with amino acids uniformly labeled with the non-radioactive dense isotopes ^{15}N, ^{13}C, and ^{2}H. Newly synthesized proteins will be of higher density than pre-existing proteins and can be separated using velocity gradient centrifugation. The density of most proteins range from 1.34-1.38 g/cm^3. The total substitution of amino acids containing all three heavy isotopes results in a maximal density increase of approximately 8% (0.109 g/cm^3).

The velocity of sedimentation of a particle is directly related to the difference between the particle density (d_p) and the density of the gradient (d_g). Since the density of most proteins is approximately 1.36 g/cm^3, an 8% increase would yield protein with a density of 1.47 g/cm^3. If a gradient is constructed such that the average density is 1.2 g/ml, the difference in density between the dense protein and the gradient (d_p-d_g) is 0.27 while the value for the protein of normal density is 0.16. Therefore, under the correct gradient conditions an 8% increase in the density of protein can result in a 1.7-fold increase (0.27/0.16) in the velocity of sedimentation. To achieve an average gradient density of 1.2 g/ml, sucrose gradients are constructed with D_2O instead of H_2O. The newly synthesized protein of high density can be separated from pre-existing protein of normal density by velocity sedimentation and the respective populations of dense and normal receptor quantitated using L-[^{125}I]T3 (25).

Figure 1 illustrates the time course of dense amino acid labeling of thyroid hormone receptor in GH_1 cells. Cells were first incubated without hormone in medium containing dense amino acids for 1.5, 5, or 20 hours (25). One hour before harvesting the cells, 5 nM L-[^{125}]T3 was added to the medium to quantitate receptor levels. Labeled dense and normal receptors were then extracted from isolated nuclei with 0.4 M KCl and separated by sedimentation in 17-32% D_2O-sucrose gradients.

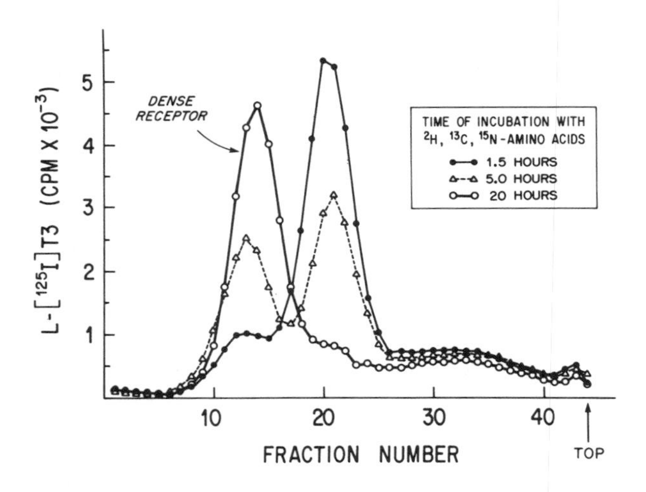

Figure 1. Separation of normal and dense thyroid hormone receptors in sucrose-D_2O gradients. GH_1 cells were cultured in medium without thyroid hormone for a total of 43 hours. Normal medium was replaced with dense medium 1.5, 5, or 20 hours prior to harvesting the cells. The dense medium was supplemented with 5 nM L-[^{125}I]T3 1 hour before harvesting the cells to identify the receptor. Nuclei were isolated and hormone-receptor complexes were extracted with 0.4 M KCl. The extracts were sedimented in sucrose-D_2O gradients as described in the text. The direction of sedimentation is from right to left. From Raaka, B.M., and Samuels, H.H (25).

After 1.5 hours in dense amino acid medium approximately 90% of the receptors are of normal density and a more rapidly sedimenting species corresponding to newly synthesized dense receptor is identified (fraction 12). After 5 hours in dense medium approximately equal amounts of normal and dense receptor are present. Lastly, after 20 hours approximately 95% of the receptor is of the dense form. By measuring the kinetics of disappearance of receptor of normal density the receptor half-life can be determined (32). In cells cultured without thyroid hormone exposure, except for the one hour incubation with L-[^{125}I]T3 to occupy receptor, the half-life ($t_{1/2}$) was found to be 4.7 ± 0.5 hours (25).

The effect of L-T3 on reducing receptor levels during the first 15 hours of incubation was shown to result from a rapid effect of hormone on decreasing the synthetic rate of receptor (25). Reduction of receptor caused by thyroid hormone is rapidly reversed after removal of hormone by cells (25). Over a 20 hour period receptor repopulates the nucleus to the level identified in cells cultured chronically without hormone. The recovery of receptor is due to an increase in receptor synthesis to the rate which occurs in cells cultured without hormone. Therefore, during both the onset of and the recovery from thyroid hormone-mediated receptor reduction, changes in nuclear receptor levels are primarily determined by rapid alterations in receptor synthesis which is controlled by thyroid hormone. Although the mechanism of L-T3 regulation of receptor levels has not yet been defined, all studies indicate that the reduction of receptor is mediated by an interaction of iodothyronine with nuclear associated receptor (23,25). This observation suggests that the thyroid hormone-receptor complex may act to lower receptor levels by decreasing the expression of the receptor gene which would result in the decrease in steady state amounts of thyroid hormone receptor mRNA.

REGULATION OF GROWTH HORMONE GENE EXPRESSION BY THYROID HORMONE

In GH_1 and GC cells, L-T3 and other iodothyronines stimulate an increase in growth hormone synthesis and mRNA accumulation in which the steady state level of induction parallels the extent of receptor occupancy by hormone (7,10). To more precisely compare the relationship between the level of L-T3-receptor complexes and the biologic response, relative growth hormone gene transcription rates were determined using isolated nuclei (33). GC cells were incubated with L-T3 for varying times and the nuclei were then isolated and incubated with [α-^{32}P]UTP at 29°C for 1 hour _in vitro_ (12). Labeled nuclear RNA was then isolated and hybridized to nitrocellulose filter immobilized rat growth hormone cDNA (34) to determine the level of radiolabeled growth hormone gene transcripts (12). This procedure results in _in vitro_ elongation of gene transcripts initiated in the intact cell. Since no initiation occurs _in vitro_, this approach provides an estimate of the instantaneous rate of growth hormone gene transcription at the time of nuclear isolation (33).

Figure 2 compares the kinetics of binding of L-[^{125}I]T3 to receptor and the kinetics of stimulation of growth hormone gene transcription (12). Maximal growth hormone gene transcription rates occurred 1 hour after L-T3 incubation and at 30 minutes the transcription rate was 71% of the maximal level. The level of L-T3-receptor complexes at 30 minutes was 74% of the maximal value observed at 1-2 hours of incubation. Further incubation resulted in a decrease in the level of L-T3-receptor complexes and a parallel decrease in the rate of growth hormone gene transcription, suggesting that the level of hormone-receptor complexes is rate limiting for this response. In other experiments we have estimated the half-life of growth hormone mRNA to be about 40 hours which does not appear to altered by L-T3 incubation (12).

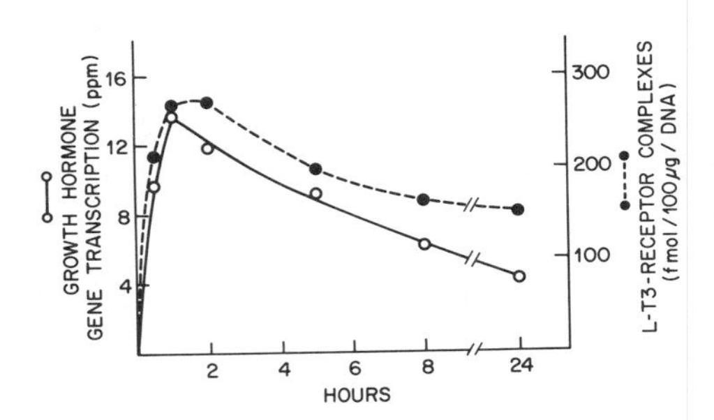

Figure 2. Comparison of the kinetics of L-T3-receptor binding and receptor abundance to changes in the rate of transcription of the growth hormone gene. GC cells were inoculated and cultured in 175 cm^2 flasks for 3 days and were then incubated with medium containing 10% (v/v) hormone depleted fetal calf serum for an additional 48 hours. The cells were then incubated with 5 nM L-T3 for the times indicated. The nuclei were then isolated and transcribed *in vitro* with [α-^{32}P]UTP, and the [^{32}P]RNA was isolated, purified, and hybridized to nitrocellulose filters containing immobilized plasmid with an 802 nucleotide growth hormone cDNA insert (12,34). The results are expressed as ppm (o) of the total [^{32}P]RNA input in the hybridizations. The levels of L-T3-receptor complexes (o) were determined by incubating parallel flasks with 5 nM L-[^{125}I]T3 for the times indicated. From Yaffe and Samuels, (12).

Our studies (12) indicate that: 1) stimulation of growth hormone gene transcription by L-T3 is directly proportional at any point in time to the level of L-T3-receptor complexes; 2) stabilization of growth hormone mRNA does not appear to play an important role in stimulation of growth hormone synthesis by thyroid hormone; and 3) the kinetics of stimulation of the synthesis of growth hormone by L-T3 can be accounted for by transcriptional control and a growth hormone mRNA half-life of about 40 hours. Furthermore, these studies (12) indicate that thyroid hormone can rapidly stimulate transcription of the growth hormone gene with no significant lag time between the occupancy of receptor and the transcriptional response (Figure 2). These observations support the notion that the thyroid hormone-receptor complex interacts with DNA sequences in the environs of the growth hormone gene to elicit direct control of transcription.

5'-FLANKING DNA SEQUENCES OF THE GROWTH HORMONE GENE MEDIATE THYROID HORMONE STIMULATION OF GENE EXPRESSION

The rat growth hormone gene, cloned in the Eco RI-Hind III sites of pBR322 (prGHeh-5.8), consists of 2.1 kb of DNA encompassing five exons, four introns, and 1.8 kb of 5'-flanking and 1.9 kb of 3'-flanking DNA sequences (35,36). Co-transformation of mouse Ltk$^-$ cells with the cloned rat growth hormone and the herpes simplex virus thymidine kinase genes yields stable transformants with growth hormone mRNA which is shorter than that of GC or GH_3 cells (37,38). This appears to reflect aberrant transcriptional initiation in the second intron of the growth hormone gene (38) and no stimulation resulted from thyroid hormone incubation (38). Introduction of rat growth hormone gene sequences into fibroblastic cells using a retroviral vector also resulted in no documented stimulation by thyroid hormone (39).

Recent gene transfer studies indicate that cell-specific *trans*-acting factors and 5'-flanking DNA sequences play an important role in regulating the expression of genes in specific cell types (40-42). Therefore, GC cells would appear to be ideal for growth hormone gene transfer studies. To analyze the regulatory role of the 5'-flanking region of the rat growth hormone gene we have transfected GC cells with a chimeric gene (pGH-xgpt) which consists of 1.8 kb of the 5'-flanking region of the rat growth hormone gene ligated to *E. coli* DNA containing the structural gene for xanthine-guanine phosphoribosyl transferase (XGPT). pGH-xgpt yields stable transformants with relatively high frequency which demonstrate regulated expression of the XGPT gene by thyroid hormone (43). These studies support the notion that 5'-flanking sequences of the rat growth hormone gene contain a DNA control element which can mediate regulated expression by thyroid hormone (43).

Chimeric Plasmid Construction

Figure 3 outlines the construction of pGH-xgpt (43). The rat growth hormone gene (prGHeh-5.8) (36), obtained through the generosity of John D. Baxter, University of California San Francisco Medical Center, was cleaved with XhoI and BamHI. This gives a fragment cleaved at +8, relative to the cap site of the growth hormone gene, which contains 1.8 kb of the 5'-flanking region contiguous through sequences from the EcoRI site to the Bam HI site of pBR322. pSV2-gpt, a construct developed by Mulligan and Berg (44), was cleaved with HindIII and BamHI. This yields a fragment of about 1.9 kb containing the XGPT gene as well as the SV40 termination and polyadenylation signals but without the SV40 early promoter and enhancer region (44,45). This fragment was ligated to the BamHI cohesive end of pBR322 contiguous with the 5' region of the rat growth hormone gene. The XhoI and HindIII cohesive ends were blunt-ended with the Klenow fragment of *E. coli* DNA polymerase and the plasmid (pGH-xgpt) was circularized by blunt end ligation with T4 DNA ligase. The filled-in fragment from the XhoI site (CTCGA), which extends to +11, contributes a deoxyadenosine residue to the filled-in HindIII site (AGCTT). This regenerates the the HindIII restriction sequence (AAGCTT) which begins at +11 and is cleaved between +11 and +12.

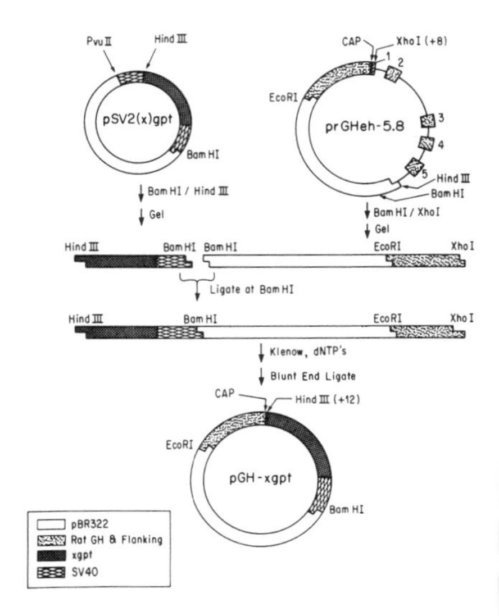

Figure 3. Construction of pGH-xgpt. Details are given above in the text. From Casanova et al. (43).

Development of Stable Transformants of GC Cells

GC cells were transfected with pGH-xgpt using $CaPO_4$ precipitation (47,48) and stable transformants were selected with medium containing mycophenolic acid, xanthine and L-T3 (43). pGH-xgpt yielded stable transformants with a frequency of about one colony/30,000 cells. No stable transformants developed from cultures transfected with pBR322. In addition, no stable transformants were obtained with pSV2-gpt when 150,000-300,000 cells were transfected. However, based on transfection of a greater number of cells, we estimate that pSV2-gpt transforms GC cells with a frequency of about one colony/10^6 cells. Furthermore, these transformants replicate at very low rates in selective medium compared to stable transformants generated by pGH-xgpt. Control transfection experiments with pSV2-gpt using cultured H4 hepatoma cells yielded many stable transformed clones suggesting the the SV40 early enhancer and/or promoter in pSV2-gpt does not function efficiently in GC cells (43).

All clones derived from pGH-xgpt demonstrated L-T3 regulation of XGPT enzyme activity. This might be expected if the XGPT gene was under control of a thyroid hormone regulatory element in the 5' region of the growth hormone gene since L-T3 was present in the selection medium. The results presented here were performed with one of the stable transformed lines and are similar to that observed using cells derived from colonies which were pooled before amplification in selective medium. The level of thyroid hormone nuclear receptors in the stable transformed cell line, determined as previously described (12), was 320 fmol/100 ug DNA (about 20,000 receptors/cell nucleus) which is in the same range reported for untransfected GC cells (12).

Stimulation of XGPT Enzyme Levels by Thyroid Hormone in Stable Transformants of GC cells

Transfected cells, maintained in selective medium containing 2.5 nM L-T3, and untransfected (control) GC cells were incubated with nonselective DHAP medium supplemented with 10% (v/v) thyroid hormone depleted calf serum (7,12). Thirty h later the media was replaced with the same medium and 5 nM L-T3 was added to half of the cultures. After 24 h the cells were refed with the same media. Twenty h later aliquots of media were saved for determination of growth hormone by radioimmunoassay (7) and the cells were harvested for quantitation of cell protein, XGPT enzyme activity, and for cytoplasmic RNA purification (12). XGPT levels are determined by measuring the enzymatic formation of [^{14}C]xanthosine-5'-monophosphate ([^{14}C]XMP) from [^{14}C]xanthine and 5-phosphoribosyl-1-pyrophosphate. Figure 4 illustrates that only lysates of the transfected cells synthesized [^{14}C]XMP and this was stimulated 5-fold by L-T3 incubation.

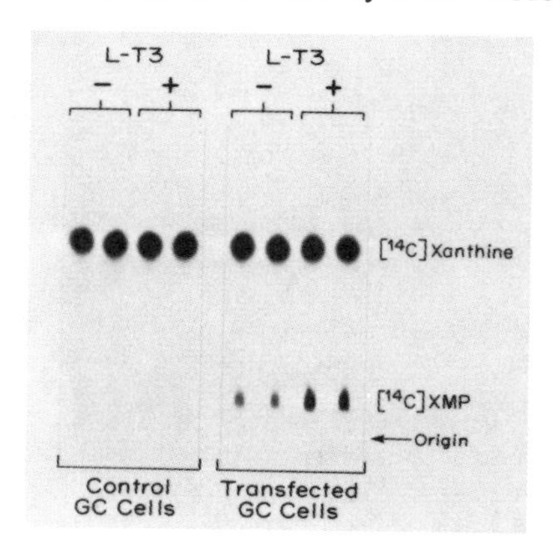

Figure 4. Stimulation of XGPT enzyme levels by L-T3 in stable transformants of GC cells transfected with pGH-xgpt. Transfected and untransfected (control) GC cells were cultured in 75 cm^2 flasks ± 5 nM L-T3 and assayed for XGPT enzyme levels as described (43). Ten ul of cell lysate containing 10 ug of protein from duplicate flasks were incubated with 10 ul of buffer containing [^{14}C]xanthine + 5-phosphoribosyl-1-pyrophosphate for 30 minutes at 37°C. Five ul samples from the assays were chromatographed on PEI-cellulose thin layer plates which were then dried and autoradiographed at -80°C for 15 hours using Kodak X-Omat AR film. From Casanova et al. (43).

Stimulation of XGPT mRNA levels by L-T3 and S1 Nuclease Mapping

The levels of XGPT mRNA are shown in Figure 5. L-T3 stimulated a marked increase (about 30-fold) in the levels of XGPT mRNA in the transfected cells which migrates primarily as a 1.8 kb species. Virtually no XGPT mRNA was detected in transfected cells cultured without L-T3, but it was detected after extending the autoradiographic time from 15 h to 80 h. Two very minor species (one greater than 1.8 kb and one at about 1.3 kb) were also stimulated by L-T3. No hybridization signal was detected in untransfected GC cells even after extremely long autoradiographic times.

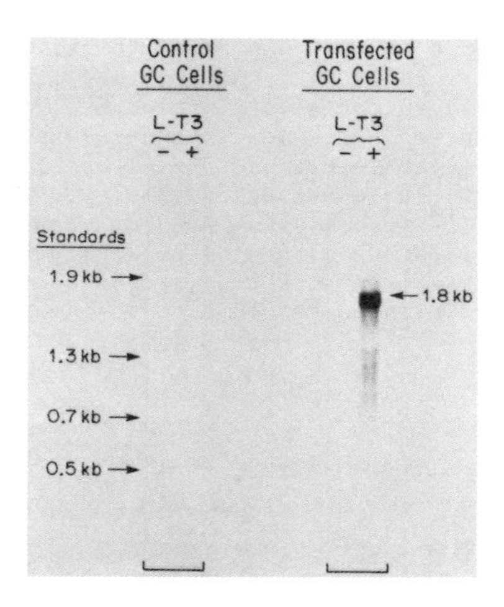

Figure 5. Stimulation of XGPT mRNA levels by L-T3. Twenty ug of total cytoplasmic RNA from the experiment in Figure 4 was electrophoresed in a 1.25% agarose-formaldehyde gel. RNA was transferred to nitrocellulose and the blot was hybridized to nick translated pSV2-gpt (3×10^8 dpm/ug DNA) and autoradiographed for 15 h at $-80^{\circ}C$ using Kodak X-Omat AR film and a Lighting Plus intensifying screen (Du Pont). Without L-T3 (-); With L-T3 (+). From Casanova et al. (43).

The length of the bacterial DNA is 1,057 bp (49) and the distance from the end of the bacterial DNA to the SV40 polyadenylation signal is about 750 nucleotides (44,45). Therefore, excluding the length of any poly(A^+) tail, an XGPT mRNA initiated at the cap site of the growth hormone gene would be about 1.8 kb which is in agreement with the size observed (Figure 5). The site of initiation of the mRNA was analyzed by S1 nuclease analysis (46,50) (Figure 6). pGH-xgpt was digested with Asp718. This gives a DNA fragment of about 650 nucleotides which includes: 312 of 5'-flanking DNA to the cap site of the growth hormone gene; 11 including the cap site to the HindIII restriction site; and 322 to the 5' overhang in the bacterial DNA. This fragment was end labeled with T4 polynucleotide kinase and used for the S1 nuclease analysis (43,46,50). An XGPT mRNA initiated at the cap site of the growth hormone gene would protect a fragment of 333 nucleotides (11 nucleotides including the cap site of the growth hormone gene and 322 nucleotides of E. coli DNA) (43). Figure 6 indicates that cytoplasmic RNA from transfected cells incubated with L-T3 protects only one DNA fragment of about 333 nucleotides indicating that initiation begins at the cap site of the growth hormone gene. No protected DNA fragments were identified using RNA derived from transfected cells cultured without L-T3 (43).

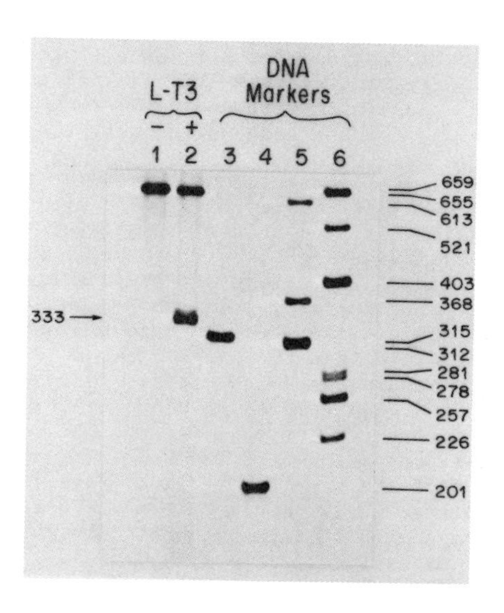

Figure 6. S1 nuclease analysis of XGPT mRNA. Twenty-five ug of total cytoplasmic RNA from transfected cells cultured ± 5 nM L-T3 was hybridized to the ^{32}P-end labeled fragment and treated with S1 nuclease as described in the text. DNA size markers consisted of polynucleotide kinase-labeled fragments from restriction enzyme digests of pBR322 or pSV2-gpt as described below. Lane 1, without L-T3; Lane 2, with L-T3; Lane 3, Hind III-Asp718 digest of pSV2-gpt (318 nucleotides); Lane 4, Hind III-Bgl II digest of pSV2-gpt (201 nucleotides); Lane 5, Taq I digest of pBR322; Lane 6, Alu I digest of pBR322. Autoradiography was at -80°C for 4 h using Kodak X-Omat AR film and a Lighting Plus intensifying screen. The band at the top of lanes 1 and 2 reflects rehybridization of the labeled DNA probe. From Casanova et al. (43).

DISCUSSION

Although thyroid hormone has been shown to influence the expression of a number of rat genes (1), the growth hormone gene is one of only a few genes which have been cloned and sequenced (35,36). Therefore, with appropriate gene transfer studies, it should be possible to delineate the location and the sequence of the DNA element(s) which mediate thyroid hormone regulation of growth hormone gene expression. We utilized growth hormone producing cultured GC cells for our analysis since cell-specific factors appear to play an important role in regulation of gene expression in specific cell types (40-42). Although it would be of interest to use the entire rat growth hormone gene for transfer studies, it would be difficult to distinguish between expression of the transferred and the endogenous growth hormone gene. Therefore, we developed a chimeric plasmid (pGH-xgpt) to explore whether 5'-flanking sequences of the rat growth hormone gene mediate regulated expression of the gene by thyroid hormone. In pGH-xgpt, expression of the XGPT gene would be under control of the growth hormone gene 5'-flanking region and could reveal whether this region contains a thyroid hormone responsive control element.

pGH-xgpt yields stable transformants of GC cells with relatively high frequency and L-T3 markedly stimulates XGPT mRNA levels (Figure 5). Furthermore, the mRNA is initiated at the cap site of the growth hormone gene (Figure 6) indicating correct hormonal regulation of transcriptional initiation. The extent of L-T3 stimulation of XGPT mRNA is much greater (about 30-fold) than the 5-fold stimulation of XGPT enzyme activity. This difference can be explained if XGPT mRNA has a much shorter half-life than the enzyme. In this study, the transfected cells were maintained in selective medium with 2.5 nM L-T3 up to the time the medium was replaced with nonselective medium containing hormone-depleted calf serum. Half of the cells were maintained without L-T3 for 74 h while the remaining cells received L-T3 for the last 44 h of the experiment. Therefore, XGPT enzyme levels in hormone-depleted medium would reflect L-T3 stimulation prior to the medium exchange and, with a longer half-life than XGPT mRNA, would not decrease as markedly over the 74 h period.

DNA sequences 3' to the start site of transcription have been shown to influence the expression of the human β-globin (51), chick thymidine kinase (52), and the rearranged immunoglobulin heavy chain (42) genes. However, 5'-flanking DNA sequences have been shown to be important for steroid hormone regulation of the genes for chick ovalbumin and lysozyme (53,54) and human metallothionein II_A (55) and corresponds to the steroid hormone receptor binding regions of these genes (53-55). A high affinity glucocorticoid receptor binding domain has recently been identified in the first intron of the human growth hormone gene (56) suggesting that hormone regulatory elements may exist within intragenic sites in the growth hormone gene family.

Evidence from several laboratories indicates that thyroid hormone stimulates growth hormone mRNA accumulation primarily by enhancing the rate of transcription of the gene (5,11,12). However, it has recently been suggested that stabilization of growth hormone mRNA by L-T3 is the major cause of growth hormone mRNA accumulation in GH_3 and GC cells (57,58). In this study we document that 5'-flanking DNA of the rat growth hormone gene (including 11 nucleotides 3' of the cap site) contains a control element which mediates regulated expression by thyroid hormone. Regulation of a heterologous gene further supports the notion that thyroid hormone mediates expression by transcriptional control mechanisms and not mRNA stabilization. There are a number of convenient restriction endonuclease sites in the 5'-flanking region of the growth hormone gene (-550, Pst I; -312, Asp718; and -236, Bgl II) (34,36). This should allow for the construction of deletion mutants to localize the borders and identify the DNA sequence of the thyroid hormone responsive element in the 5'-flanking region of the rat growth hormone gene. These studies, coupled with an analysis of the interaction of the thyroid hormone receptor with DNA sequences in the 5'-flanking region, should provide additional insights into the molecular basis of transcriptional regulation by the thyroid hormone-receptor complex.

ACKNOWLEDGEMENT

We would like to thank Mary McCarthy for expert secretarial assistance. This research was supported by Grants AM16636 and AM21566 from the National Institutes of Health.

REFERENCES:

1. Molecular Basis of Thyroid Hormone Action. (Oppenheimer, J.H., and Samuels, H.H., eds.) Academic Press, N.Y., 1983.
2. Wolff, E.C., and Wolff, J. The mechanism of action of the thyroid hormones. in The Thyroid Gland (Pitt-Rivers, R., and Trotter, W.R., eds.) Butterworth and Co., Ltd., London. 1: 237, 1964.
3. Hervas, F., Morreale de Escobar, G. and Escobar Del Ray, F.: Rapid effects of single small doses of L-thyroxine and triiodo-L-thyronine on growth hormone, as studied in the rat by radioimmunoassay. Endocrinology 97:91, 1975.
4. Dobner, P.R., Kawasaki, E.W., Yu, L.-Y., and Bancroft, F.C.: Thyroid or glucocorticoid hormone induces pre-growth-hormone mRNA and its probable nuclear precursor in rat pituitary cells. Proc. Natl. Acad. Sci. U.S.A. 78:2230, 1981.
5. Evans, R.M., Birnberg, N.C., and Rosenfeld, M.G.: Glucocorticoid and thyroid hormones transcriptionally regulate growth hormone gene expression. Proc. Natl. Acad. Sci. U.S.A. 79:7659, 1982.
6. Martial, J.A., Baxter, J.D., Goodman, H.M and Seeburg, P.H.: Regulation of growth hormone messenger RNA by thyroid and glucocorticoid hormones. Proc. Natl. Acad. Sci. U.S.A. 74:1816, 1977.
7. Samuels, H.H., Stanley, F., and Shapiro, L.E.: Dose-dependent depletion of nuclear receptors by L-triiodothyronine: Evidence for a role in the induction of growth hormone synthesis in cultured GH_1 cells. Proc. Natl. Acad. Sci. U.S.A. 73:3877, 1976.
8. Samuels, H.H., Stanley, F., and Shapiro, L.E.: Control of growth hormone synthesis in cultured GH_1 cells by 3,5,3'-triiodo-L-thyronine and glucocorticoid agonists and antagonists: studies on the independent and synergistic regulation of the growth hormone response. Biochemistry 18:715, 1979.

9. Seo, H., Vassart, G., Brocas, H., and Refetoff, S.: Triiodothyronine stimulates specifically growth hormone mRNA in rat pituitary tumor cells. Proc. Natl. Acad. Sci. U.S.A. 74:2054, 1977.
10. Shaprio, L.E., Samuels, H.H. and Yaffe, B.M.: Thyroid and glucocorticoid hormones synergistically control growth hormone mRNA in cultured GH_1 cells. Proc. Natl. Acad. Sci. U.S.A. 75:45, 1978.
11. Spindler, S.R., Mellon, S.H. and Baxter, J.D.: Growth hormone gene transcription is regulated by thyroid and glucocorticoid hormones in cultured rat pituitary tumor cells. J. Biol. Chem. 257:11627, 1982.
12. Yaffe, B.M., and Samuels, H.H.: Hormonal regulation of the growth hormone gene: relationship of the rate of transcription to the level of nuclear thyroid hormone-receptor complexes. J. Biol. Chem. 259:6284, 1984.
13. Magnuson, M.A., and Nikodem, V.M.: Molecular cloning of a cDNA sequence for rat malic enzyme: direct evidence for induction in vivo of rat liver malic enzyme by thyroid hormone. J. Biol. Chem. 258:12712, 1983.
14. Siddiqui, U.A., Goldflam, T., and Goodridge, A.G.: Nutritional and hormonal regulation of the translational levels of malic enzyme and albumin mRNA's in avian liver cells in vivo and in culture. J. Biol. Chem. 256:4544, 1981.
15. Towle, H.C., Mariash, C.N., Schwartz, H.L., and Oppenheimer, J.H.: Quantitation of rat liver messenger ribonucleic acid for malic enzyme induction by thyroid hormone. Biochemistry 20:3486, 1981.
16. Carr, F.E., Jump, D.B., and Oppenheimer, J.H.: Distribution of thyroid hormone-responsive translated products in rat liver polysome and postribosomal ribonucleoprotein populations. Endocrinology 115:1737, 1984.
17. Oppenheimer, J.H.: The nuclear receptor-triiodothyronine complex: relationship to thyroid hormone distribution, metabolism, and biological action. In Molecular Basis of Thyroid Hormone Action. (Oppenheimer, J.H., and Samuels, H.H., eds.) Academic Press, New York. pp 1-34, 1983.
18. Samuels, H.H.: Identification and characterization of thyroid hormone receptors and action using cell culture techniques. in: Molecular Basis of Thyroid Hormone Action. (Oppenheimer, J.H., and Samuels, H.H., eds.) Academic Press, New York pp 35-64, 1983.
19. Latham, K.R., Ring, J.C. and Baxter, J.D.: Solubilized nuclear "receptors" for thyroid hormones: physical characteristics and binding properties, evidence for multiple forms. J. Biol. Chem. 251:7388, 1976.
20. Perlman, A.J., Stanley, F., and Samuels, H.H.: Thyroid hormone nuclear receptor: evidence for multimeric organization in chromatin. J. Biol. Chem. 257:930, 1982.
21. Koerner, D., Schwartz, H.L, Surks, M.I., Oppenheimer, J.H. and Jorgensen, E.C.: Binding of selected iodothyronine analogues to receptor sites of isolated rat hepatic nuclei. J. Biol. Chem. 250:6417, 1975.

22. Samuels, H.H., and Tsai, J.S.: Thyroid hormone action in cell culture: demonstration of nuclear receptors in intact cells and isolated nuclei. Proc. Natl. Acad. Sci. U.S.A. 70:3488, 1973.
23. Samuels, H.H., Stanley, F., and Casanova, J.: Relationship of receptor affinity to the modulation of thyroid hormone nuclear receptor levels and growth hormone synthesis by L-triiodothyronine and iodothyronine analogs in cultured GH_1 cells. J. Clin. Invest. 63:1229, 1979.
24. Schwartz, H.L., Trence, D., Oppenheimer, J.H., Jiang, N.S. and Jump, D.B.: Distribution and metabolism of L- and D-triiodothyronine (T3) in the rat: preferential accumulation of L-T3 by hepatic and cardiac nuclei as a probable explanation of the differential biological potency of T3 enantiomers. Endocrinology 113:1236, 1983.
25. Raaka, B.M., and Samuels, H.H.: Regulation of thyroid hormone nuclear receptor levels in GH_1 Cells by 3,5,3'-triiodo-L-thyronine: use of dense amino acid labeling to determine the influence of hormone on the receptor half-life and the rate of appearance of newly synthesized receptor. J. Biol. Chem. 256:6883, 1981.
26. Oppenheimer, J.H., Schwartz, H.L., and Surks, M.I.: Tissue Differences in the concentration of triiodothyronine nuclear binding sites in the rat: liver, kidney, pituitary, heart, brain, spleen, and testes. Endocrinology 95:897, 1974.
27. Samuels, H.H., and Tsai, J.S.: Thyroid hormone action: demonstration of similar receptors in isolated nuclei of rat liver and cultured GH_1 cells. J. Clin. Invest. 53:656, 1974.
28. Casanova, J., Horowitz, Z.D., Copp, R.P., McIntyre, W.R., Pascual, A., and Samuels, H.H.: Photoaffinity labeling of thyroid hormone nuclear receptors: influence of n-butyrate and analysis of the half-lives of the 57,000 and 47,000 molecular weight receptor forms. J. Biol. Chem. 259:12084, 1984.
29. MacLeod, K.M., and Baxter, J.D.: Chromatin receptors for thyroid hormones: interaction of the solubilized proteins with DNA. J. Biol. Chem. 251:7380, 1976.
30. Samuels, H.H., Perlman, A.J., Raaka, B.M., and Stanley, F.: Organization of the thyroid hormone receptor in chromatin. Recent Progr. Hormone Res. 38:557, 1982.
31. Pascual, A., Casanova, J., and Samuels, H.H.: Photoaffinity labeling of thyroid hormone nuclear receptors in intact cells. J. Biol. Chem. 257:9640, 1982.
32. Schimke, R.T.: Methods for analysis of enzyme synthesis and degradation in animal tissues. Methods Enzymol. 40:241, 1975.
33. McKnight, G.S., and Palmiter, R.D.: Transcriptional regulation of the ovalbumin and conalbumin genes by steroid hormones in chick oviduct. J. Biol. Chem. 254:9050, 1979.
34. Seeburg, P.H., Shine, J., Martial, J.A., Baxter, J.D., and Goodman, H.M.: Nucleotide sequence and amplification in bacteria of the structural gene for rat growth hormone. Nature 270:486, 1977.

35. Page, G.S., Smith, S. and Goodman, H.M.: DNA sequence of the rat growth hormone gene: location of the 5' terminus of the growth hormone mRNA and identification of an internal transposon-like element. Nucleic Acids Res. 9:2087, 1981.
36. Barta, A., Richards, R.I., Baxter, J.D., and Shine, J.: Primary structure and evolution of rat growth hormone gene. Proc. Natl. Acad. Sci. U.S.A. 78:4867, 1981.
37. Moore, D.D., Walker, M.D., Diamond, D.J., Conkling, M.A., and Goodman, H.M.: Structure, expression, and evolution of growth hormone genes. Recent Prog. Hormone Res. 38:197, 1982.
38. Karin, M., Eberhadt, N.L., Mellon, S.H., Malich, N., Richards, R.I., Slater, E.P., Barta, A., Martial, J.A., Baxter, J.D., and Cathala, G.: Expression and hormonal regulation of the rat growth hormone gene in transfected mouse cells. DNA 3:147, 1984.
39. Miller, A.D., Ong, E.S., Rosenfeld, M.G., Verma, I.M. and Evans, R.M.: Infectious and selectable retrovirus containing an inducible rat growth hormone minigene. Science 225:993, 1984.
40. Walker, M.D., Edlund, T., Boulet, A.M. and Rutter, W.J.: Cell-specific expression controlled by the 5'-flanking region of insulin and chymotrypsin genes. Nature 306:557, 1983.
41. Episkopou, V., Murphy, A.J.M., and Efstratiadis, A.: Cell-specified expression of a selectable hybrid gene. Proc. Natl. Acad. Sci. U.S.A. 81:4657, 1984.
42. Gillies, S.D., Morrison, S.L., Oi, V.T., and Tonegawa, S.: A tissue-specific transcription enhancer element is located in the major intron of a rearranged immunoglobulin heavy chain gene. Cell 33:717, 1983.
43. Casanova, J., Copp, R.P., Janocko, L., and Samuels, H.H.: 5'-Flanking DNA of the rat growth hormone gene mediates regulated expression by thyroid hormone. J. Biol. Chem. 260:11744, 1985.
44. Mulligan, R.C., and Berg, P.: Expression of a bacterial gene in mammalian cells. Science 209:1422, 1980.
45. Mulligan, R.C. and Berg, P.: Factors governing the expression of a bacterial gene in mammalian cells. Mol. Cell. Biol. 1:449, 1981.
46. Maniatis, T., Fritsch, E.F., and Sambrook, J.: Molecular Cloning: A Laboratory Manual, Cold Spring Harbor Laboratory, N.Y. 1982.
47. Loyter, A., Scangos, G.A., and Ruddle, F.H.: Mechanisms of DNA uptake by mammalian cells: Fate of exogenously added DNA monitored by the use of fluorescent dyes. Proc. Natl. Acad. Sci. U.S.A. 79:422, 1982.
48. Shen, Y.M., Hirschhorn, R.R., Mercer, W.E., Surmacz, E., Tsutsui, Y., Soprano, K.J., and Baserga, R.: Gene transfer: DNA microinjection compared with DNA transfection with a very high efficiency. Mol. Cell. Biol. 2:1145, 1982.
49. Richardson, K.K., Fostel, J., and Skopek, T.R.: Nucleotide sequence of the xanthine guanine phosphoribosyl transferase gene of E. coli. Nucleic Acids Res. 11:8809, 1983.
50. Weaver, R.F. and Weissmann, C.: Mapping of RNA by a modification of the Berk-Sharp procedure: the 5' termini of 15 S β-globin mRNA precursor and mature 10 S β-globin mRNA have identical map coordinates. Nucleic Acids Res. 7:1175, 1979.

51. Charnay, P., Treisman, R., Mellon, P., Chao, M., Axel, R. and Maniatis, T.: Differences in Human α- and β-globin gene expression in mouse erythroleukemia cells: The role of intragenic sequences. Cell 38:251, 1984.
52. Merrill, G.F., Hauschka, S.D., and McKnight, S.L.: tk enzyme expression in differentiating muscle cells is regulated through an internal segment of the cellular tk gene. Mol. Cell. Biol. 4:1777, 1984.
53. Dean, D.C, Knoll, B.J., Riser, M.E. and O'Malley, B.W.: A 5'-flanking sequence essential for progesterone regulation of an ovalbumin fusion gene. Nature 305:551, 1983.
54. Renkawitz, R., Schutz, G., von der Ahe, D., and Beato, M.: Sequences in the promoter region of the chicken lysozyme gene required for steroid regulation and receptor binding. Cell 37:503, 1984.
55. Karin, M., Haslinger, A., Holtgreve, H., Richards, R.I., Krauter, P., Westphal, H.M., and Beato, M.: Characterization of DNA sequences through which cadmium and glucocorticoid hormones induce human metallothionein-II_A gene. Nature 308:513, 1984.
56. Moore, D.D., Marks, A.R., Buckley, D.I., Kapler, G., Payvar, F., and Goodman, H.M.: The first intron of the human growth hormone gene contains a binding site for glucocorticoid receptor. Proc. Natl. Acad. Sci. U.S.A. 82:699, 1985.
57. Nyborg, J.K., Nguyen, A.P., and Spindler, S.J.: Relationship between thyroid and glucocorticoid hormone receptor occupancy, growth hormone gene transcription, and mRNA accumulation. J. Biol. Chem. 259:12337, 1984.
58. Diamond, D.J., and Goodman, H.M.: Regulation of growth hormone messenger RNA synthesis by dexamethasone and triiodothyronine: transcriptional rate and mRNA stability changes in pituitary tumor cells. J. Mol. Biol. 181:41, 1985.

20. DNA Sequences Required for Steroid Hormone Regulation of Transcription

M. Pfahl,[a] *H. Ponta,*[b] *P. Skroch,*[b] *A. Cato,*[b] *and B. Groner*[c]

[a]La Jolla Cancer Research Foundation, 10901 North Torrey Pines Road, La Jolla, CA 92037; [b]Kernforschungszentrum Karlsruhe, Institute for Genetics and Toxicology, D-7500 Karlsruhe, Federal Republic of Germany; [c]Ludwig Institute for Cancer Research, Inselspital, 3010 Bern, Switzerland

INTRODUCTION

In the last few years our understanding of the mechanisms of steroid hormone action has greatly advanced through the development of purification schemes for several steroid receptors and through the cloning of hormone-responsive genes. It is now clear that steroid hormone receptors, when complexed with their steroid hormone, can recognize and bind to specific DNA sequences (1-7). In mouse mammary tumor virus (MMTV), some of these DNA sequences could be directly correlated with the sequences necessary for hormone-responsive induction of RNA transcription (8-9). The DNA sequences recognized in the MMTV long terminal repeat (LTR) regions have recently been shown to belong to a class of DNA regulatory signals termed enhancers (10).

The MMTV enhancer responds to the activated glucocorticoid receptor (GR) complex, i.e., it responds to the outside signal of the glucocorticoid hormone. In this paper we further analyze the sequence requirements for this glucocorticoid responsive enhancer (GRE) and provide evidence that the GRE or associated sequence can respond to a number of steroid hormones including glucocorticoids, progestins, estrogens, and androgens.

RESULTS

Localization of glucocorticoid receptor binding sites in the MMTV LTR

A number of different *in vitro* DNA binding assays have been used to demonstrate that cloned MMTV proviral DNA contains specific binding sites for the activated glucocorticoid receptor (3,4,11,12). The approaches included the nitrocellulose filter binding assay (13), a DNA cellulose competition assay (14) and DNA footprinting (15). Results obtained with all three techniques pointed to an approximately 140 nucleotide region located between -200 and -60 upstream of the MMTV LTR cap site as the main GR binding site (8,9,16,17). We will describe here one set of experiments in which we use a number of deletion fragments in the DNA cellulose competition assay to demonstrate that the MMTV GRE contains multiple GR binding sites.

In the DNA cellulose competition assay, ^{3}H-labeled hormone-receptor complex is mixed with a DNA-cellulose suspension, where the cellulose-bound DNA is nonspecific (calf thymus) DNA. The amount of hormone-receptor complex bound to DNA-cellulose can be easily determined by pelleting the DNA-cellulose and counting the radioactivity associated with it. If free DNA is mixed with DNA-cellulose, the free DNA competes with the DNA cellulose for labeled steroid-receptor complex. DNA fragments containing specific binding sites for the glucocorticoid-receptor complex compete more strongly than nonspecific DNA fragments (3). This assay permits the use of unpurified receptor, which is of special advantage when receptor mutants or glucocorticoid antagonist-receptor complexes are to be investigated, and also if receptors from different species are to be compared.

To obtain deletions extending to various degrees into the GRE, we cloned a 550 bp MMTV LTR fragment (3), comprising LTR sequences up to -265 from the cap site, into the Pst I site of plasmid pBR322 and constructed deletions extending from the 5' end into this fragment (1). The deletion end-points were determined by DNA sequencing. The results from the test of these deletion fragments in the DNA cellulose competition assay are shown in Table 1. Specific binding was observed for deletions extending to -139 and -100 but not for a deletion extending to +26. If we calculate the increase in affinity for one hypothetical site per fragment (see ref. 3), we observe a decrease in the affinity of the -139 fragment compared to the original 550 bp fragment, and we see a further decrease in affinity when the deletion extends to -100 with respect to the RNA cap site. Deletion fragment +26 shows only nonspecific binding. These results confirmed results obtained with other deletions (8). The findings indicate that the GRE contained in the 500 bp LTR fragment consists of multiple binding sites or that large portions of the GRE can be deleted without greatly affecting the binding of GR to the GRE. The system presented represents a relatively simple approach towards the analysis of specific steroid receptor DNA binding regions.

Mutations introduced into the glucocorticoid receptor binding sites strongly reduce the hormonal responsiveness

Deletion analyses and footprinting experiments pointed out two protein binding areas for the glucocorticoid receptor complex defined in the GRE as indicated in Figure 2A between positions -71 to -124 and -163 to -191 (8,17). In these receptor binding regions, the hexanucleotide sequence 5'-TGTTCT-3' is found four times. The guanosine residues of these hexanucleotides are protected by the receptor protein from methylation by dimethylsulfate (18). This hexanucleotide motif is also described in other glucocorticoid regulated genes, e.g., the chicken lysozyme gene, and, with a slight variation - the substitution of the 3' T by a C - in the human metallothionein IIA gene (6). We decided to test the biological significance of each of these motifs in the GRE of the MMTV LTR by substituting different hexanucleotides. In addition, we substituted the "enhancer consensus" sequences 5'-GTGGTTT-3' (8,36) at -147 and -132.

The derivation of the mutant genes is outlined in Figure 1. A Bam HI-EcoRI restriction fragment of the MMTV LTR was cloned into the pEMBL9$^+$ vector. The Bam HI site is a synthetic linker site introduced into the MMTV LTR at position -236 with respect to the transcriptional start (19). It contains the GRE, the promoter, and the RNA start site. Single-stranded DNA was isolated from transformed E. coli K12 71/18 bacteria. Synthetic oligomers (24mer and 27mer) spanning the region to be mutated (* which were a gift from Pierre Chambon, Strasbourg, and Ernst Winnacker, Munich) were synthesized. They are homologous to the wild-type LTR sequences except for the hexanucleotide to be mutated. In place of this hexanucleotide, we introduced a restriction enzyme recognition site. The synthetic oligomers and the sequencing primer for the pEMBL9$^+$ vector were hybridized to the pEMBL9$^+$ vector with the LTR insert (20). The primed second strand was elongated with DNA polymerase, and the gap between the 5' end of the synthetic oligomer and the newly synthesized DNA starting at the sequencing primer was closed with T4 DNA ligase. The restriction fragment containing the GRE and the MMTV promoter with a mismatch at the site to be mutated, was excised and isolated. It was introduced in front of the HSV tk gene (RI-Pvn II fragment of 1.6 kb length) (21) which had been cloned into the SP6 vector. The resulting plasmid DNA was methylated with purified DAM-methylase *in vitro* to prevent repair synthesis of the heteroduplex. Transformation of bacteria led to two types of progeny plasmid: one with the wild-type GRE and the other with the mutated GRE. They were separated in a second round of transfections and identified by restriction enzyme digestion. We have been successful in isolating and analyzing four different mutants. The two at position -96 and position -130 have not yet been analyzed.

The plasmid DNAs were introduced into mouse LTk$^-$ cells by calcium-phosphate precipitation, and tk$^+$ colonies were grown to mass cultures. The response to dexamethasone treatment was evaluated 4 hrs after administration of the steroid hormone by the S1 mapping technique. The transcription from the authentic MMTV promoter leads to a protected DNA fragment of 104 nucleotides, indicated by an arrow. The intensity of this band is a direct measure for the cellular concentration of transcripts.

All four mutants are affected in their response to dexamethasone. Three of them still show inducibility by dexamethasone (Fig. 2B), but the level of induction is much reduced when compared to that of the wild-type. These mutations are located around positions -145, -116, and -80. They affect the proximal binding domain and one of the presumptive enhancer sequences. The fourth mutant differs from the others in that it does not respond to dexamethasone at all (Fig. 2B). This mutant is located at position -173, and therefore affects the distal receptor binding area. Thus, the sequence of all of the hexanucleotides tested so far has to be kept unchanged for a strong hormonal response. The sequence in the distal protected area (-172) seems to be an absolute requirement for hormone regulation, whereas the hexanucleotides in the proximal protected area (-80 and -116) as well as the "core enhancer sequence" at -145 potentiate the induction by the steroid hormone. Not all sequences within the GRE seem to be absolutely required for the hormonal induction. Internal deletions and substitutions have been introduced around position -107 (22,23) and, to a limited extent, can be accommodated without functional consequence.

Induction of MMTV transcripts by glucocorticoids, progesterone, estrogen, and androgen in human mammary tumor cells

The high incidence of mammary tumors in the female GR mouse strain is due to the expression of the MMTV provirus gene associated with the Mtv-2 locus (24-26). Breeding females of this strain of mice develop pregnancy-dependent mammary tumors which appear under the hormonal influence of pregnancy and regress after parturition (27). Comparable hormone-dependent mammary tumors can be induced in ovariectomized GR mice with 17-ethynyl-19-nortestosterone, a compound with progestational activity coupled with a relatively low estrogenicity (28).

Although these experiments indicate a role for ovarian hormones in murine mammary tumorigenesis, most of the molecular genetics studies carried out so far have been with glucocorticoids. We therefore set out to investigate the role in MMTV gene expression played by steroid hormones other than glucocorticoid. We constructed two chimeric plasmids made up of the MMTV LTR region and the bacterial neomycin resistance gene (neo). Plasmid pLneo was constructed by substituting the promoter region of the plasmid pSV2neo (29) with a Pvu II/Hind III fragment of GR40 (30) bearing the right LTR from GR40. Plasmid pL2neo was constructed by inserting the Bgl II fragment of GR40 containing the right LTR into pLneo in the correct orientation to yield a provirus-like arrangement where the two LTRs flank the neo gene (Fig. 3). These two constructs were transfected into the human mammary adenocarcinoma cell lines MCF7 and T47D which have receptors for all four classes of steroid hormones (31) - namely, glucocorticoids, progesterone, estrogen, and androgen. After transfection, single or pooled clones of these cells were selected with the antibiotic G418 (29). S1-nuclease mapping was carried out with total cellular RNA from these clones after having been treated with or without steroid hormones. The probe used in these studies was a 2.3 kb Sal I/Bgl II fragment which should protect a 360 bp piece of DNA from S1-nuclease digestion, if mRNA transcripts are to be correctly initiated from the MMTV promoter (Fig. 3).

We observed in these studies that not only was the correct MMTV transcriptional start site utilized (see arrows in Fig. 4), but also the amount of mRNA initiating at this site was considerably increased by treatment of the cells with the synthetic glucocorticoid (dexamethasone), 17 β-estradiol, dihydrotestosterone, or by the progestins, R5020, or medroxyprogesterone acetate (Fig. 4 and unpublished data). In some experiments, a fragment of about 315 bp was observed, indicating the initiation of mRNA transcript within the neo gene region (Fig. 4B). However, since this fragment was not reproducibly observed, we take it to be the product of an artefact generated by the S1-nuclease mapping procedure. The magnitude of induction observed after treatment by the various hormones is quantitated in Table 2 after densitometric scanning of the correct 360 bp band in autoradiographs following gel electrophoresis.

A combined treatment of the cells with dexamethasone and R5020 did not show the reported anti-glucocorticoid effect of progestin (32), indicating that the R5020 effect is mediated by the progesterone receptor (results not shown). Similarly, it is unlikely that the induction by R5020 is mediated by androgen receptor, as androgens themselves do not cause an increase in MMTV promoter activity in the population of MCF7 cells we used. It is possible that these cells may have functionally defective androgen receptor (Table 1). Many lines of evidence show that the reported increase in MMTV transcription by the various hormones occurs through their corresponding receptors. Some of this evidence comes from studies with specific anti-hormones or from the dose-response curves of induction by the hormones (33, and unpublished data). Another piece of evidence comes from studies in which R5020 has been shown to be an ineffective inducer in transfected rat XC cells. This result is perhaps not surprising, as XC cells have no progesterone receptor but do contain glucocorticoid receptor which responds to dexamethasone treatment (33-34). Further evidence comes from *in vitro* studies with a partially purified progesterone receptor in which *it has* been shown that the progesterone receptor can bind to defined nucleotide sequences in the MMTV LTR region, overlapping sequences bound by the glucocorticoid receptor (35). Indeed, constructions bearing these sites show both glucocorticoid and progesterone inducibility when transferred into cells with functional receptors for both hormones (33). Although it is not yet clear which sequences mediate the androgen and estrogen responses, the finding that all four classes of steroid hormones induce the transcriptional activity of the MMTV genome has interesting physiological and evolutionary implications.

CONCLUSIONS

The activated GR binds to specific regions in the MMTV GRE. When these regions are partially deleted, only a partial decrease in binding activity is observed. These results are most consistent with multiple independent GR binding sites in the GRE. Our findings from the *in vitro* mutagenesis study, that the replacement of specific short sequences in the GRE results in a drastic decrease of hormone responsiveness, are therefore somewhat surprising. One has to assume that i) sequences which are outside the major recognition sites of the GR, like the enhancer consensus sequence at -148, are essential for the function of the GRE and ii) that the GRE requires more than just receptor binding for its function. This leads us to propose that the hormone receptor induces a specific DNA structure or topology when interacting with sequences in the GRE, and that this specific DNA structure is optimally induced when all receptor sites are occupied. Evidence for such specific DNA structures has been reported (1,37).

The response of the MMTV LTR to four different classes of steroid hormones makes it possible to hypothesize that the response of a gene to a particular steroid hormone is primarily dependent on the presence of the specific steroid in the cell and its functional receptor. It is, however, also possible that all hormone responsive enhancers share a relatively high homology in nucleotide sequence - since they respond to a specific class of proteins - and that some of these enhancers are sufficiently recognized by all steroid hormone receptors whereas others respond mostly to a specific steroid hormone receptor. While it appears most likely that the MMTV GRE responds also to progestins, further experiments are necessary to determine which sequences in the LTR respond to estrogens and androgens.

ACKNOWLEDGEMENTS

This work was supported in part by grants from NSF (PCM8403355) and NIH (AM35083) to M.P.

REFERENCES

1. Pfahl, M.: In Biochemical Actions of Hormones (G. Litwack, editor), Academic Press, New York. Vol. XIII (in press).
2. Payvar, F., G.L. Firestone, S.R. Ross, V.L. Chandler, O. Wrange, J. Carlstedt-Duke, J.A. Gustafsson, and K.R. Yamamoto: J. Cell. Biochem. 19:241, 1982.
3. Pfahl, M.: Cell 31:475.
4. Govindan, M.V., E. Spiess, and J. Majors: Proc. Natl. Acad. Sci. USA 79:5157, 1982.
5. Karin, M., A. Haslinger, H. Holtgreve, R.I. Richards, P. Krauter, H.M. Westphal, and M. Beato: Nature (London) 308:513, 1984.
6. Renkawitz, R., G. Schutz, D. von der Ahe, and M. Beato: Cell 37:503, 1984.
7. Scheidereit, C., S. Geisse, H.M. Westphal, and M. Beato: Nature (London) 304:749, 1983.
8. Pfahl, M., D. McGinnis, M. Hendricks, B. Groner, and N.E. Hynes: Science 222:1341, 1983.
9. Chandler, V.L., B.A. Maler, and K.R. Yamamoto: Cell 33:489, 1983.
10. Ponta, H., N. Kennedy, P. Skroch, N.E. Hynes, and B. Groner: Proc. Natl. Acad. Sci. USA 82:1020, 1984.
11. Payvar, F., O. Wrange, J. Carlstedt-Duke, S. Okret., J.A. Gustafsson, and K. Yamamoto: Proc. Natl. Acad. Sci. USA 78:6628, 1981.
12. Geisse, S., C. Scheidereit, H.M. Westphal, N.E. Hynes, B. Groner, and M. Beato: EMBO J 1:1613, 1982.
13. Riggs, A.D., H. Suzuki, and S. Bourgeois: J. Mol. Biol. 48:67, 1970.
14. Simons, S.S., Jr.: Biochim. Biophys. Acta 496:339, 1977.
15. Galas, D., and A. Schmitz: Nucl. Acids Res 5:3157, 1978.
16. Payvar, F., D. DeFranco, G.L. Firestone, B. Edgar, O. Wrange, S. Okret, J.-A. Gustafsson, and K.R. Yamamoto: Cell 35:381, 1983.
17. Scheidereit, C., S. Geisse, H.M. Westphal, and M. Beato: Nature 304:749, 1983.
18. Scheidereit, C., and M. Beato: Proc. Natl. Acad. Sci. USA 81:3029, 1984.
19. Hynes, N.E., A. Van Ooyen, N. Kennedy, Herrlich, P., H. Ponta, and B. Groner: Proc. Natl. Acad. Sci USA 80:3637, 1983.
20. Zeller, M.J., and M. Smith: DNA 3:479, 1984.
21. Wilkie, N.M., J.B. Clements, W. Boll, N. Mantei, D. Lonsdale, and C. Weissman: Nucl. Acids Res. 7:859, 1979.

22. Majors, J., and H.E. Varmus: Proc. Natl. Acad. Sci. USA 80:5866, 1983.
23. Lee, F., C.V. Hall, G.M. Ringold, Dobson, D.E., J. Luk, and P.E. Jekob: Nucl. Acids Res. 12:4191, 1984.
24. Van Nie, R., and J. de Moes: Int. J. Cancer 20:588, 1977.
25. Michalides, R., L. van Deemter, R. Nusse, and R. van Nie: Proc. Natl. Acad. Sci. USA 75:2368, 1978.
26. Michalides, R., R. van Nie, R. Nusse, N.E. Hynes, and B. Groner: Cell 23:165, 1978.
27. Van Nie, R., and A. Dux: J. Natl. Cancer Inst. 46:885, 1971.
28. Van Nie, R., and J. Hilgers: J. Natl. Cancer Inst. 56:27, 1976.
29. Southern, P.J., and P. Berg: J. Mol. Appl. Genet. 1:327, 1982.
30. Kennedy, N., G. Knedlitschek, B. Groner, N.E. Hynes, P. Herrlich, R. Michalides, and A.J.J. van Ooyen: Nature 295:622, 1982.
31. Horwitz, K.B., D.T. Zava, A.K. Thilagar, E.V. Jensen, and W.L. McGuire: Cancer Res. 38:2434, 1978.
32. Rousseau, G.G., J.D. Baxter, and G.M. Tomkins: J. Mol. Biol. 67:99, 1972.
33. Ponta, H., N. Kennedy, P. Herrlich, N.E. Hynes, and B. Groner: J. Gen. Virol. 64:567, 1983.
34. Cato, A.C.B., R. Miksicek, G. Schutz, and M. Beato: Science (submitted), 1985.
35. von der Ahe, D., S. Janick, C. Scheidereit, R. Renkawitz, G. Schutz, and M. Beato: Nature 313:706, 1985.
36. Weiher, H., M. Konig, and P. Gruss: Science 219:626, 1983.
37. Zaret, K.S., and K.R. Yamamoto: Cell 38:29, 1984.

TABLE 1. Receptor Binding Characteristics of MMTV LTR Deletion Fragments

DNA fragment tested (bp)	μg DNA for 1/2 max. competition	Increase in Affinity*	
		for total fragment	for one specific site/fragment
.55 Kb (550)	.56	8.4	4,600
-139 (430)	.61	7.2	3,100
-100 (390)	.65	6	2,340
+26 (265)	4.7	0	0
calf thymus	4.7	0	0

* The increase is the fold affinity minus one. For the calculation of the increase in affinity for one specific site per fragment, the assumption has been made that each bp of a DNA fragment can start a new binding site (neglecting end effects). The increase in affinity for one specific site per fragment is therefore the observed increase for the total fragment times the number of bp per fragment (3).

TABLE 2. The Magnitude of Induction of MMTV Transcription by Steroid Hormones

Hormone / Cell Line	Dex -/+	R5020 -/+	DHT -/+	E_2 -/+
T47D				
clone 4	7.9	10.5	n.d.	n.d.
clone 10	8.5	9.7	10.9	n.d.
MCF7				
pooled clones	4.2	3.3	1.1	3.9

The intensities of the 360 bp band in the autoradiographs (Fig. 2) from samples treated with hormone were scanned and compared to samples without hormone. The markers represent arbitrary units. Abbreviations: Dex, Dexamethasone(10^{-6} M); R5020, 17,21-dimethyl-pregna-4,9(10)-diene-3, 20-dione (3 x 10^{-8} M); DHT, dihydrotestosterone (10^{-8} M); E_2, 17-estradiol; n.d., not determined.

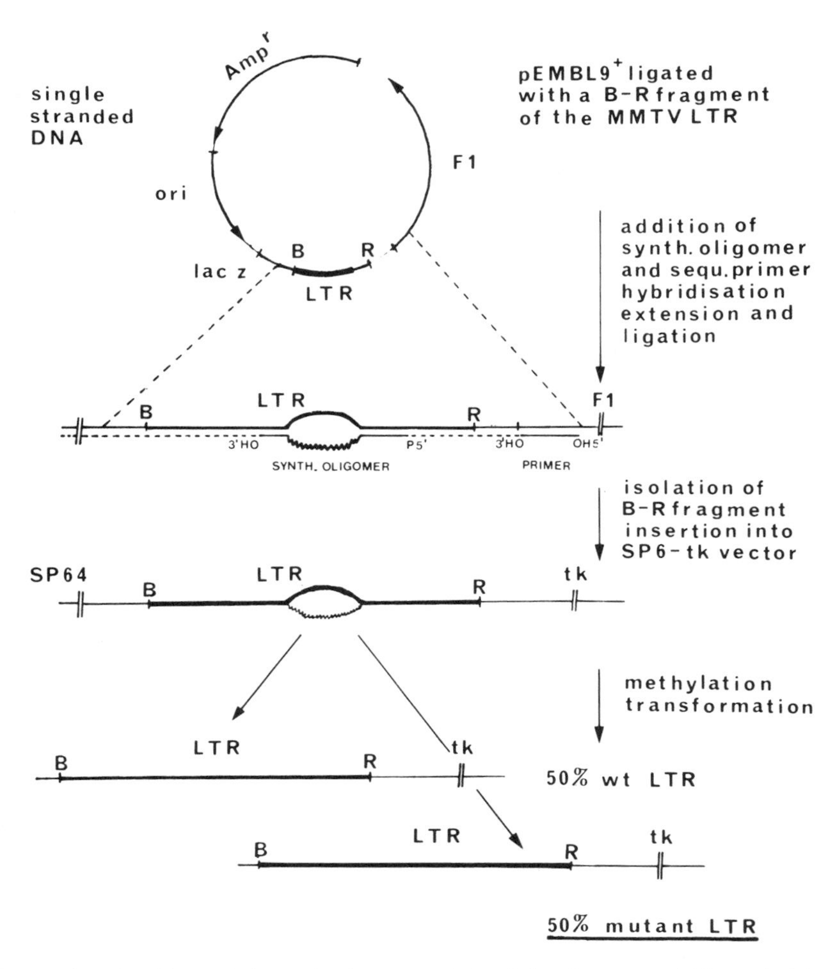

Figure 1. Outline of the strategy used to introduce specific mutations into the GRE of MMTV.

A single-stranded DNA of the GRE was obtained by cloning a Bam HI-Eco RI fragment into the pEMBL9$^+$ vector. Synthetic oligonucleotides spanning the site to be mutated were annealed with the vector DNA. Elongation of primers restored the double-stranded Bam HI-Eco RI fragment as a heteroduplex molecule with a mismatch at a specific sequence. This restriction fragment was recloned to obtain the wild-type and mutated double-stranded sequence and reinserted into the LTR-tk construct. The sequences mutated in this manner are indicated in Figure 2A.

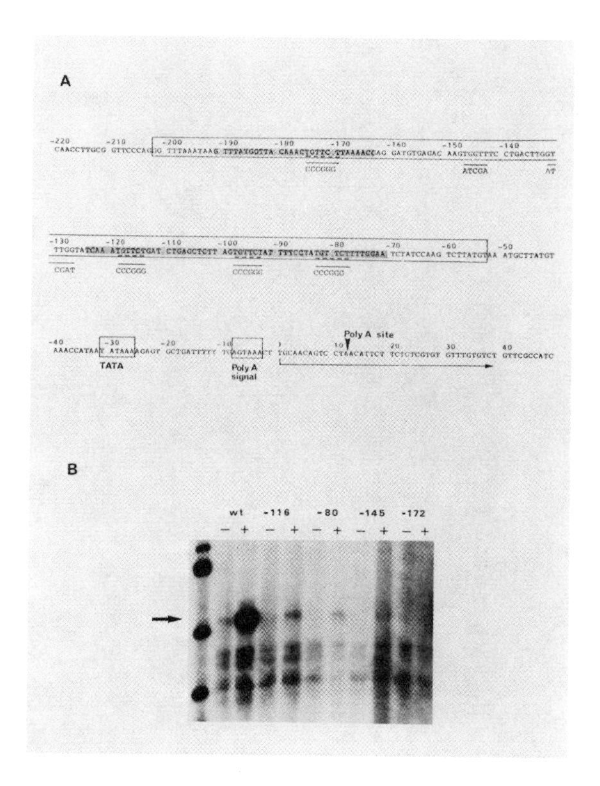

Figure 2. Effect of mutations on the hormone responsiveness of the GRE

A) Partial nucleotide sequences of the MMTV LTR. The RNA initiation site is indicated by an arrow. The nucleotides preceding the RNA initiation site are shown and several features are highlighted: The TATA box at position -30, the GRE at position -52 to -202, the sequence identified as glucocorticoid receptor binding sites (shaded) at position -71 to -124 and -162 to -191, the TGT TCT motifs at positions -78 to -83, -93 to -98, -113 to -119, and -170 to -175, and the possible enhancer core sequences (underlined) at positions -147 to -143 and -132 to -127. B) Single-stranded nuclease protection assay of RNA derived from cells transfected with LTR-tk constructs. The RNA was hybridized to a single stranded DNA fragment (Bam HI-Eco RI) labelled at the 5' end of the Eco RI site. The hybrids were digested with S1 nuclease and the protected DNA fragment of 106 nucleotides was visualized by gel electrophoresis and autoradiography. The arrow indicates the position of the protected fragment, indicative of RNA initiated at the MMTV LTR start site. wt: wild-type construct transfected into L cells and grown with (+) and without (-) 10^{-6} M dexamethasone. -116, -80, -145, -172 denotes mutated constructs transfected into L cells. The nucleic acid substitutions of these constructions are indicated in A.

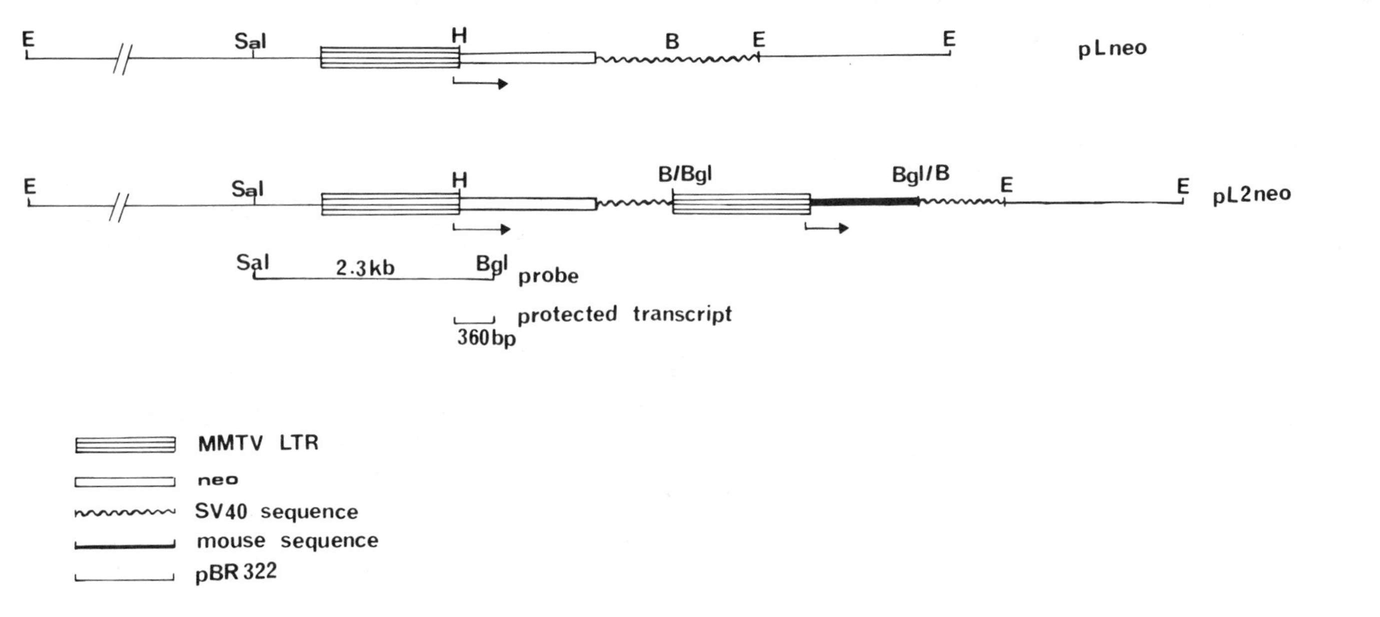

Figure 3. The plasmid constructs pLneo and pL2neo.

The open bars represent the neomycin-resistance gene sequences (8), the striped bars represent the MMTV LTR sequences, the wavy line represents the SV40 sequences, the shaded bar, mouse sequence, and the horizontal line, pBR322 sequence. Symbols: E = Eco RI, Sal = Sal I, H = Hind III, B = Bam HI, Bgl = Bgl II. The 2.3 kb Sal I/Bgl II fragment used as hybridization probe in the S1-nuclease mapping technique is indicated, and the position of the 360 bp fragment is also shown.

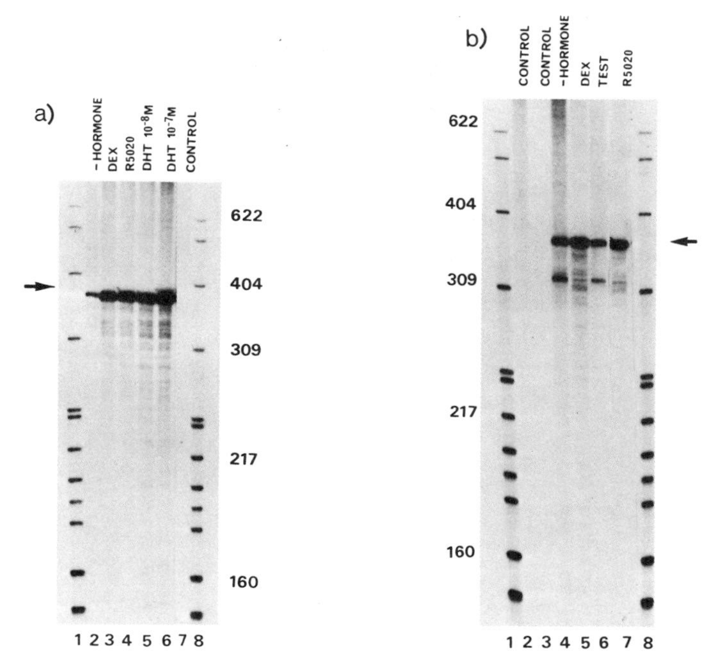

Figure 4. Steroid hormones increase the amount of mRNA initiating at the MMTV promoter in transfected T47D and MCF7 cells.

A) Single-strand specific nuclease protection assay showing mRNA initiated at the MMTV promoter in T47D cells transfected with pL2neo. The RNA from transfected cells was hybridized to a single-stranded DNA fragment (Sal I-Bgl II in Figure 3) of 2.3 kb which was labelled at the 5' end of the Bgl II site. The hybrids were digested with S1 nuclease, and the protected DNA fragment of 360 nucleotides was visualized by gel electrophoresis and autoradiography. The arrow shows the position of the protected fragment, indicative of RNA initiated at the MMTV LTR start site. The RNAs for these assays were derived from a transfected cell clone treated with the following hormones: Lanes 2-6: cellular RNA from transfected T47D cells treated without hormone (2); with 10^{-6} M dexamethasone (3); with 3 x 10^{-8} M R5020 (4); with 10^{-8} M DHT (5); or 10^{-7} M DHT (6). Lanes 1 and 8: Hpa II digested pBR322 marker DNA. Lane 7: yeast RNA. B) Single-strand specific nuclease protection assay showing mRNA initiated at the MMTV promoter in MCF7 cells transfected with pLneo. The RNAs for these studies (from pooled, transfected MCF7 cell clones) were derived from cells treated with the following hormones: Lanes 2-7: cellular RNA from non-transfected MCF7 cells (2); RNA from MCF7 cells treated with 10^{-6} M dexamethasone (3); transfected MCF7 without hormone (4); with 10^{-6} M dexamethasone (5); with 10^{-7} testosterone (6); or with 3 x 10^{-8} M R5020 (7). Lanes 1 and 8: Hpa II digested pBR322 marker DNA.

21. The Hormone Response Element of the MMTV LTR: A Complex Regulatory Region

M. Cordingley, H. Richard-Foy, A. Lichtler, and G. L. Hager

Hormone Action and Oncogenesis Section, LEC, National Cancer Institute, Bethesda, MA 20892

INTRODUCTION

Regulation of gene expression by steroid hormones is a fundamental mechanism of intercellular communication in multicellular organisms. The mouse mammary tumor virus (MMTV) has emerged in the past five years as a rich model system for the study of gene regulation by glucocorticoids at the transcriptional level. It was first demonstrated that regulation in this system was primarily at the level of transcription in 1979 (1,2). The application of recombinant DNA and transfection technologies to demonstrate transfer of regulation from MMTV to genes not normally induced by steroids by our group (3,4) and Lee et al (5) in 1981 opened the door to a rapid accumulation of information concerning basic mechanisms of steroid hormone action. We now know that the cis-acting sequences which mediate hormone action at the MMTV promoter are located between 75 and 200 base pairs upstream from the cap site for transcription initiation (6,7). We know that these signals can be separated from the minimal LTR (long terminal repeat) promoter, and confer regulation on a heterologous promoter when juxtaposed (8-10). Finally, and importantly, we know that this cis-acting sequence corresponds to sites at which the glucocorticoid receptor bind in vitro (11-14). The initial framework of this regulatory system has therefore been established.

RESULTS

Interaction of Hormone Response Element with Enhancers

In our original experiments identifying the MMTV LTR as the locus of hormone action (3,4), we observed that the MMTV promoter was very inefficient in the host cells (NIH 3T3) employed in those experiments. To produce a more conveniently assayable LTR construction, we added an active transcriptional enhancer element from the Harvey murine sarcoma virus (HaMuSV) LTR to the MMTV LTR promoter (15). Since the MMTV LTR was serving as a weak promoter in these cells and the HaMuSV enhancer was known to be a strong activating element, we expected these constructions to be expressed constitutively, with very little if any inducibility. Unexpectedly, the MMTV LTR-Ha enhancer fusions manifested strong inducibility (15). Induction ratios were as high as fifty-fold in a focus formation assay, and ranged from five to ten-fold in steady state RNA levels.

Of great interest was the mechanism by which induction was preserved in the enhancer-LTR fusions. When the enhancer LTR cassette was used to drive ras gene expression in a focus formation assay, the number of foci recovered with the enhancer LTR cassette in the presence of hormone was comparable to that found with the enhancer alone, driving its native, non-regulated promoter. Focus formation activity with the Ha-enhancer MMTV LTR cassette in the absence of hormone, however, was approximately fifty-fold depressed, suggesting that the LTR was actually serving more as a negative influence on transcription in the absence of hormone, rather than a hormone inducible positive element.

These results must be interpreted with care, because the focus forming activity of a given construction is expected to be subject to a threshold effect, which will depend on the amount of ras gene product required to achieve phenotypic transformation. The experiments were therefore extended with chimeras in which enhancer-LTR cassettes were linked to the CAT (chloramphenicol acetyltransferase) gene, and the accumulation of CAT activity measured in transient expresssion assays, a more reliable test of transcription activity (15). Here again, there was a depression of CAT expression in the absence of hormone compared to the constitutive level from the enhancer driven noninducible promoter. The results from the focus formation assay, therefore, were correctly interpreted as a transcriptional effect.

The results from these experiments suggested that a negative transcriptional element was present in the MMTV LTR. To identify such a potential element, deletion analysis was carried out with the MMTV LTR driving the ras gene in the focus formation assay, and with the CAT gene in the transient expression assay, in both cases with a constitutive enhancer activating the promoter.

This series of experiments (16) indicated a more complex transcriptional structure for the MMTV LTR. Deletion of sequences in the LTR-ras fusions between approximately -350 and -100 with respect to the MMTV cap site gave rise to chimeras with equal focus forming activity both in the presence and absence of hormone, while the unaltered LTR showed a fifty-fold lower transforming activity in the absence of hormone. That is, assuming the focus forming assay was responding primarily to transcriptional activity at the MMTV promoter, a region was removed in this deletion series that was responsible for a repressive effect. A similar deletion series carried out in chimeras with the enhancer-LTR cassette driving the CAT gene confirmed a negative acting element in the same region, but also revealed a hormone responsive positive effect. Deletion of the −350 to −100 region in these experiments resulted both in a higher constitutive level of CAT activity in the transient expression assay in the absence of hormone, and a decreased level of activity in the presence of hormone. Since the transient expression assay is presumably a more accurate indicator of transcription activity than the focus forming assay, we assume that the hormone responsive positive element was not detected in the transformation assay because of threshold effects.

Of several models that could be advanced to explain these complex results, two relatively straightforward possibilities will be considered here. One could propose that the LTR contains both a cis-acting negative element and a cis-acting hormone responsive positive element. The increase in constitutive expression (minus hormone) would then be explained as the deletion of the negative element, permitting marked activation by the strong HaMuSV enhancer at the MMTV promoter, with the effect of the weaker hormone responsive element masked by the predominant enhancer effect, particularly in the transformation assay with the ras fusions. This model predicts two elements, a hormone independent negative or repressive element, and a hormone dependent positive element. Alternatively, one could propose that the effects are mediated entirely by a single, hormone responsive element. This element must then be postulated to exert both a cis-acting negative effect in the absence of hormone, and a cis-acting positive effect upon induction.

It is likely that these models could not be distinguished by the initial series of low resolution deletion experiments (16), since relatively large blocks of DNA were removed. To address this question directly, we modified the oligonucleotide-directed mutagenesis technique to permit the precise alteration of small blocks (1-15 base pairs) of sequence (17). In these experiments, a single stranded vector was engineered in m13 phage with the MMTV LTR driving the ras oncogene. Unnecessary sequences from the original LTR-ras cassette (3,15) were first removed to permit insertion in the size limited m13mp8 phage molecule. Appropriate unique restriction sites were then created in the LTR region of the molecule to facilitate deletion of discrete blocks of sequence from the LTR region. Heteroduplexing of these deletions with full length single stranded molecules creates substrates which are single stranded only in the region to be mutagenized. In addition, the full length strand of the heteroduplex is prepared from dam- bacteria, and the deleted strand from dam+ host, so that the molecule is hemizygous with respect to methylation. After targeting the gap with an appropriate mismatched oligonucleotide, filling in with Klenow polymerase, and ligation, recombinant molecules are selected on a restricting host. Phage with the mutant strand are therefore enriched in the progeny. This procedure, referred to as "oligo scanning mutagenesis," results in 5-10% of the progeny containing mutant molecules, even when the mismatch is a great as 10-12 base pairs (17). When this construction is used to transfect NIH3T3 cells, and focus formation by the ras gene is monitored, approximately 200 times as many foci are recovered when hormone is present in the medium during the transfection experiment (Lichtler and Hager, unpublished observations). This represents a very powerful assay for the initial screening of mutants, since focus formation can be monitored directly without the necessity of any biochemical procedures.

The region of the LTR that contains the hormone receptor binding sites (-75 to -190) was arbitrarily divided in 10 bp blocks for mutagenic replacement. These mutants were then tested for both their constitutive response and their induction properties. The results (Lichtler and Hager, unpublished observations) were most compatible with the second model. Replacement mutants in which one of the receptor binding sites had been significantly altered led to an impaired hormone response, and an increased constitutive level of activity. Mutants in which both receptor binding sites were altered manifested a complete loss of induction and a high constitutive level of activity. That is, both the enhancer blocking effect and the hormone response effect were simultaneously eliminated.

While this last set of data is consistent with the second model proposed above, it does not prove the model. In particular, it is possible that there are in fact two kinds of elements, and the mutants we have tested thus far, while of high resolution, fortuitously impair both elements at the same time. What is clear from the experimental evidence thus far is that the hormone response region is complex, with multiple functional elements.

Tissue Specificity of the MMTV LTR

The MMTV LTR exhibits considerable variability with respect to cell type in the efficiency with which it promotes transcription initiation. Murine mammary cell lines isolated from MMTV induced tumors usually express large amounts of MMTV RNA, as great as three percent of cell RNA (18). The low transfection efficiency observed when we originally fused the LTR to a heterologous gene (3) was therefore quite surprising, as discussed above. These experiments suggested that NIH3T3 cells are a poor host for MMTV transcription.

The actual rates of initiation at specific LTR promoters are difficult to determine in tumor cell lines because of the multiple copy numbers. This question was addressed more directly in the experiments with the LTR fused both to the ras gene (transformation assay) and to the CAT gene (transient expression), since the observed signal in both these approaches is averaged over a large number of molecules, and position effect is effectively removed. Whereas expression of CAT driven from the MMTV promoter was undetectable in NIH3T3 cells, inducible transcription was easily measured in 34i-cl101, a cell line derived from an MMTV induced mammary tumor, and presumably more representative of the natural host cell for MMTV expression (19). These experiments suggest an element of tissue specificity for the MMTV LTR reminiscent of that recently described for several gene systems.

In studies of MMTV induced murine mammary tumors where one provirus has been acquired, it has been found that the genome is usually, if not always, inserted in the vicinity of one of two (int-1)(20) (int-2)(21), and possibly a third (R. Callahan, personal communication) cellular genes. In the cases of int-1 and int-2, newly induced cellular transcripts have been correlated with these insertional events. These events are most readily explained (20,21) in terms of transcriptional activation in cis; that is, the MMTV genome must contain an element responsible for the induction of these cellular genes when integrated in their vicinity. It seems likely that the same element is responsible for insertional activation in the case of the mammary tumor loci, and cell type specificity in the transfection experiments (15), although in neither case has the location of the active element yet been mapped.

What relationship does this cis-acting, "enhancer like" element bear to hormone inducibility? Chandler et. al. (9), supported by Ponta et al (10) have argued that the hormone responsive element (HRE) is in fact a glucocorticoid dependent enhancer. Some properties of the HRE (orientation independence and action at a distance) are indeed characteristic of enhancers. Certain other properties, however, are difficult to explain under this model. Glucocorticoid receptors are widely distributed in mammalian cell types. They are quite abundant in the NIH3T3 cells, for example, where we found the MMTV LTR to be a very weak promoter. The one known protein to act at the MMTV LTR, therefore, is not limiting in cells where the promoter is relatively inactive, albeit inducible. Furthermore, the cis-acting repressive effect discussed above is not consistent with the enhancer model, since enhancers with different specificities do not interfere with each other, even in cells where one is completely inactive (W. Schaffner, personal communication).

We believe that the steroid hormone response is fundamentally different in its mode of action. We suggest that the hormone response region in the MMTV LTR is complex, serving both to repress transcription at adjacent promoters in the absence of activated hormone receptor complex, and to activate transcription when the receptor complex is mobilized by ligand. Whether these properties reside in one definable sequence element, or whether they represent separate and distinct elements remains to be seen. Results with the oligo-scanning high resolution mutants discussed above suggest both activities reside in the same element, but this point requires further investigation.

Chromatin Structure of the MMTV LTR

The intricate nucleoprotein structural organization of mammalian DNA is likely to play some role in the regulation of expression. The higher order folding found in condensed chromatin is certainly incompatible with active transcription. Whether the local assembly of DNA control sequences into nucleosome structures critically affects the interaction of those sequences with soluble transcription factors is not yet clear.

We have investigated the chromatin structure of the MMTV LTR, focusing initially on the spacing of nucleosomes with respect to known control regions in the molecule. Using the indirect end-labelling technique of Wu (22) and Georgiev (23), in conjunction with partial micrococcal nuclease digestions, we find that nucleosomes are organized in a phased array on the LTR (24), with approximately 190 base pair spacing. The first nucleosome in this array appears to include both sets of sequences that interact preferentially with hormone receptor in vitro (11-14). It is probable, therefore, that the sequences with which receptor interacts in the nucleus have considerable tertiary structure.

The nucleoprotein organization of the LTR sequences could bear important ramifications for the mechanism of transcriptional regulation. The nonrandom localization of nucleosomes indicates, for example, that one face of the DNA helix interacts with nucleosome cores over a considerable portion of the regulatory region; this face would almost certainly be unavailable for interaction with soluble proteins. Higher order juxtapositioning of the helix could also bring regions at considerable distance on the primary sequence into close proximity.

These considerations have led us to begin attempts to study the hormone regulatory element in its native chromatin state. Chimeras with the LTR mobilized on the episomal bovine papilloma virus (BPV) vector appear to provide a promising system for these studies (25). The phasing experiments discussed above were carried out in this system. In addition, we have been able to show recently that hormone receptor interacts specifically with LTR chromatin in vivo (26).

The Hormone Regulatory Element - Mechanism of Action

At least three kinds of activities must now be explained in a comprehensive model of hormone induction by glucocorticoid at the MMTV LTR. First and foremost, binding of activated receptor complex results in a higher rate of transcription initiation at promoters in the vicinity. Second, in the absence of bound receptor complex, the hormone response region can inhibit the activity of cis-acting constitutive enhancers in the vicinity. Third, the complete LTR promoter structure exhibits considerable cell type specificity with respect to efficiency of promotion. It must also be remembered that the same hormone, and probably the same activated receptor complex, can repress expression of some genes, as in the case of the proopiomelanocortin locus (27). It is now known that down regulation at this locus is primarily a transcriptional effect (Drouin, personal communication).

Two general classes of models can be advanced to explain the mechanism by which the negative effect occurs. One possibility is the existence of a classical repressor molecule, acting at some site in the vicinity of the hormone response element. Indirect evidence for this model has already been obtained through the utilization of protein synthesis inhibitors (Cordingley and Hager, unpublished observations). It has been found that inhibition of protein synthesis results in a rapid increase in the rate of transcription initiated at the MMTV promoter, as measured by transcription "run on" experiments in isolated nuclei. Although inhibition of synthesis of a labile repressor is the most obvious explanation of this effect, there are other possible interpretations. The complex metabolism of the hormone receptor complex might be involved. If this complex were stabilized in the absence of protein synthesis, a super induction may result. This model would not explain the effect found in the absence of hormone, unless a small amount of receptor is bound to the LTR even in the absence of induction.

A second possible model for the negative effect emerges from the observations on chromatin structure of the LTR discussed above. Sequences localized on the surface of a nucleosome could be considerably sequestered from soluble transcription factors.

Either a single hormone responsive element must be responsible for each of the activities discussed, or the regulatory region must contain multiple elements, each potentially interacting with cognate diffusable regulators. Mutational analysis carried out to date has not resolved multiple elements, but the studies have been designed primarily to test only for hormone responsiveness. With increasing awareness of the multiple effects, future experimentation will be designed to explore the mechanism of action in terms of this greater complexity.

REFERENCES

1. Young, H.A., T.Y. Shih, E.M. Scolnick, and W.P. Parks: Steroid induction of mouse mammary tumor virus: effect upon synthesis and degradation of viral RNA. J Virol 21:139, 1977.

2. Ringold, G.M., K.R. Yamamoto, J.M. Bishop, and H.E. Varmus: Glucocorticoid-stimulated accumulation of mouse mammary tumor virus RNA: increased rate of synthesis of viral RNA. Proc Natl Acad Sci USA 74:2879, 1977.

3. Huang, A.L., M.C. Ostrowski, D. Berard, and G.L. Hager: Glucocorticoid regulation of the Ha-MuSV p21 gene conferred by sequences from mouse mammary tumor virus. Cell 27:245, 1981.

4. Hager, G.L.: Expression of a viral oncogene under control of the mouse mammary tumor virus promoter: A new system for the study of glucocorticoid regulation. Prog Nucl Acid Res Mol Biol 29:193, 1983.

5. Lee, F., R. Mulligan, P. Berg, and G. Ringold: Glucocorticoids regulate expression of dihydrofolate reductase cDNA in mouse mammary tumour virus chimeric plasmids. Nature 294:228, 1981.

6. Pfahl, M., D. McGinnis, M. Hendricks, B. Groner, and N.E. Hynes: Correlation of glucocorticoid receptor binding sites on MMTV proviral DNA with hormone inducible transcription. Science 222:1341, 1983.

7. Lee, F., C.V. Hall, G.M. Ringold, D.E. Dobson, J. Luh, and P.E. Jacob: Functional analysis of the steroid hormone control region of mouse mammary tumor virus. Nucleic Acids Res 12:4191, 1984.

8. Majors, J., and H.E. Varmus: A small region of the mouse mammary tumor virus long terminal repeat confers glucocorticoid hormone regulation on a linked heterologous gene. Proc Natl Acad Sci USA 80:5866, 1983.

9. Chandler, V.L., B.A. Maler, and K.R. Yamamoto: DNA sequences bound specifically by glucocorticoid receptor in vitro render a heterologous promoter hormone responsive in vivo. Cell 33:489, 1983.

10. Ponta, H., N. Kennedy, P. Skroch, N.E. Hynes, and B. Groner: Hormonal response region in the mouse mammary tumor virus long terminal repeat can be dissociated from the proviral promoter and has enhancer properties. Proc Natl Acad Sci USA 82:1020, 1985.

11. Scheidereit, C., S. Geisse, H.M. Westphal and M. Beato: The glucocorticoid receptor binds to defined nucleotide sequences near the promoter of mouse mammary tumour virus. Nature 304:749, 1983.

12. Payvar, F., D. DeFranco, G.L. Firestone, B. Edgar, O. Wrange, S. Okret, J.A. Gustafsson, and K.R. Yamamoto: Sequence-specific binding of glucocorticoid receptor to MTV DNA at sites within and upstream of the transcribed region. Cell 35:381, 1983.

13. Pfahl, M.: Specific binding of the glucocorticoid-receptor complex to the mouse mammary tumor proviral promoter region. Cell 31:475, 1982.

14. Miller, P.A., M.C. Ostrowski, G.L. Hager, and S.S. Simons Jr: Covalent and noncovalent receptor-glucocorticoid complexes preferentially bind to the same regions of the long terminal repeat of murine mammary tumor virus proviral DNA. Biochemistry 23:6883, 1984.

15. Ostrowski, M.C., A.L. Huang, M. Kessel, R.G. Wolford, and G.L. Hager: Modulation of enhancer activity by the hormone responsive regulatory element from mouse mammary tumor virus. EMBO J 3:1891, 1984.

16. Kessel, M., G. Khoury, A. Lichtler, M.C. Ostrowski, and G.L. Hager: The mouse mammary tumor virus LTR contains negative and positive regulatory elements. (submitted to Nucleic Acids Res)

17. Lichtler, A., and G.L. Hager: Multiple mismatch mutants induced by oligonucleotide site directed mutagenesis. (manuscript submitted to Gene)

18. Dudley, J.P., and H.E. Varmus: Purification and translation of murine mammary tumor virus mRNA's. J Virol 39:207, 1981.

19. Parks, W.P., E.S. Hubbell, R.J. Goldberg, F.J. O'Neill, and E.M. Scolnick: High frequency variation in mammary tumor virus expression in cell culture. Cell 8:87, 1976.

20. Nusse, R., and H.E. Varmus: Many tumors induced by the mouse mammary tumor virus contain a provirus integrated in the same region of the host genome. Cell 31:99, 1982.

21. Peters, G ., S. Brookes, R. Smith, and C. Dickson: Tumorigenesis by mouse mammary tumor virus: evidence for a common region for provirus integration in mammary tumors. Cell 33:369, 1983.

22. Wu, C.: The 5' ends of Drosophila heat shock genes in chromatin are hypersensitive to DNase I. Nature 286:854, 1980.

23. Georgiev, G.P., V.V. Bakayev, S.A. Nedospasov, S.V. Razin, and V.L. Mantieva: Studies on structure and function of chromatin. Mol Cell Biochem 40:29, 1981.

24. Richard-Foy, H, M.C. Ostrowski, and G.L. Hager: Nucleosomes are phased on the LTR of mouse mammary tumor virus. (submitted to Mol Cell Biol).

25. Ostrowski, M.C., H. Richard-Foy, R.G. Wolford, D.S. Berard, and G.L. Hager: Glucocorticoid regulation of transcription at an amplified, episomal promoter. Mol Cell Biol 3:2045, 1983.

26. Ostrowski, M.C., and G.L. Hager: Specific binding of glucocorticoid receptor to chromatin. (submitted to Mol Cell Biol).

27. Roberts, J.L., C.L. Chen, J.H. Eberwine, M.J. Evinger, C. Gee, E. Herbert, and B.S. Schachter: Glucocorticoid regulation of proopiomelanocortin gene expression in rodent pituitary. Recent Prog Horm Res 38:227, 1982.

22. Applications for Glucocorticoid-Dependent Mouse Mammary Tumor Virus LTR Chimeras in Conditional Gene Regulation and Mutant Selection Studies

B. Groner,[a] *H. Ponta,*[b] *R. Ball,*[a] *and M. Pfahl*[c]

[a]Ludwig Institute for Cancer Research, Inselspital, 3010 Bern, Switzerland; [b]Kernforschungszentrum Karlsruhe, Institute for Genetics and Toxicology, D-7500 Karlsruhe, Federal Republic of Germany; [c]La Jolla Cancer Research Foundation, 10901 North Torrey Pines Road, La Jolla, CA 92037

INTRODUCTION

Although glucocorticoid induction of gene transcription is a multifaceted process, a few of the interacting components have been identified in considerable detail. Molecular studies on the glucocorticoid receptor molecule (14,28) and on DNA sequences required for glucocorticoid induction of transcriptional activity (21) are most advanced. Investigations into the transcriptional control sequences of the proviral DNA of mouse mammary tumor virus (MMTV) have yielded a precise delimitation of a regulatory element responsible for glucocorticoid hormone induction of transcription (22). This element is located within the long terminal repeat (LTR) of MMTV and contains specific DNA sequences which interact with the glucocorticoid receptor (10,19,25). The interaction of the receptor with these DNA sequences is an essential prerequisite for the regulation of transcription (7). The hormonal effect originally observed in the transcription of viral RNA can be conferred to heterologous genes either by replacing a promoter with the MMTV LTR promoter sequence (9,10,13) or by supplementing a promoter with the hormonal response element (HRE) (22). This element is a purely regulatory element without coding potential or transcribed sequences. It can confer hormonal induction onto the heterologous genes over distances of more than 1 kb and functions when it is placed either 5' or 3' of the regulated gene in both of the possible orientations (22). The HRE is therefore functionally related to enhancer sequences (8). The dependence of the HRE function on glucocorticoid hormone allows its description as a conditional enhancer.

We have made use of both aspects (the possibility to confer hormonal inducibility to heterologous genes and the functional similarity to enhancer sequences) to employ the MMTV LTR in the study of biological processes. The variation of the concentration of a single component is potentially desirable for the study of many complex multicomponent interactions. Here we describe three biological systems in which the conditional gene expression can be used as a tool in the investigations of cellular transformation, antigen recognition by T lymphocytes and mutant selection in enhancer binding proteins.

RESULTS AND DISCUSSION

Hormonal Regulation of Oncogene Expression

Oncogenes were initially identified as the regions within the genomes of retroviruses which enabled these viruses to transform cells in culture or to cause tumors in animals (2). Oncogenes have also been detected independently from viral associations after gene transfer of DNA derived from tumor cells (31). The cellular oncogenes are able to morphologically transform normal NIH/3T3 fibroblasts. They act dominantly in these acceptor cells, i.e., the presence and expression of the activated cellular oncogenes is sufficient to cause NIH/3T3 cells to become tumorigenic, grow in soft agar and grow under low serum conditions (4,33). Although all of these cellular properties can be attributed to the action of oncogenes,the pleiotropic phenotype of the transformed cells is mediated through the effect of the oncogene product on the cellular genetic program. Conditional oncogenes, i.e., chimeric constructs in which the expression of oncogenes is subjected to the glucocorticoid regulated MMTV promoter, have been utilized to investigate the biochemical action of oncogenes and their effect on the cellular metabolism.

a) MMTV LTR-ras

One of the earliest experiments demonstrating the possibility to confer hormonal regulation of gene transcription from the MMTV proviral gene to a chimeric MMTV LTR construct, made use of the v-ras coding region (9). The region of Harvey muring sarcoma virus which codes for the p21 transforming protein was linked to the MMTV LTR and the hybrid gene was introduced into NIH/3T3 cells. The levels of transcript as well as the levels of p21 protein were shown to be regulated by glucocorticoid hormone and hormone dependent changes in the phenotype of the transfected cells were observed.
More recently the activated human cellular H-ras gene was subjected to the MMTV-LTR (R. Jaggi, B. Salmons, manuscript in preparation). The construct was introduced into NIH/3T3 cells and the stably transfected cell clones were characterized in the absence and presence of glucocorticoid hormone. The clones showed a morphologically transformed phenotype only in the presence of hormone, formed large colonies in soft agar and induced tumors in nude mice within ten days. Cell clones containing the LTR-Hras proto-oncogene appear morphologically normal in medium with and without dexamethasone, form only small colonies in soft agar in the presence of dexamethasone and are not tumorigenic. The activated p21 protein accumulates rapidly in MMTV LTR-Hras transfected cells and reaches a maximum about five hours after hormone addition. The level of LTR-Hras mRNA and p21 protein is down-regulated after 8 to 25 hours and 24 to 42 hours,

respectively. No, or only a moderate, down-regulation was observed in cells containing the LTR-Hras proto-oncogene. The mechanism of regulation seems to be at the transcriptional level, i.e., oncogene product accumulation provides a signal for inhibition of MMTV LTR transcription. A drastic accumulation of cellular ornithine decarboxylase mRNA and c-myc mRNA was observed as a secondary effect upon induction of the MMTV LTR-Hras gene.

b) MMTV LTR-mos

Introduction of a MMTV LTR-mos construct into NIH/3T3 cells resulted in transfectants which are able to modulate the level of mos encoded p37 (15). Addition of dexamethasone to these cells leads to the induction of p37 and morphological transformation. This process is reversible and only a very low level of p37 is required to confer a transformed phenotype onto the NIH/3T3 cells. The low levels of p37 observed are sufficient for initiation and maintenance of transformation. Transformation-specific phenotypic parameters have been examined as a function of p37 levels and soft agar growth seems to require higher levels of p37 than focus formation in monolayers.

c) MMTV LTR-src

A MMTV LTR-src construct has been used to modulate the levels of the v-src encoded pp60 phosphoprotein in transfected rat-2 cells (11). These studies allowed the assessment of the contributions of quantitative (increase in cellular concentration) and qualitative (differences in amino acid sequence between v-src and c-src) parameters in the transformation process by the pp60 gene product. A distinct threshold level of pp60 v-src was discovered which is required for transformation. The mere presence of pp60v-src is not a sufficient prerequisite and even high levels of pp60 c-src are not capable of cellular transformation. These studies demonstrated that only the concurrence of elevated expression and the amino acid differences found in v-src can cause neoplastic transformation of rat-2 cells. Associated cellular protein phosphorylation differences (pp36), found in RSV infected cells, are not necessarily observed in the transition from the normal to the transformed state in MMTV LTR-src transfected cells.

d) MMTV LTR-polyoma Middle T

Established rat fibroblasts were transfected with a MMTV LTR- mT construct and the extent of mT expression could be regulated over a 100-fold range (24). Morphological transformation on monolayers and anchorage independent growth in soft agar required increasing levels of mT expression. Only those cells expressing high enough mT levels to promote soft agar growth could form tumors in syngeneic rats. The important functional parameter for the involvement of mT in the transformation process seems to be the level of mT phosphorylated by pp60 c-src and not the total amount of mT.

e) MMTV LTR - c myc

The recombination of the MMTV LTR with the coding region of the c-myc gene and introduction of this construct into Balb/c - 3T3 cells was used to determine whether the c-myc gene product functions as an intracellular mediator of the mitogenic response to PDGF (1). Upon

induction of the c-myc RNA levels with glucocorticoid hormone it was found that the probability of cell division in the absence of PDGF and the sensitivity to the mitogenic activity of EGF are increased. The c-myc product has therefore a growth promoting activity in Balb/c -3T3 cells.
These examples demonstrate the usefulness of conditional gene expression in the study of oncogenes. A precise and reproducible regulation of expression can be achieved in transfected cells dependent on glucocorticoid hormone in the growth medium. Further attenuation of the oncogene expression levels might become possible through suboptimal hormone concentrations (see below) and allow a further study of the effect of oncogene product levels on the various transformation parameters _in vitro_. Finally, the _in vitro_ recombination with the MMTV LTR allows a conditional expression of any oncogene, thereby expanding the framework of studies which are up to now dependent on the spontaneous occurrence of temperature sensitive oncogenes (32).

Expression of the Major Histocompatibility Antigen H-2L^d under the Control of the MMTV LTR

We have applied the MMTV LTR mediated conditional gene expression to a major histocompatibility antigen gene (H-2L^d) (23). The MHC class I molecules serve as recognition elements for T cells in the detection of foreign antigens. Lysis by cytotoxic T cells is dependent upon recognition of foreign antigens in conjunction with class I histocompatibility antigens (35). This process probably plays a role also in the elimination of spontaneously altered cell variants which arise _in vivo_. A variety of carcinoma cells have reduced levels of class I molecules at their surface, whereas augmentation of class I gene expression in, for instance, Ad-12 transformed cells leads to a reduction of their tumorigenicity. The ability to modulate the cell surface expression of the H-2L^d antigen by growth medium parameters might provide an excellent prerequisite for studying the foreign antigen presenting cell and T cell interaction in vitro.

a) A MMTV LTR-H-2L^d Chimeric Construct

To manipulate the expression of the H-2L^d antigen, the promoter region of the H-2L^d gene was replaced by the MMTV LTR. Figure 1 shows the strategy for the construction of the MMTV LTR H-2L^d chimera. The hormonally inducible MMTV promoter is contained in a fragment of the LTR comprising the entire U3 region (1194 nucleotides), the RNA initiation site designated +1, and 106 nucleotides of the R and U5 region (6). A restriction fragment delimited by a Bgl II site at the 3' border of the LTR (Figure 1A) and a HpaII site downstream of the RNA initiation site were used. The HpaII site was converted by synthetic linker ligation into a BamHI site (Figure 1C). The MMTV LTR promoter region was linked with a partially Bam HI digested H-2L^d gene fragment (5) which is truncated at a Bam HI site located about 15 nucleotides 5' of the translation initiation codon (Figure 1B). This partial Bam HI fragment (composed of 1.9 and 2.4 kb fragments) contains the entire coding region of the H-2L^d gene, but no promoter element. The transcriptional control sequences are contributed by the MMTV LTR (Figure 1C).

b) Cell Surface Expression of LTR H-2L^{d} Transfected Cells is Inducible by Glucocorticoid Hormone

Two plasmids containing the authentic H-2L^{d} gene with its own promoter region (Figure 1B) and the recombined MMTV LTR H-2L^{d} (Figure 1C) were introduced into L cells by transfection procedures. Mouse L cells are derived from C3H mice and express the K haplotype (H-2K^{k} and H-2D^{k}) antigens. The effect of dexamethasone on the expression of the transfected H-2L^{d} and the MMTV LTR H-2L^{d} as well as on the endogenous H-2K^{k} genes was determined (Figure 2). The two H-2 antigens were detected by specific monoclonal antibodies and quantitated by cell surface indirect immunofluorescence. The expression of the endogenous H-2K^{k} gene was unaffected by dexamethasone treatment (Figure 2A) and no endogenous H-2L^{d} expression could be detected (2B). L cells transfected with the authentic H-2L^{d} gene express high levels of H-2L^{d} antigen (2D). This level of expression is unresponsive to dexamethasone, but appears higher than H-2K^{k} expression in untransfected cells (compare 2A with 2D). Transfection with the H-2L^{d} gene and expression of H-2L^{d} antigen causes a decrease in the cell surface expression of the endogenous H-2K^{k}, irrespective of dexamethasone treatment (compare 2A with 2C). Such a reduction in endogenous H-2 expression after exogenous H-2L^{d} transfection has been reported. L cells transfected with MMTV LTR H-2L^{d} DNA expressed H-2L^{d} also in the absence of dexamethasone. They responded to dexamethasone treatment with about a three-fold increase in fluorescence intensity (2F). Concomitantly the cell surface expression of endogenous H-2K^{k} decreased after dexamethasone treatment (2E). This decrease could be due to competition for non-covalently associated β-2microglobulin. The association of β-2 microglobulin with H-2 antigen has been proposed as being required for cell surface expression of class I major histocompatibility antigens (12). These results show that the MMTV LTR H-2L^{d} construct is able to respond to dexamethasone not only on the transcriptional level, but that the cell surface concentration of the histocompatibility antigen can be modulated by hormone.

Selection of Glucocorticoid Receptor Mutants in a Rat Fibroblast Cell Line : a General Scheme for the Detection of Mutants in Enhancer Binding Proteins

a) Strategy of the Selection

Mutants affecting the glucocorticoid receptor gene or the function of the receptor protein usually do not alter the phenotype of cultured mammalian cells sufficiently for an efficient phenotypic selection. Exceptions are found in certain lymphoid cell lines which are growth inhibited or killed by glucocorticoid hormones (16,17,18). The inability to select these mutants could be overcome if a suitable selectable marker could be introduced into mammalian cells which responds to glucocorticoids. A suitable gene could be defined as a gene which responds to physiological doses of hormone by the production of a mRNA and a protein product which confers concentration dependent growth properties on the cell. The thymidine kinase (tk) gene is useful for this purpose. For and against the expression of the tk gene can be selected by addition of HAT (hypoxanthine, aminopterin and thymidine) or BrdU (bromodeoxyuridine) to the growth medium, respectively (29). We have therefore introduced a glucocorticoid-

responsive MMTV LTR-tk gene construct into rat-2 cells. The gene construct used in this study contains the MMTV LTR sequences up to position -236 (p Δ236, 10). It has been shown that this MMTV LTR fragment is able to confer hormonal responsiveness to linked heterologous gene promoters. The MMTV LTR fragment was linked to a truncated tk gene which retained part of its regulatory sequences up to position -78 (10,20). Rat-2 cells are derived from rat embryo fibroblasts. They are thymidine kinase deficient and contain normal amounts of GR (M. Pfahl, unpublished results). Glucocorticoids exert a profound inhibitory effect on the proliferation of L cells (mouse fibroblasts), but we did not observe growth inhibition by glucocorticoids on rat-2 cells. A slightly altered cellular morphology, however, was caused by the presence of hormone in the growth medium.
Rat-2 cells were transfected with plasmid Δ236 (MMTV LTR-tk) and HAT resistant tk^+ clones were selected in medium containing 10^{-6} M dexamethasone. Ten clones - dependent on dexamethasone for growth in HAT - were isolated. Removal of dexamethasone from HAT medium resulted in rapid cell death. All clones grew well in medium containing 60 µg ml of bromodeoxyuridine (BrdU) in the absence of dexamethasone. BrdU allows only the growth of tk^- cells. Dexamethasone inhibited the growth in medium with BrdU, as expected. Figure 3 shows the influence of dexamethasone on the growth of one transfected cell clone (13-13-1) in selective medium. A minimum of 2 x 10^{-8} M dexamethasone is required to kill the cells in BrdU medium, whereas 2.5 x 10^{-9} M dexamethasone is required for growth in HAT medium. These growth characteristics demonstrate a fine tuning of hormonal responsiveness of the chimeric promoter. It allows only a very low constitutive level of expression in the absence of hormone and it is very differentially inducible. The extent of induction is a function of the glucocortiocid concentration in the growth medium. The cellular clone 13-13-1 can be used to obtain mutants deficient in GR function by selection for the tk^- phenotype, i.e., by selection for cells which are no longer expressing the tk gene upon dexamethasone stimulation.

b) Mutagenesis and Receptor Mutant Selection

We assume that rat-2 cells are diploid for the GR gene. Spontaneous mutations affecting both GR gene alleles should occur at a very low frequency of about 10^{-12}. After mutagenesis the frequency might be increased to 10^{-8} (3). We decided on a scheme in which mutagenized cells are initially selected for mutants with a defect in a single GR allele. After a second exposure to mutagen we select cells in which both alleles are affected. The advantage of this approach is that the frequency of mutation in one allele can be expected to be 10^{-4} to 10^{-5} as previously observed (3,16). The mutagen N-methyl-N-nitro-N-nitroso guanidine (NG) was used to mutagenize 10^7 cells of clone 13-13-1. Twenty-four hours after mutagenesis, cells were transferred to medium containing BrdU and 2 x 10^{-8} M dexamethasone. Under these conditions non-mutagenized 13-13-1 cells do not survive (Figure 3). We can assume, however, that cells which have a two-fold reduced amount of receptor due to the inactivation of one GR gene allele might be able to grow. 13-13-1 cells can grow in BrdU and 10^{-8} M dexamethasone

(Figure 3). Cells were maintained in BrdU and 2 x 10^{-8} M dexamethasone for 10 days. Viable cells were then grown in 10^{-6} M dexamethasone and HAT for one week to exclude mutants in the structural part of the tk gene and the hormone element (HRE) of the MMTV LTR. Surviving cells were again mutagenized and completely GR-negative (r^-/r^-) clones were selected in BrdU containing 10^{-6} M dexamethasone. After two weeks of selection colonies were isolated and characterized.

c) Analysis of Hormonal Response Mutants

Sixty clones were grown in DME medium in the presence and absence of dexamethasone. The wild type cells respond to dexamethasone with a rounded morphology. In the absence of dexamethasone they are elongated, of typical fibroblast morphology. The dexamethasone-responsive tk^- cell clones and the dexamethasone-nonresponsive tk^- clones can therefore be distinguished phenotypically. The dexamethasone-responsive clones could be mutants in the tk gene, mutants in the HRE sequence or mutants in which specific transcription factors are affected. Seven of the 60 clones appeared to be dexamethasone-nonresponsive. To further characterize the cell clones, nuclear transfer tests of the steroid receptor complex and hormone binding studies were carried out. The procedures have been described by Pfahl et al. (17). The nuclear transfer was measured at five different hormone concentrations. The results are summarized in Table I. Three of the non-responsive mutants (R6, R32a, R32) show a decreased nuclear transfer and one clone (R2) reveals a slightly increased nuclear transfer. Steroid binding activity is normal in five of the clones and reduced two- to three-fold in clones R32 and R16. From these data it appears most likely that the tk^- phenotype in clones R6, R32a, and R32 results from non-functional glucocortiocid receptor. This conclusion is being supported by transfection experiments in which p 236 was reintroduced into clones R6, R32a and R32. Tk^+ clones were obtained at a reduced frequency and all clones were constitutively expressing the tk gene in the absence of hormone. The other cell clones need to be characterized further to determine whether their tk^- phenotype is due to lesions in the glucocorticoid receptor gene.

d) Implications of the Functional Similarity between Glucocorticoid Receptor and Enhancer Binding Proteins

The selection scheme described above enabled us to obtain GR mutants in a cell in which such mutants could previously not be selected. The mutants isolated fit into the classes of nt^- (R6, R32a, R32) and possibly nt^i (R2) (16,26,27). It has been suggested that the nt^- and nt^i phenotypes are due to alterations in the DNA-binding domain of the GR (16,34). A defect in the DNA-binding could results in the nt^- phenotype (reduced nuclear transfer). This might be due to a reduced DNA affinity and a lesser retention in the nucleus. In addition, it has been shown that the binding to specific DNA sequences can be impaired for some nt^- receptors (19). The nt^i phenotype probably signals increased DNA affinity of the GR. So far we have only detected receptor mutants which appear to be due to a defect in the DNA binding of the receptor. Our selection scheme was not designed to

detect mutations which lack hormone binding activity. The selection was in part carried out at hormone concentrations more than two orders of magnitude above the binding constant. Point mutations which results in a reduction of hormone binding by the receptor by a factor of 10 to 50 might not have been detected in the second part of the selection scheme. Mutations which reduce the affinity of the receptor for the hormone even further might be rare. The selection scheme most likely detects mutations in which the DNA binding of the receptor complex is reduced five- to ten-fold.
The functional resemblance of the GR with enhancer binding proteins has been pointed out (22). The principles of this selection scheme might therefore be applicable for the isolation of mutants of other enhancer binding proteins. These could be represented by tissue-specific proteins conferring cell type specificity to enhancer function or more generally occurring proteins such as the SV40 enhancer binding protein. The scheme might furthermore be applicable to the selection of mutations in other steroid receptor genes. This possibility is based on the observation that MMTV LTR contains sequence elements which mediate responsiveness to steroid hormone receptors for progesterone, estrogen and testosterone (Pfahl, et al., this volume). The availability of mutants in enhancer binding proteins might facilitate their characterization and isolation. The combination of these mutants with gene transfer procedures employing total genomic DNA could provide an approach to the molecular cloning of enhancer binding protein genes. In the case of the above described mutants it might be possible to transfect GR minus cell clones with human genomic DNA from cells expressing GR and select for GR gene acquisition by the tk^+ phenotype. This methodology, initially based on the knowledge of the hormone response element (HRE) of the MMTV LTR, could become valuable in the general study of enhancer action.

REFERENCES

1. Armelin, H.A., M.C.S. Armelin, K. Kelly, T. Stewart, P. Leder, B.H. Cochran, and C.D. Stiles: Functional role of c-myc in mitogenic response to platelet-derived growth factor. Nature 310:655, 1984.
2. Bishop, J.M.: Viral oncogenes. Cell 42:23, 1985.
3. Bourgeois, S., and R.F. Newby: Diploid and haploid states of the glucocorticoid receptor gene of mouse lymphoid cell lines. Cell 11:423, 1977.
4. Cooper, G.M.: Cellular transforming genes. Science 218:801, 1982.
5. Evans, G.A., D.H. Margulies, R. Daniel Camerini-Otero, K. Ozato, and J.G. Seidman: Structure and expression of a mouse major histocompatibility antigen gene, H-$2L^d$. Proc Natl Acad Sci USA 79:1994, 1982.
6. Fasel, N., K. Pearson, E. Buetti, and H. Diggelmann: The region of mouse mammary tumor virus DNA containing the long terminal repeat includes a long coding sequence and signals for hormonally regulated transcription. The EMBO J 1:3, 1982.
7. Groner, B., N.E. Hynes, U. Rahmsdorf, and H. Ponta: Transcription initiation of transfected mouse mammary tumor virus LTR DNA is regulated by glucocorticoid hormones. Nucleic Acids Res 11:4713, 1983.
8. Gruss, P.: Magic enhancers. DNA 3:1, 1984.
9. Huang, A.L., M.C. Ostrowski, D. Berard, and G.C. Hager: Glucocorticoid regulation of the Ha Mu SV p21 gene conferred by sequences from mouse mammary tumor virus. Cell 27:245, 1981.
10. Hynes, N.E., A. van Ooyen, N. Kennedy, P. Herrlich, H. Ponta, and B. Groner: Subfragments of the long terminal repeat cause glucocorticoid responsive expression of mouse mammary tumor virus and of an adjacent gene. Proc Natl Acad Sci USA 80:3637, 1983.
11. Jakobovits, E.B., J.E. Majors, and H.E. Varmus: Hormonal regulation of the Rous Sarcoma Virus src gene via a heterologous promoter defines a threshold dose for cellular transformation. Cell 38:757, 1984.
12. Kamarck, M.E., J.A. Barbosa, and F.H. Ruddle: Somatic cell genetic analysis of HLA-A, B, C and human β_2-microglobulin expression. Somatic Cell Genetics 8:385, 1982.
13. Majors, J., and H.E. Varmus: A small region of the mouse mammary tumor virus long terminal repeat confers glucocorticoid homrone regulation on a linked heterologous gene. Proc Natl Acad Sci USA 80:5866, 1983.
14. Miesfeld, R., S. Okret, A.C. Wikström, Oe. Wrange, J.-A. Gustafsson, and K.R. Yamamoto: Characterisation of a steroid hormone receptor gene and mRNA in wild-type and mutant cells. Nature 312:779, 1984.
15. Pappkoff, J., and G.M. Ringold: Use of the mouse mammary tumor virus long terminal repeat to promote steroid-inducible expression of v-mos. J Virol 52:420, 1984.
16. Pfahl, M., R.J. Kelleher, and S. Bourgeois: General features of steroid resistance in lymphoid cell lines. Molec Cellul Endocrinology 10:193, 1978.

17. Pfahl, M., T. Sandros, and S. Bourgeois: Interaction of glucocorticoid receptors from lymphoid cell lines with their nuclear acceptor sites. Molec Cellul Endocrinology 10:175, 1978.
18. Pfahl, M., and S. Bourgeois: Analysis of steroid resistance in lymphoid cell hybrids. Somatic Cell Genetics 6:63, 1980.
19. Pfahl, M.: Specific binding of the glucocorticoid-receptor complex to the mouse mammary tumor proviral promoter region. Cell 31:475, 1982.
20. Pfahl, M., D. McGinnis, M. Hendricks, B. Groner, and N.E. Hynes: Correlation of glucocorticoid receptor binding sites on MMTV proviral DNA with hormone inducible transcription. Science 222:1341, 1983.
21. Ponta, H., W.H. Gunzburg, B. Salmons, B. Groner, and P. Herrlich: Mouse mammary tumor virus: a proviral gene contributes to the understanding of eukaryotic gene expression and mammary tumorigenesis. J Gen Virol 66:931, 1985a.
22. Ponta, H., N. Kennedy, P. Skroch, N.E. Hynes, and B. Groner: Hormonal response region in the mouse mammary tumor virus long terminal repeat can be dissociated from the proviral promoter and has enhancer properties. Proc Natl Acad Sci USA 82:1020, 1985b.
23. Ponta, H., R. Ball, M. Steinmetz, and B. Groner: Hormonal regulation of cell surface expression of the major histocompatibility antigen $H\text{-}2L^d$ in transfected cells. EMBO J 4:13, 1985c.
24. Raptis, L., H. Lamfrom, and T.L. Benjamin: Regulation of cellular phentotype and expression of polyomavirus middle T antigen in rat fibroblasts. Molec Cellul Biol 5:2476, 1985.
25. Scheidereit, C., and W. Beato: Contacts between hormone receptor and DNA double helix within a glucocorticoid regulatory element of mouse mammary tumor virus. Proc Natl Acad Sci USA 81:3029, 1984.
26. Schmidt, T.J., J.M. Harmon, and B.E. Thompson: Activation labile glucocorticoid receptor complexes of a steroid resistant variant of CEM-C7 human lymphoid cells. Nature 286:507, 1980.
27. Sibley, C.H., and G.M. Tomkins: Isolation of lymphoma cell variants resistant to killing by glucocorticoids. Cell 2:213, 1974.
28. Svec, F.: Glucocorticoid receptor regulation. Life Sci 36:2359, 1985.
29. Szybalska, E.H., and W. Szybalski: Genetics of human cell lines, IV. DNA mediated heritable transformation of a biochemical trait. Proc Natl Acad Sci USA 48:2026, 1962.
30. Tanaka, K., K.J. Isselbacher, G. Khoury, and G. Jay: Reversal of oncogenesis by the expression of a major histocompatibility complex class I gene. Science 228:26, 1985.
31. Varmus, H.E.: The molecular genetics of cellular oncogenes. Ann Rev Genet 18:553, 1984.
32. Weber, M.J., and R.R. Friis: Dissociation of transformation parameters using temperature-conditional mutants of Rous Sarcoma Virus. Cell 16:25, 1979.
33. Weinberg, R.A.: Oncogenes of spontaneous and chemically induced tumors. Adv Cancer Res 36:149, 1982.
34. Yamamoto, K.R. and B.M. Alberts: Steroid receptors: elements for modulation of eukaryotic transcription. Ann Rev Biochem 45:721, 1976.

35. Zinkernagel, R.M., and P.C. Doherty: MHC restricted cytotoxic T cells: studies on the biological role of polymorphic major transplantation antigens determining T cell restriction specificity, function and responsiveness. Adv Immunol 27:51, 1979.

Construction of a H-2L^d gene with a MMTV promoter

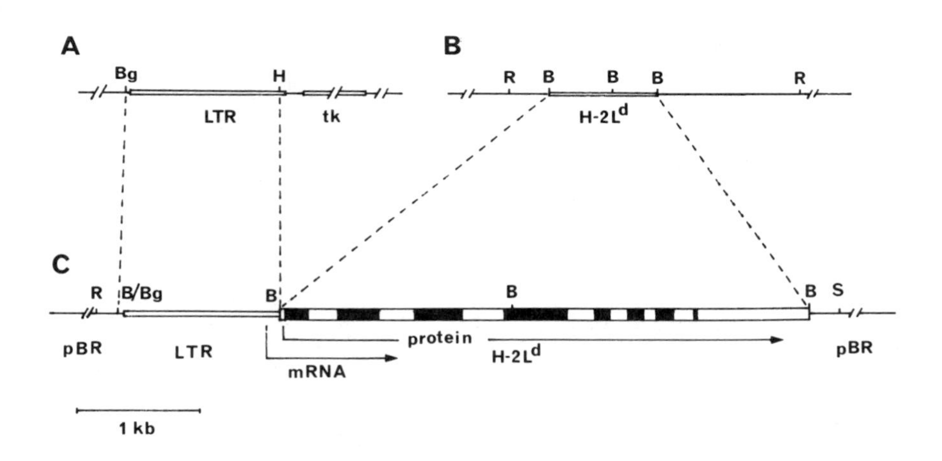

Figure I. Construction of the plasmid MMTV LTR H-2L^d. The plasmid MMTV LTR H-2L^d (C) was constructed from restriction fragments contained in plasmids 2.6 (A) and pH-2L^d (B). The MMTV LTR is contributed by a 1.2 kb Bgl II-Hpa II fragment (A). The protein coding region of H-2L^d is contributed by a 4.3 partial Bam HI fragment (B). These fragments were joined upon conversion of the Hpa II site into a BamHI site. The mRNA initiation site (contributed by the LTR fragment) and the translational start site (contributed by the H-2L fragment) are indicated by arrows in (C). Exons are indicated in black. B: Bam HI, Bg: Bgl II, H: Hpa II, R: EcoRI, S: Sal I.

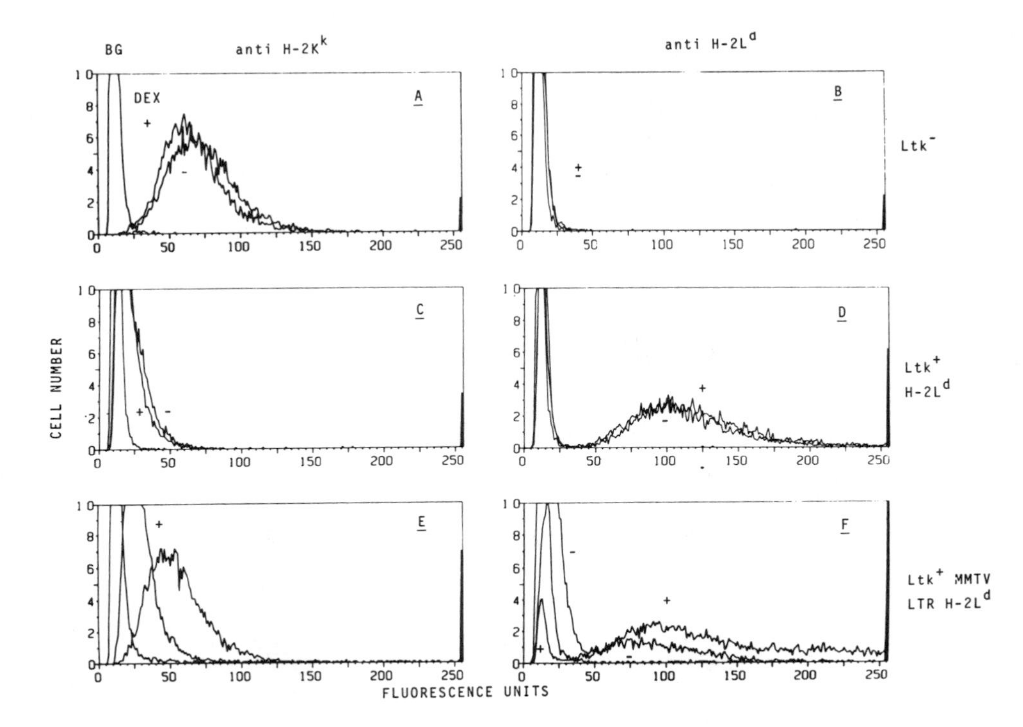

Figure 2. Effect of dexamethasone on the cell surface expression of major histocompatibility antigens. Ltk^- cells (panels A, B), pH-$2L^d$ transfected Ltk^- cells (C, D) and MMTV LTR H-2L transfected Ltk^- cells (E, F) were pretreated for 24 hours with 10^{-6} M dexa-methasone (+ dex) before staining with anti H-2 K^k antibody (16-3-22 S) (A, C, E) or anti H-$2L^d$ antibody (28-14-8 S) (B, D, F). Background staining with a control antibody (P3X63 Ag8) is indicated by the superimposed left most peak (BG). The cytofluorographic analysis of the fluorescence intensity of 10^4 cell/sample is shown using an identical electronic amplification throughout. Dexamethasone had no effect on cell size as determined by narrow angle light scatter.

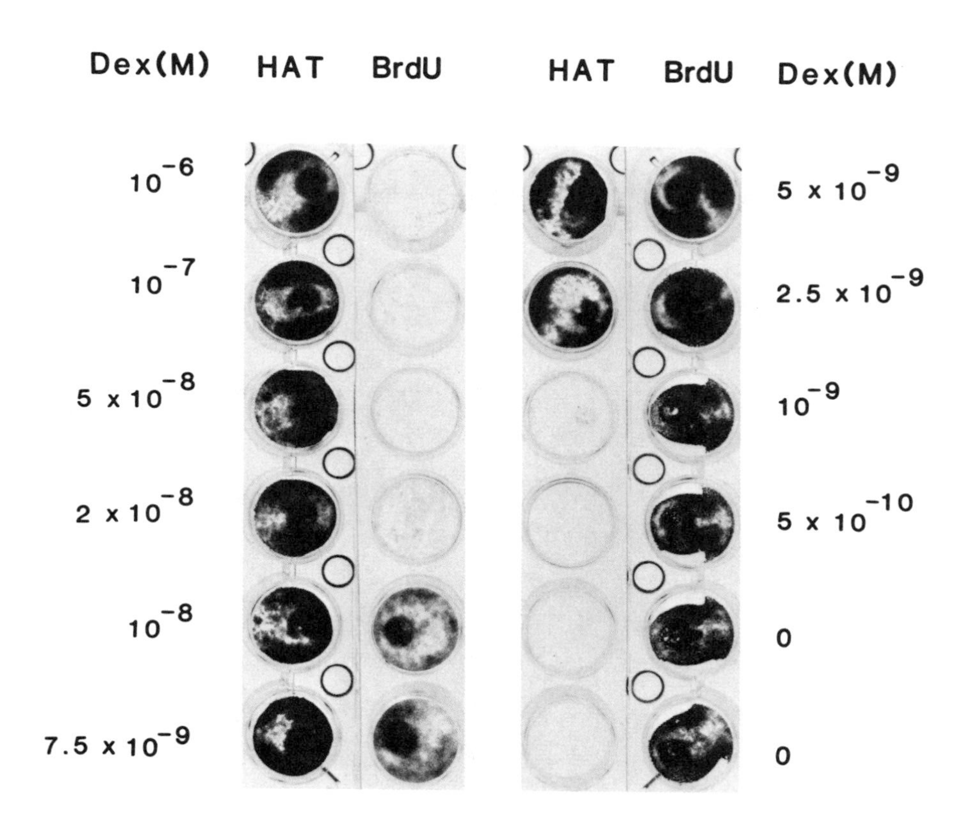

Figure 3. Effect of dexamethasone on the growth of cells from clone 13-13-1. These cells were transfected with DNA from the MMTV LTR -tk construct (p**Δ**236) and grown in medium containing HAT and BrdU in addition to the indicated concentration of dexamethasone. 10^4 cells were plated per 1 ml well and grown for one week. Cells were fixed, stained and photographed. White areas in the corners of some intensely stained wells result from overgrowth and detachment of cells.

TABLE I. Characteristics of the Glucocorticoid Receptors in Mutant Clones

Mutant (receptor)	% Nuclear Transfer	Kd (TA) ($M \times 10^9$)
wild-type	84	2.5
R-6	53	2.4
R-2	92	2.8
R-32a	56	3.5
R-33	86	
R-34	74	2.8
R-32	62	6.8
R-16	78	9.4

23. Specific DNA-Binding Proteins and DNA Sequences Involved in Steroid Hormone Regulation of Gene Expression

T. Spelsberg, J. Hora, M. Horton, A. Goldberger, B. Littlefield, R. Seelke, and H. Toyoda

Department of Cell Biology, Section of Biochemistry, Mayo Clinic and Graduate School of Medicine, Rochester, MN 55905

MECHANISM OF ACTION OF STEROID HORMONES

Steroid hormones circulate in the blood and are taken up by target cells via complexes with intracellular binding proteins termed receptors, that are hormone and tissue specific (1). Each receptor binds its specific steroid with very high affinity, having an equilibrium dissociation constant (K_d) in the range of 10^{-9} to 10^{-10} M. Once bound by their specific steroid hormones, the steroid receptors undergo a conformational change which allows them to bind with high affinity to sites on chromatin, termed nuclear acceptor sites (2,3). There are estimated 5,000 to 10,000 of these sites expressed with an equal number not expressed ("masked") in intact chromatin. The result of the binding to nuclear acceptor sites is an alteration of gene transcription (3-5) or, in some cases, gene expression (6) as measured by the changing levels of specific RNAs and proteins in that target tissue. Each steroid regulates specific effects on the RNA and protein profiles. The chronology of the above mechanism of action after injection of radiolabelled steroid is as follows: Steroid-receptor complex formation (1 minute), nuclear acceptor sites (2 minutes), effects on RNA synthesis (10 to 30 minutes), and finally the changing protein profiles via changes in protein synthesis and protein turnover (1 to 6 hours) (2,3). Thus steroid receptors represent one of the first identified intracellular gene regulatory proteins. The receptor molecules themselves are regulated by the presence or absence of the steroid molecule.

NUCLEAR ACCEPTOR SITES FOR STEROID RECEPTORS

The nuclear acceptor sites for steroid receptors have been identified as either specific DNA sequences or a complex of DNA and protein. Only studies on the nuclear acceptor sites, containing the specific DNA sequences bound by the specific acceptor proteins, will be described in this chapter. The other class of sites, involving specific DNA sequences alone, have been shown to adjoin the structural genes they regulate (7,8,9). In some studies, these DNA domains have been shown to be required for glucocorticoid effects on the genes. Whether or not the steroid receptors actually bind to these DNA sequences, however, is still questionable because many properties reported for the steroid receptor binding to these sites indicate a non-specific interaction. For example, none of the studies identifying these specific DNA binding sequences for steroid receptors have included studies using native chromatin, which is the native state of DNA in cell nuclei. Further, the data presented to date suggest that these sequences show only a 2 fold enhanced binding over the whole genomic DNA with a lack of saturable binding (9-15). Moreover, they fail to display receptor specificity (14). Lastly, with regard to the glucocorticoid receptor in rat liver, the exact DNA sequence involved in the binding of steroid receptors is controversial (15). It is possible that these particular DNA sequences may actually be involved in binding of "other" regulatory proteins which may play a role in the steroid action on gene transcription.

On the other hand, the acceptor sites involving DNA-acceptor protein complexes, display a high affinity, high capacity binding, which is both saturable and steroid receptor specific (2,16,17). These protein-DNA acceptor sites have been identified for the avian oviduct progesterone receptor (2,17), and estrogen receptor (18), the cow and rabbit uterine estrogen receptor (19,20), the guinea pig uterine progesterone receptor (21), the sheep brain estrogen and progesterone receptor (22), the rat testicular androgen receptors (23), and finally the glucocorticoid receptors (24) in rat liver and in a human leukemia cell line (25). In the chick oviduct model system, it has been suggested that the acceptor protein-DNA complex serves as the initial binding site (acceptor site) for the avian oviduct progesterone receptor (PRov) via the receptor (B) subunit. The latter is initially bound by the receptor subunit (A) to form a dimer. Quickly thereafter the receptor subunit (A) dissociates from the B subunit and binds to DNA sequences neighboring the ovalbumin (and other) structural genes to alter transcription (17,26,27). To support this model, the A subunit has been used to show specific binding to DNA sequences adjacent to the ovalbumin gene (9) while the B species has been shown to bind the protein-DNA sequences (M. Horton, J. Hora, B. Littlefield and T. Spelsberg, in preparation).

COMPOSITION OF THE NUCLEAR ACCEPTOR SITES OF THE PRov

Studies in this laboratory have shown that the nuclear acceptor sites for the PRov (the initial nuclear binding site) appear to involve specific acceptor proteins (3,17,28) and possibly specific acceptor DNA sequences (28,29). The acceptor proteins can be reannealed to whole hen DNA to generate reconstituted acceptor sites (28), which display the same patterns and specific binding of PRov as the native acceptor sites. Thus, using the reconstitution approach, a specificity for both specific proteins (17,28) and specific DNA sequences (28,29) have been strongly intimated. The acceptor proteins have been characterized as small molecular weight, hydrophobic proteins of two classes, one with a pI in the range of 5.5 to 6.0 and

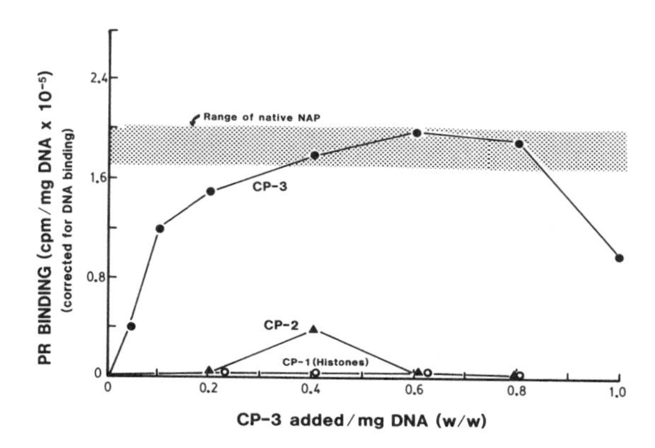

Figure 1. Recovery of acceptor activity as a function of the ratio of protein to DNA in the reconstitution assay and the chromatin protein fraction used. The CP-1, CP-2, and CP-3 protein fractions were reconstituted to hen DNA at varying concentrations as shown in the abscissa and described elsewhere (28). Reconstituted NAPs with CP-1 (○), CP-2 (▲), and CP-3 (●) were analyzed for DNA recovery and acceptor activity (40). The mean of four replicate [^{3}H]PR binding assays in one typical experiment are shown. The hatched area represents the range of [^{3}H]PR binding to native NAP (undissociated) obtained in these studies. Values are corrected for DNA binding. (Data was reproduced with permission from T.C. Spelsberg et al., Biochemistry 23:5103, 1984).

one with a pI in the range of 6.5 to 7.0 (2,3,16,17). The evidence that specific DNA sequences are involved in the acceptor sites is presented in the next section.

SPECIFIC PROTEINS AND DNA SEQUENCES INVOLVED IN THE NUCLEAR ACCEPTOR SITES FOR PRov

Figure 1 shows the results of PRov binding to reconstituted acceptor sites consisting of various fractions of chromosomal proteins reannealled to hen genomic DNA. Only one fraction, i.e. chromosomal protein fraction 3 (CP-3), is capable of reconstituting the nuclear acceptor sites. As can be seen, the number of PR binding sites which can be reconstituted is limited, even when an excess of acceptor protein is reannealed to the hen DNA. Enriched acceptor protein preparations display the same patterns. These results suggest that a limited number of specific DNA sequences in the avian genome contribute to the regeneration of acceptor sites. Figure 2 shows the results of extensive studies on the species specificity with regard to the source of the DNA required to regenerate the PRov acceptor sites. In Panel A, genomic DNAs of many animals can substitute for hen DNA to regenerate PRov acceptor sites when reannealed with the avian oviduct

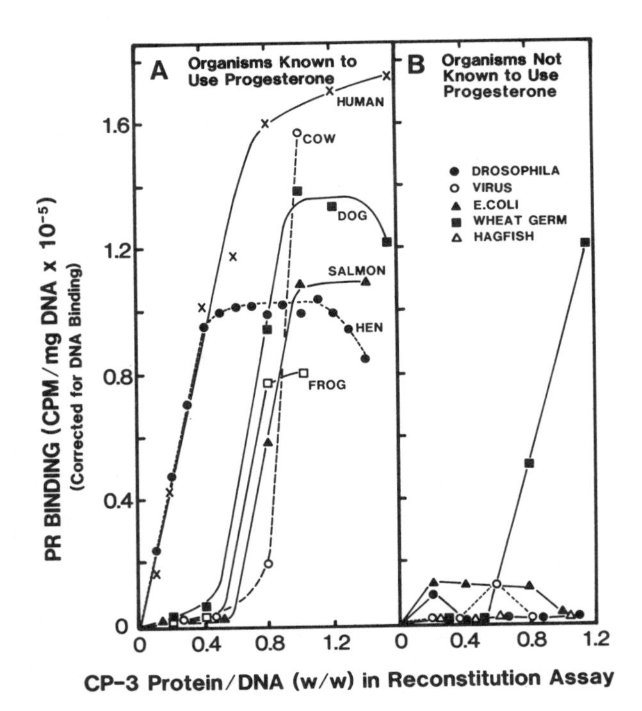

Figure 2. Binding of isolated chick oviduct $[^3H]$PR to reconstituted heterologous NAPs composed of hen oviduct CP-3 protein and DNA from various sources. The NAPs were reconstituted using isolated hen oviduct CP-3 chromatin fraction and the pure DNA from a variety of sources as described elsewhere (29). Reconstituted NAPs were then resuspended in a Buffer containing 4 mM Tris-HCl + 0.2 mM EDTA, pH 7.5 and bound with $[^3H]$PR by the streptomycin method as described elsewhere (16,41). The binding assay was the same as described elsewhere (28). The mean of triplicate analysis of PR binding to the reconstituted NAPs in the same experiment are shown. The binding values for each NAP are corrected for PR binding to the respective pure DNA (the same pure DNA used in the reconstituted NAP). The data is divided by NAPs containing DNA from organisms known to utilize the steroid progesterone (Panel A), and from organisms not utilizing progesterone (Panel B). The following symbols define the source of the DNA in the reconstituted NAPs: Panel A -- (X) human uteri; (○) cow thymus (calf); (■) dog spleen; (●) hen spleen; (▲) salmon sperm; and (□) frog liver. Panel B -- (●) Drosophila larvae; (▲) E. coli; (○) Charon 4A virus; (△) hagfish; and (■) wheat germ. (Data was reproduced with permission from H Toyoda, et al., P.N.A.S. (US), in press).

acceptor protein (fraction CP-3). Interestingly, these organisms utilize progesterone as a steroid. The hen and human DNAs require less acceptor protein to regenerate a maximal level of PRov acceptor sites than do the DNAs from other animals. In contrast, the genomic DNAs from organisms which do not utilize progesterone fail to regenerate PRov acceptor sites when reconstituted with avian oviduct acceptor protein. These data support the idea that specific DNA sequences play a role in the PRov acceptor sites as a complex with the acceptor proteins, and are, in part, species specific.

EVOLUTIONARY CONSERVATION OF THE NUCLEAR ACCEPTOR SITES FOR PRov

Since the avian oviduct acceptor proteins for PRov can be reannealed to the genomic DNAs of various animals, e.g. human, dog, cow, frog, etc. (Figure 2), the PRov acceptor sites, including the acceptor protein and acceptor DNA sequences appear to be conserved during biological evolution. This possibility is supported by 1) steroids (e.g. progesterone) as well as steroid receptors in general have also been conserved during evolution (30,31); 2) monoclonal antibodies against PRov acceptor proteins (complexed with the specific acceptor site sequences) not only inhibit PRov binding to hen oviduct acceptor sites but also inhibit PRov binding to the acceptor sites from the cow uterus (32); and finally 3) the results in Figure 2 show a correlation between the organisms whose genomic DNAs can regenerate PRov acceptor sites with avian oviduct acceptor protein, and organisms known to utilize progesterone and its receptor. In contrast, the genomic DNAs from organisms not using progesterone nor its receptor, fail to regenerate acceptor sites (29).

EVIDENCE THAT THE ACCEPTOR PROTEIN-DNA SEQUENCES FOR PRov DO NOT ADJOIN THE OVALBUMIN GENE

In the estrogen withdrawn chicken, progesterone regulates the synthesis of ovalbumin (33,34). Thus, it was of interest to determine whether or not the acceptor site DNA sequences reside in or near this gene. Figure 3 shows the results of PR binding to the PRov acceptor sites prepared by reconstituting chick oviduct acceptor protein (CP-3) to cloned genomic ovalbumin gene. The ovalbumin clone (POV-12) was a gift from Dr. Bert O'Malley, Baylor College of Medicine, Houston, Texas. This gene represents the natural gene and contains the natural introns including flanking sequences of 500-3,500 base pairs (35). As shown in Figure 3, this reconstituted protein-ovalbumin gene DNA complex does not display the specific PRov acceptor activity whereas the reconstituted complex containing whole hen genomic DNA does display the acceptor activity. Thus, these acceptor protein-DNA sequences (i.e. the PRov acceptor sites) do not appear to reside in or near the ovalbumin gene. This conclusion is supported by the fact that the native PRov acceptor sites are resistant to DNase activity (16,36,37). This property is the opposite of that found for transcriptionally active regions of chromatin (38) but is similar to that reported for acceptor sites of other steroid receptors which have been localized in the nuclear matrix (39,40).

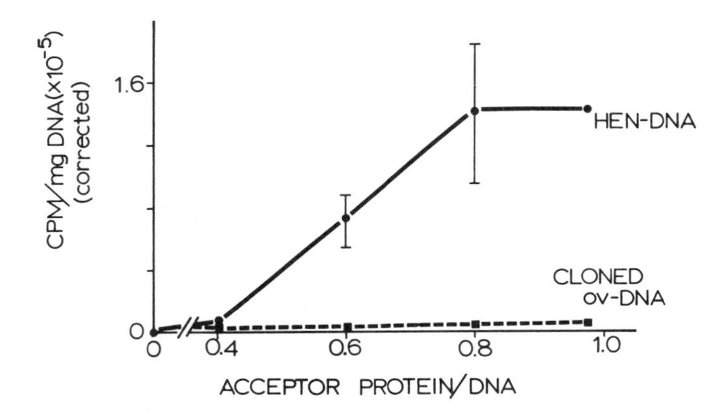

Figure 3. Reconstitution of PRov acceptor sites using avian oviduct chromatin proteins and cloned ovalbumin gene. The acceptor protein from hen oviduct chromatin was reconstituted with (●—●) whole genomic hen DNA and (■-■) cloned genomic ovalbumin gene (POV-12). POV-12 is a plasmid constructed by inserting the entire ovalbumin gene plus 3.5 kb of flanking 5' and 0.5 kb of flanking 3' DNA into the parent plasmid, pBR322 (35). The latter was a gift from Dr. Bert O'Malley, Dept. of Cell Biology, Baylor College of Medicine, Houston, Texas. The reconstitution and the cell free binding assays are described elsewhere (28,41). The extent of PRov binding is plotted against the ratio of protein to DNA added to the reconstitution assay.

FUTURE DIRECTION

The role of the acceptor proteins and acceptor DNA sequences in the nuclear binding of PRov is not clear. Whether the PRov binds directly to the acceptor protein (bound to DNA) or to DNA whose structure is perturbed by the DNA acceptor protein complex is unknown. The cis position (distance) of these acceptor sites with regard to structural genes affected by the progesterone is presently under study. If the PRov acceptor sites are found to be markedly distant from the progesterone regulated genes, it is possible that regulatory genes may exist to serve as intermediates in the process of steroid regulation of gene transcription. These regulatory genes could reside distant from the structural genes and, by neighboring acceptor sites, could be regulated by the PRov. Using the cloned acceptor site DNA sequences as a probe, the "speculated" regulatory sequences neighboring the acceptor sites should be retrievable from a library of chicken genomic DNA.

ACKNOWLEDGEMENTS

This work was supported by grants from the NICHDD (HD9140 and HD16075), training grants HD07108 (J.H.) and CA90441 (M.H. and A.G.), and the Mayo Foundation.

1. Jensen, E.V. and DeSombre, E.R. Mechanism of action of the female sex hormones. Ann Rev Biochem 41:203, 1972.
2. Thrall, C., Webster, R.A. and Spelsberg, T.C. "Steroid receptor interaction with chromatin". In: The Cell Nucleus, Vol. VI, Part 1, Harris Busch (ed.), Academic Press, New York, pp. 461, 1978.
3. Spelsberg, T.C. "Chemical characterization of nuclear acceptors for the avian progesterone receptor". In: Biochemical Actions of Hormones, G. Litwack (ed.), Academic Press, Vol. 9, pp. 141, 1982.
4. O'Malley, B.W. and Means, A.R. Female steroid hormones and target cell nuclei: The effects of steroid hormones on target cell nuclei are of major importance in the induction of new cell functions. Science 183:610, 1974.
5. Breathnach, R. and Chambon, P. Organization and expression of eucaryotic split genes coding for proteins. Ann Rev Biochem 50:349, 1981.
6. Moore, J.T., Norvitch, M.E., Wieben, E.D. and Veneziale, C.M. Expression of a secretory protein gene during androgen-induced cell growth. J Biol Chem 259:14750, 1984.
7. Payvar, F., Wrange, O., Carlstedt-Duke, J., Okret, S., Gustafsson, J.A. and Yamamato, K. Purified glucocorticoid receptors bind selectively in vitro to a cloned DNA fragment whose transcription is regulated by glucocorticoids in vivo. Proc Natl Acad Sci 78:6628, 1981.
8. Chandler, V.L., Maler, B.A. and Yamamoto, K.R. DNA sequences bound specifically by glucocorticoid receptor in vitro render a heterologous promoter hormone responsive in vivo. Cell 33:489, 1983.
9. Compton, J.G., Schrader, W.T. and O'Malley, B.W. Selective binding of chicken progesterone receptor A subunit to a DNA fragment containing ovalbumin gene sequences. Biochem Biophys Res Comm 105:96, 1982.
10. Bailley, A., Atger, M., Atger, P., Cerbon, M.A., Alizon, M., Vu Hai, M.T., Logeat, F. and Milgrom, E. The rabbit uteroglobin gene. Structure and interaction with the progesterone receptor. J Biol Chem 258:10384, 1983.
11. Mulvihil, E.R., LePennec, J.P. and Chambon, P. Chicken oviduct progesterone receptor: Location of specific regions of high affinity binding in cloned DNA fragments of hormone-responsive genes. Cell 28:621, 1982.
12. Dean, D.C., Knoll, B.J., Riser, M.E. and O'Malley, B.W. A 5'-flanking sequence essential for progesterone regulation of an ovalbumin fusion gene. Nature 305:551, 1983.
13. Payvar, F., DeFranco, D., Firestone, G.L., Edgar, B., Wrange, O., Okret, S., Gustafsson, J.A. and Yamamoto, K.R. Sequence specific binding of glucocorticoid receptor to MTV DNA at sites within and upstream of the transcribed region. Cell 35:381, 1983.
14. Miller, P.A., Ostrowski, M.C., Hager, G.L. and Simons, S.S., Jr. Covalent and non-covalent receptor-glucocorticoid complexes preferentially bind to the same regions of the long terminal repeat of murine mammary tumor virus proviral DNA. Biochemistry 23:6883, 1984.

15. deAhe, D., Janich, S., Scheidert, C., Renkawitz, R., Schutz, G. and Beato, M. Glucocorticoid and progesterone receptors bind to the same sites in two hormonally regulated promoters. Nature 313:706, 1985.
16. Spelsberg, T.C. "Role of Nonhistone Chromatin Proteins and Specific DNA Sequences in the Nuclear Binding of Steroid Receptors". In: Chromosomal Nonhistone Proteins - Vol. I Biology, L.S. Hnilica (ed.), CRC Press, Inc., Cleveland, pp. 47, 1983.
17. Spelsberg, T.C., Littlefield, B.A., Seelke, R., Martin-Dani, G., Toyoda, H., Boyd-Leinen, P., Thrall, C. and Kon, O.L. Role of specific chromosomal proteins and DNA sequences in the nuclear binding sites for steroid receptors. Recent Prog Hormone Res 39:463, 1983.
18. Ruh, T.S. and Spelsberg, T.C. Acceptor sites for the estrogen receptor in hen oviduct chromatin. Biochem J 210:905, 1982.
19. Ruh, T.S., Ross, P., Jr., Wood, D.M. and Keene, J.L. The binding of [^{3}H]oestradiol receptor complexes to calf uterine chromatin. Biochem J 200:133, 1981.
20. Singh, R.K., Ruh, M.F. and Ruh, T.S. Binding of [^{3}H] estradiol and [^{3}H]H1285 receptor complexes to rabbit uterine chromatin. Biochim Biophys Acta 800:33, 1984.
21. Cobb, A. and Leavitt, W.W. Characterization of nuclear binding sites for different forms of the progesterone receptor (abstract). 67th Annual Meeting of the Endocrine Society June 19-21 (Baltimore, MD), p. 83, No. 330.
22. Perry, B.N. and Lopez, A. The binding of ^{3}H-labelled oestradiol- and progesterone-receptor complexes to hypothalmic chromatin of male and female sheep. Biochem J 176:873, 1978.
23. Klyzsejko-Stefanowicz, L., Chui, J.F., Tsai, Y.H. and Hnilica, L.S. Acceptor proteins in rat androgenic tissue chromatin. Proc Natl Acad Sci 73:1954, 1976.
24. Hamana, K. and Iwai, K. Glucocorticoid receptor complex binds to nonhistone protein and DNA in rat liver chromatin. J Biochem (Tokyo) 83:279, 1978.
25. Ruh, M.F., Singh, R.K., Bellone, C.J. and Ruh, T.S. Binding of [^{3}H]triamcinolone acetonide-receptor complexes to chromatin from B-cell leukemia line, BCL. Biochim Biophys Acta 844:24, 1985.
26. Schrader, W.T., Heuer, S.S. and O'Malley, B.W. Progesterone receptors of chick oviduct: Identification of 6S receptor dimers. Biol Reprod 12:134, 1975.
27. Buller, R.E., Schwartz, R.J., Schrader, W.T. and O'Malley, B.W. Progesterone-binding components of chick oviduct. In vitro effect of receptor subunits on gene transcription. J Biol Chem 251:5178, 1976.
28. Spelsberg, T.C., Gosse, B., Littlefield, B., Toyoda, H. and Seelke, R. Reçonstitution of native-like nuclear acceptor sites of the avian oviduct progesterone receptor: Evidence for involvement of specific chromatin proteins and specific DNA sequences. Biochemistry 23:5103, 1984.

29. Toyoda, T., Seelke, R., Littlefield, B.A. and Spelsberg, T. Evidence for specific DNA sequences in the nuclear acceptor sites of the avian oviduct progesterone receptor. Proc Natl Acad Sci USA, 1984, in press.
30. Greene, G.L., Closs, L.E., Fleming, H., DeSombre, E.R. and Jensen, E.V. Antibodies to estrogen receptor: Immunochemical similarity of estrophibin from various animal species. Proc Natl Acad Sci USA 74:3681, 1977.
31. Greene, G.L., Nolan, C., Engler, J.P. and Jensen, E.V. Monoclonal antibodies to human estrogen receptor (hybridomas, estrophilin, affinity chromatography, breast cancer). Proc Natl Acad Sci USA 77:5115.
32. Goldberger, A., Horton, M. and Spelsberg, T.C. Monoclonal antibodies which inhibit binding of avian oviduct progesterone receptor to native hen oviduct nuclear acceptor sites (Abstract) of the 67th Annual Meeting of the Endocrine Society, June 19-21 (Baltimore, MD), No. 5, p. 29.
33. O'Malley, B.W., Spelsberg, T.C., Schrader, W.T., Chytil, F. and Steggles, A.W. Mechanism of interaction of hormone-receptor complex with the genome of a eucaryotic target cell. Nature 235:141, 1972.
34. Palmiter, R.D., Moore, P.B., Mulvihill, E.R., and Emtage, S. A significant lag in the induction of ovalbumin messenger RNA by steroid hormones: A receptor translocation hypothesis. Cell 8:557, 1976.
35. Lai, E.C., Woo, S.L.C., Bordelon-Riser, M.E., Fraser, T.H. and O'Malley, B.W. Ovalbumin is synthesized in mouse cells transformed with the natural chicken ovalbumin gene. Proc Natl Acad Sci USA 72:244, 1980.
36. Hora, J. Horton, M.C., and Spelsberg, T.C. (in preparation).
37. Toyoda, H. and Spelsberg, T.C. DNA sequence specific interactions with acceptor proteins of the avian oviduct progesterone receptors. Abstract of the 63rd Annual Meeting for the Endocrine Society, Cincinnati, Ohio, No. 857, p. 297.
38. Igo-Kemenes, T., Horz, W. and Zachau, H.G. Chromatin. Ann Rev Biochem 51:89, 1982.
39. Barrack, E.R. and Coffey, D.S. The specific binding of estrogens and androgens to the nuclear matrix of sex hormone responsive tissues. J Biol Chem 255:7265, 1980.
40. Barrack, E.R. and Coffey, D.S. Biological properties of the nuclear matrix: Steroid hormone binding. Recent Prog Hormone Res 38:133, 1982.
41. Spelsberg, T.C. A rapid method for analysis of ligand binding for DNA and soluble nucleoproteins using streptomycin: Application to steroid receptor ligands. Biochemistry 22:13, 1983.

24. Interaction of Steroid Hormone Receptors with DNA

M. Beato

Institut für Molekularbiologie und Tumorforschung, Philipps-Universität, 3550 Marburg, Federal Republic of Germany

INTRODUCTION

Regulation of gene expression in prokaryotes is often accomplished by virtue of the interaction of regulatory proteins with DNA elements close to the promoters of the regulated genes. In general, substrates, metabolic products or signal molecules either directly bind to the regulatory proteins and modulate their function, or alternatively they regulate the concentration of secondary signal molecules which in turn bind to the regulatory proteins and influence their activity. An example of a directly acting metabolic signal is the repression of the *lac* operon by lactose analogues. An example of a system responding to a secondary signal molecule is the activation of several bacterial operons through binding of cAMP to its receptor protein, itself a regulatory protein.

The general principle of gene regulation through regulatory proteins with a double specificity, e.g. for a physiological ligand and for defined DNA sequences, is also found in eukaryotes. In yeast, for instance, transcription of the *GAL1* and *GAL10* genes is induced by galactose through its interaction with the *GAL4* gene product, a regulatory protein that recognizes upstream activating sequences located between the promoters for *GAL1* and *GAL10* (1).

In animals cells one of the best characterized systems to study gene regulation is the modulation of transcription by steroid hormones. Since the discovery by Clever and Karlson (2) that the steroid hormone ecdysoneis able to induce the appearance of puffs in giant chromosomomes of *Chironomus tentans*, this field has attracted the attention of many biochemists interested in the molecular mechanisms of gene regulation in eukaryotes.

RESULTS

The hormone receptors

Steroid hormones influence the expression of specific genes by virtue of their interaction with so-called receptor molecules. These are soluble proteins with high affinity and specificity for the corresponding hormone. Although the intracellular localization of the unoccupied receptors is a matter of controversy, there is a general agreement that after binding the steroid, the complex of hormone and receptor is localized to the cell nucleus where it is tightly bound to chromatin. The nature of the chromatin components that interact with the receptor *in vivo* has not been definitively established, but it is known that the steroid hormone receptors are DNA binding proteins.

Experiments with crude cytosol or with partially enriched receptors prepared at low temperature in the absence of salt have shown that binding of the hormone receptors to DNA-cellulose requires pretreatment of the steroid-receptor complex either at elevated temperatures (above 20°C) or at high ionic strength (3). These treatments bring about an ill-defined conformational change of the complex, that has been called "activation" or "transformation", and can be prevented by sodium molybdate and analogous compounds. In addition it is known that the steroid hormone receptors are phosphoproteins, and that changes in the state of phosphorylation could be related to the ability of the receptor to bind both steroid and DNA (3).

Progress in the techniques of receptor purification and the availability of antisera and monoclonal antibodies have allowed one to develop relatively precise models of receptor structure. In principle, all steroid hormone receptors studied so far are composed of at least three structural domains that appear to have a functional correlate as they can be separated by mild digestion with proteolytic enzymes. Two of these domains, the one responsible for steroid binding and that responsible for interaction with DNA, are functionally well defined, and can be isolated together as a single polypeptide of Mr 40 kDa (4,5). The third domain has only been identified due to its antigenicity; most of the antisera and monoclonal antibodies obtained so far are directed against epitopes located in this third receptor domain. A functional significance for this part of the receptor molecule has been concluded from the analysis of the *nt*^{i} variants of glucocorticoid resistant mouse lymphoma cells. This variant contains a glucocorticoid receptor of Mr 40 kDa, that binds steroid and DNA, but has lost the antigenic domain (6). Obviously this form of the receptor is not able to mediate the cellular response to the hormone.

Binding sites for the glucocorticoid receptor in regulated genes

The first succesful identification of a specific interaction between a steroid hormone receptor and defined DNA sequences of a hormonally regulated gene was accomplished with the glucocorticoid receptor and the mouse mammary tumor virus (MMTV). Transcription of the MMTV proviral DNA is known to be induced by glucocorticoid hormones in a variety of cell lines (7). In gene transfer experiments with cloned MMTV proviral genomes, the DNA sequences relevant for glucocorticoid regulation were localized to the so-called long terminal repeat (LTR) region (8,9). DNA binding experiments with purified and activated rat liver glucocorticoid receptor led to the identification of several binding sites for the steroid-receptor complex located between 70 and 190 base pairs (bp) upstream of the transcription start point (10,11). Common to all these sites is the hexanucleotide motif 5'-TGTTCT-3', that was later shown to be specifically contacted by the receptor at the N7 position of the guanines in both strands (12).

In the meantime , several other binding sites for the glucocorticoid receptor have been identified in regulated genes of various species. The nucleotide sequence of 20 of these binding sites is shown in Figure 1. Ten of these binding sites have been analyzed by DNaseI footprinting and methylation protection studies (Fig. 1A), while the other ten have been defined with less precise methods: either by DNaseI footprinting alone (HMTIIAII, CVITII and MSVI), by exonuclease III footprinting (HGHI, rTOI, rTOII, COVI and COVII), or by the nitrocelluose filter binding assay (MSVII and MSVIII). In the case of MTVI, HMTIIAI, MRib, CLYSII, rTOI, rTOII and MSVI, there are functional data available suggesting that the receptor binding sequences do act as glucocorticoid regulatory elements in gene transfer experiments, indicating that the corresponding DNA sequences are essential for hormonal inducibility of adjacent promoters.

A comparison of these 20 receptor binding sites allows one to derive a 16 nucleotide consensus sequence, even though the analyzed genes were cloned from different species including mouse, rat, rabbit, chicken and human. This finding underlines the evolutionary conservation of the regulatory elements for glucocorticoid hormones, and of the corresponding binding domains of the receptor protein. Within the consensus sequence the different positions exhibit different degrees of conservation, with the hexanucleotide motif as the most relevant feature of the sequence. The most highly conserved base is position 15, where we found a C in all 20 binding sites. We know that the receptor contacts the guanine in the opposite strand in

```
(A)                                   10  12  14  16
                    1 2 3 4 5 6 7 8 9   11  13  15
                    . . . . . . . . . . . . . . . .
                      *                   *
    MTV I           G G T T A C A A A C T G T T C T
                      * *     ^           *     ^
    HMT IIa I       C G G T A C A C T G T G T C C T
                        *     ^           *     ^
    RUG I           C T G T T C A C T C T G T T C T
                              ^           *     ^
    RUG II          C C G G A C A C G G A G T C C T
                              ^           *     ^
    MRib            A C T G A C A C G C T G T C C T
                              ^           *     ^
    RUG III         G T G T C A G T C T T G T T C T
                                          *     ^
    MTV IIa         T G G T A T C A A A T G T T C T
                                          *     ^
    DSC7            C G A T T T G A T C T G T T C T
                      *                   *     ^
    CLYS I          T G A C A A C T G T A G A A C A
                          ^             *       ^
    CLYS II         A A A T T C C T C T G T G G C T
                              ^                 ^
(B)
    HGH I           G G G C A C A A T G T G T C C T

    CVIT II         G G G A T C A A T G T G T T C T

    rTO I           A T G C A C A G C G A G T T C T

    rTO II          C C T T T C A T G A T G T C C T

    MSV I           G G G G A C C A T C T G T T C T

    MSV II          C T G T T C C A T C T G T T C T

    MSV III         G C T G T C T C T C T G T T C C

    COV I           G T T T T C T G C C T G T T C T

    COV II          G G T C A C G T C T T G T T C T

    HMTIIa II       C C C T C C C T C C T G T C C T

                    1 2 3 4 5 6 7 8 9   11  13  15
                    . . . . . . . . . . . . . . . .
    CONSENSUS:      G G G T A C A A T C T G T T C T
                    c t t c t   c t c g       c
                    . . . . . . . . . . . . . . . .
    (20 SITES)      15  16  16  15  14  16  18  20
                      14  15  16  13  14  19  18  18
```

Figure 1. <u>Comparison of twenty binding sites for the glucocorticoid receptor</u>. Only the DNA strand containing the hexanucleotide motif in the correct orientation is shown. Symbols are: MTVI and MTVIIa, the two main glucocorticoid receptor binding sites in the MMTV-LTR (11); HMTIIaI and HMTIIaII, the two sites in the human metallothionein IIA promoter region (13); RUGI, RUGII and RUGIII, the three sites in the rabbit uteroglobin gene region 2.6kb upstream of the transcription initiation site (14); MRib, the main binding site overlapping the transcription start point of the mouse ribosomal promoter (unpublished); DSC7, a site in the promoter region of an ecdysone responsive gene of *Drosophila* (unpublished); CLYSI and CLYSII, the two sites in the promoter region of the chicken lysozyme gene (16); HGHI, a site in the first intron of the human growth hormone gene (15); rTOI and rTOII, two sites in the upstream region of the rat tryptophan oxygenase gene (unpublished); MSVI, MSVII and MSVIII, three sites in the enhancer region of Moloney Sarcoma Virus (unpublished); COVI and COVII, two sites in the promoter region of the chicken ovalbumin gene (unpublished). The * indicate the G-residues that are contacted by the receptor, and the ^ mark C-residues which complementary G is contacted by the receptor.

all the 10 binding sites that have been studied by methylation protection (Fig. 1A). The next most conserved base is the G at position 12, that is found in 19 out of 20 binding sites, and is contacted by the receptor in 9 out of 10 binding sites analyzed so far (Fig. 1A). The two T's at positions 13 and 16 are preserved in 90% of the binding sites, whereas the T at position 11 is found in 80% of the sites. Within the hexanucleotide motif only position 14 accepts some variablity, but in 90% of the receptor binding sites there is a pyrimidine at this position.

Outside of the hexanucleotide motif the conservation of nucleotide sequence between the different binding sites is less stringent with the exception of position 6 where a C is found in 80% of the sites. We know that the G in the opposite strand is contacted by the receptor in 6 out of the 10 binding sites analyzed by methylation protection (Fig.1A). The other positions can apparently accomodate a higher degree of variability, and we know that in some cases the DNaseI footprint does not cover the complete left half of the indicated nucleotide sequence (13).

Within the 16 bp consensus sequence a certain degree of two fold rotational symmetry can be observed centered around position 9, with the structure 5'-ACANNNTCT-3' as a highly conserved element. The general structure of the consensus sequence is compatible with the receptor binding taking place by means of an initial strong interaction of a receptor monomer with the hexanucleotide motif followed by a weaker interaction of another receptor monomer - in head to head orientation - with positions 1 to 8 of the consensus sequence (see below and Fig. 2). This later interaction would be facilitated through protein-protein contacts and would therefore be less dependent on the exact nucleotide sequence of the binding site.

How could the interaction of a receptor dimer with the DNA regulatory element lead to an activation of transcription of the corresponding promoter? The position of these elements with respect of the transcription initiation site varies between 2,600 bp upstream in the case of rabbit uteroglobin (14) and 100 bp downstream in the case of the human growth hormone gene (15). Thus, whatever the induction mechanism, it must be able to act bidirectionally at a long distance from the regulated promoter. This could be accomplished by direct protein-protein interaction or alternatively by inducing a change in the conformation of the DNA that could be transmitted in both directions through relatively long distances along the DNA double helix. We have preliminary evidence supporting this latter possibility. Binding of the rat liver glucocorticoid receptor to closed circular plasmids containing the hormone regulatory element of MMTV, leads to a topological change of the plasmid DNA that can be revealed by incubation with topoisomerase I (M.Carballo and M.B., unpublished). This topological change could lead to localized alterations of DNA structure in relevant regions of the promoter that could be recognized by appropriate transcription factors. This possibility is currently being tested in our laboratory.

Relationship between the DNA binding specificity of the receptors for two different steroid hormones

It is well established that many hormonally regulated genes can be induced by more than one steroid hormone. The question arises whether each steroid hormone controls gene expression through its own independent regulatory element. To answer this question we decided to use the chick lysozyme gene, that is known to respond to all four classes of steroid hormones, including glucocorticoids and progestins (16). In microinjection experiments it has been shown that deletion of nucleotide sequences located between 160 and 208 bp upstream of the transcription start point eliminates inducibility by both dexamethasone and progesterone (16). Parallel DNA binding experiments showed that the rat liver glucocorticoid receptor and the rabbit uterus progesterone receptor bind to the same two sites within the promoter region of the chicken lysozyme gene, although with different relative affinities (16,17). Whereas the limits of the footprint are the same for both receptors in the promoter proximal binding site, in the promoter distal binding site the progesterone receptor covers a longer stretch of DNA than the glucocorticoid receptor (17). In methylation protection experiments we have recently shown that the rat liver glucocorticoid receptor and the chick oviduct progesterone receptor contact different bases within the promoter distal binding site (18). Thus, although the binding sites for the glucocorticoid and the progesterone receptors do overlap, each protein recognizes distinct features of the DNA regulatory element.

A computer graphic representation of the binding sites in the chick lysozyme promoter region is shown in Figure 2. The top line of the figure represents the strong binding site of the MMTV-LTR region between -170 and -185 (11), with a dimer of the receptor bound to it. The receptor is depicted as a 94 kDa protein composed of three domains: the A domain with the steroid binding site (the hormone is not shown for simplicity), the B domain with the DNA binding site, and the C domain with the antigenic determinants (not shown). The A and B domains can be separated by trypsin or papain digetion (5), and a minireceptor containing the A and the B domain can be easily prepared from the cytosol of frozen rat liver (4). This minireceptor exhibits a Mr of 40 kDa and binds to regulatory DNA sequences although with lower specificity than the intact 94 kDa form of the

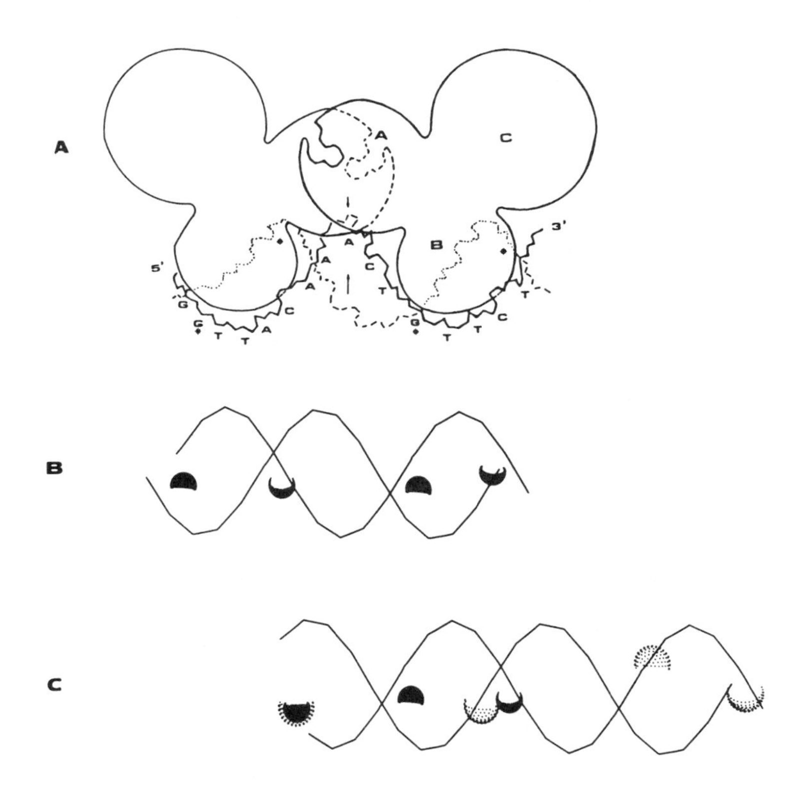

Figure 2. Models of the interaction between glucocorticoid receptor and DNA.
A. A domain model of the glucocorticoid receptor (description in the text) is shown as a complex with MTVI (Fig.1A). The * mark the Gs contacted by the receptor (12).
B. Computer graphic representation of the same DNA region shown in A, with the van der Waals spheres of the N7 atoms of the relevant guanines indicated.
C. Computer graphic model of the promoter distal binding site of the chicken lysozyme gene (16), with the contact points for the glucocorticoid receptor shown as in B, and those for the progesterone receptor indicated by lightly punctated areas (18).

receptor (11). From these, and other results (see above and Ref. 6), it is assumed that the C domain of the native receptor modulates the interaction of the B domain with DNA. A similar function has been postulated for the A domain. As in the classical model of hormone action, binding of the steroid to this domain is a prerequisite for the interaction of the B domain with DNA. This, however, has never been conclusively proven, and the recent finding of unoccupied receptor molecules within the cell nucleus (19,20) leaves the possibility open that steroid-free receptor molecules may bind to the specific DNA sequences decribed above.

The G-residues that are contacted by the receptor, as determined by methylation protection experiments are indicated by an asterisks in Figure 2A. The van der Waals spheres of the N7 atoms of these G-residues are shown in Figure 2B, which represents a computer graphic model with the backbone of the double helix outlined schematically. The promoter distal binding site of the lysozyme gene between -166 and -185 is shown in a similar type of representation in Figure 2C, with the contact points for the glucocorticoid receptor as dark half spheres, and the contact points for the progesterone receptor shown as lightly punctated spheres. It is clear that both receptors recognize different features within this regulatory element that is known to mediate biological response to both hormones (16).

A slightly different situation is found for the effect of various hormones on the expression of MMTV. In gene transfer experiments with T47D cells, a human mammary tumour cell line, the glucocorticoid regulatory element of the MMTV-LTR can also mediate progesterone induction of either the MMTV promoter or the HSV-tk-promoter (A.C.B.Cato, R.Miksicek, G. Schütz and M.B., unpublished). As we know that the progesterone receptor binds to the same sites within the LTR region that are recognized by the glucocorticoid receptor (17), in this case too both hormones act through the same regulatory element. The difference with the chick lysozyme system is that in the MMTV-LTR the receptors for both hormones yield footprints of exactly the same length (17). Unfortunately, no methylation protection data with the progesterone receptor are yet available, and therefore the features recognized by each receptor within this element can not be compared.

How can the finding of a common regulatory element for different steroid hormones be integrated in the current concepts on hormone specificty? We know that the presence of the receptor for a particular hormone is not sufficient to confer to a cell the ability to respond to the corresponding hormone. Certain cells have receptors for various steroid hormones but do not respond to any of the hormones or only to some of them. Of course, in the case of the endogenous genes the previous differentiation pathway of the cells could determine the accessibility of particular genes to hormonal regulation in general. If a particular gene, however, is susceptible to regulation by one steroid hormone, and if the different hormones act through the same regulatory element, one would expect other hormones for which the cell has appropriate receptors to be active as well. When this is not the case one would have to postulate additional regulatory events that determine the functional efficiency of the individual receptors. These events could involve either a chemical modification of the receptor (for instance, phosphorylation) or its interaction, directly or mediated through changes in DNA conformation, with cell specific factors. These factors, in addition, would have to be promoter specific, as in gene transfer experiments within the same cells only certain hormone regulatory elements are functional. The answer to these questions would be greatly facilitated by the availability of cell-free systems for the study of the events following the interaction of the receptor with the DNA regulatory elements.

SUMMARY

DNA binding experiments with purified hormone receptors combined with gene transfer experiments have been used to ilustrate the concept of hormone regulatory or reponsive element as DNA sequences that are recognized by the hormone receptor and mediate enhancement of transcription from adjacent promoters. Several examples of glucocorticoid regulatory elements are presented and the interaction of the receptor with twenty binding sites is compared. From this comparison a 16 nucleotide consensus sequence is derived

5'-GGGTACAATCTGTTCT-3'
cttct ctcg c

and a model for the interaction of a receptor dimer with the consensus sequence is presented. The molecular mechanisms responsible for the activation of transcription are discussed, with particular emphasis on the possible contribution of DNA topological changes.

A comparison of the interaction between the glucocorticoid and the progesterone receptors with the regulatory regions of the chick lysozyme gene and MMTV, shows that the same regulatory element can mediate regulation by the two hormones, although each receptor may recognize different features of the DNA molecule. This finding is discussed in the context of the prevalent ideas on hormone specificity of gene regulation.

AKNOWLEDGEMENTS

The computer graphic pictures shown in Figure 2 were prepared by Christopher Carlson and Heinz Bosshard, EMBL, Heidelberg. The experimental work summarized in this chapter has been supported by grants from the Deutsche Forschungsgemeinschaft and the Fond der Chemischen Industrie.

REFERENCES

1. Giniger, E., S.M. Varnum, and M. Ptashne: Specific DNA binding of GAL 4, a positive regulatory protein of yeast. Cell 40:767, 1985.
2. Clever, U., and P. Karlson: Induktion von Puff-Veränderungen in Speicheldrüsenchromosomen von Chironomus tentans durch Ecdyson. Exp.Cell.Res. 20:623, 1960.
3. Scmidt, T.J., and G. Litwack: Activation of the glucocorticoid-receptor complex. Physiological Reviews 62:1131, 1982.
4. Westphal, H.M., and M. Beato: Activated glucocorticoid receptor of rat liver. Purification and physical characterization. Eur.J.Biochem. 106:395, 1980.
5. Carlstedt-Duke, J., S. Okret, Ö. Wrange, and J.-A. Gustafsson: Immunological analysis of the glucocorticoid receptor: identification of a third domain separated from the steroid-binding and DNA-binding domains. Proc.Natl.Acad.Sci.USA 79:4260, 1982.
6. Westphal, H.M., K. Mugele, M. Beato, and U. Gehring: Immunochemical characterization of wild-type and variant glucocorticoid receptors by monoclonal antibodies. EMBO J 3:1493, 1984.
7. Ringold, G.M.: Glucocorticoid regulation of mouse mammary tumor virus gene expression. Biochim.Biphys.Acta 560:487, 1979.
8. Hynes, N.H., A.J.J. van Ooyen, N. Kennedy., P. Herrlich., H. Ponta, and B. Groner: Subfragments of the large terminal repeat cause glucocorticoid responsive expression of mouse mammary tumor virus and of an adjacent gene. Proc.Natl.Acad.Sci.USA 80:3637, 1983.
9. Chandler, V.L., B.A. Maler, and K.R. Yamamoto: DNA sequences bound specifically by glucocorticoid receptor in vitro render a heterologous promoter hormone responsive in vivo. Cell 33:489, 1983.
10. Payvar, F., D. DeFranco, G.L. Firestone, B. Edgar, Ö. Wrange, S. Okret, J.-A. Gustafsson, and K.R. Yamamoto: Sequence-specific binding of glucocorticoid receptor to MTV DNA at sites within and upstream of the transcribed region. Cell 35:381, 1983.
11. Scheidereit, C., S. Geisse, H.M. Westphal, and M. Beato: The glucocorticoid receptor binds to defined nucleotide sequences near the promoter of mouse mammary tumour virus. Nature 304:749, 1983.

12. Scheidereit, C., and M. Beato: Contacts between receptor and DNA double helix within a glucocorticoid regulatory element of mouse mammary tumor virus. Proc.Natl.Acad.Sci.USA 81:3029, 1984.
13. Karin, M., A. Haslinger, H. Holtgreve, R.I. Richards, P. Krauter, H.M. Westphal, and M. Beato: Characterization of DNA sequences through which cadmium and glucocorticoid hormones induce human metallothionein IA gene. Nature 308:513, 1984.
14. Cato, A.C.B., S. Geisse, M. Wenz, H.M. Westphal, and M. Beato: The nucleotide sequences recognized by the glucocorticoid receptor in the rabbit uteroglobin gene region are located far upstream from the initiation of transcription. EMBO J 3:2731, 1984.
15. Slater, E., O. Rabenau, M. Karin, J.D. Baxter, and M. Beato: Glucocortiocid receptor binding and activation of a heterologous promoter in response to dexamethasone by the first intron of the human growth hormone gene. Mol.Cell.Biol., in press.
16. Renkawitz, R., G. Schütz, D. von der Ahe, and M. Beato: Identification of hormone regulatory elements in the promoter region of the chicken lysozyme gene. Cell 37:503, 1984.
17. von der Ahe, D., S. Janich, C. Scheidereit, R. Renkawitz, G. Schütz, andM. Beato: The glucocorticoid and the progesterone receptors bind to the same sites in two hormonally regulated promoters. Nature 313:706, 1985.
18. von der Ahe, D., J-M. Renoir, T. Buchou, E-E. Baulieu, and M. Beato: The receptors for glucocorticosteroid and progesterone recognize distinct features within a DNA regulatory element. Nature, submitted.
19. King, W.J., and G.L. Greene: Monoclonal antibodies localize oestrogen receptor in the nuclei of target cells. Nature 307:745, 1984.
20. Welshons, W.V., M.E. Lieberman, and J. Gorski: Nuclear localization of unoccupied oestrogen receptors. Nature 307:747, 1984.

Index

α-difluoromethylornithine (see also DFMO), 193
Acceptors
  DNA sequences, 260, 264
  protein-DNA complex, 260
  proteins, 259-267
  site DNA, 263
Actin, 149
Adenylate cyclase, 45
Adhya, s., 45-55
Adriamycine, 184-185
Alcohol dehydrogenase gene
  promoters of, 168
Amber mutation, 81
Androgen receptors, 260
Anti-(U1) RNP, 151
Antineoplastic drug action, 183-199
Antitermination, 78, 81
Appelt, K., 57-65
Ara C, 193
1-β-D-arabionofuranosylcytosine, 193
Arabinose binding protein, 16
Aranda, A., 201-217
Autoantibodies, 149
Avian oviduct
  acceptor proteins, 263
  progesterone receptor, 260
5-aza-CR, 193
5-azacytidine, 193

β_2-microglobulin, 249
Bacteriophage
  lambda N, 77-85
  N4, 67
Ball, R., 245-258
Banjar, Z.M., 175-181
Beato, M., 269-279
Berg, J.M., 3-12
Binding
  DNA protein, Sp1, 167
  DNA proteins, 259-267
  DNA proteins in steroid hormone regulation, 259-267
  DNA specificity of steroid hormone receptors, 274-277
Binding sites for glucocorticoid receptors, 271-279
Box
  A+, 80
  A1 allele, 78
  A1 mutation, 79
  A region, 80
  A sequence, 78, 80
  C, 78

Briggs, R.C., 175-181
2-bromoacetamido-4-nitrophenol,reaction with
  lac repressor, 15
Bromodeoxyuridine, 250-251

C-myc, 247
C-src, 247
cAMP-dependent protein kinase, 45
cAMP, 269
CAP, 7, 23, 27, 45-55
Casanova, J., 201-217
CAT gene (see also chloramphenicol acetyltransferase), 234
Catabolite activator protein, 7
Catabolite gene activator protein, 23, 45-55
Cato, A., 219-231
Cell type specificity, 239-240
Chambers, J.C., 107-116
Chloramphenicol acetyltransferase gene, 234
Chromatin
  conformation, 188
  nuclear acceptor sites, 259-267
  prokaryotic, 57-65
  structure, micrococcal nuclease digestion, 238-239
  structure, nucleosome organization, 238-239
  structure, 238-239
Chromosomal proteins of PRov, 261-263
Chromosome segregation, 194
cI gene, 3, 8
Ci promoter, 9
cII protein, 46
cII protein binding affinity, 32
Cis-acting sequences, 233-239
  hormone responsive positive element, 235
  negative element, 235
Cis-control elements, 118
Cis-DDP (see also cis-diaminedichloroplatinum II), 176-177
Cis-diaminedichloroplatinum II, 176-177
Cobianchi, F., 129-146
Coliphage lambda, 77-85
Condensed chromatin, 238-239
Copp, R.P., 201-217
Cordingley, M., 233-243
2D correlated spectroscopy, 89
COSY, 89
COSY spectra, 92
Coumermycin, 69
Cro gene, 3, 79
Cro binding site, 23
Cro protein, 3, 7
Crosslinking
  DNA-protein, intercalator-induced, 185
  DNA-protein, 177, 185, 190-192

cAMP recptor protein (CRP), 45-55
  gene, 46
  gene mutations, 46
CRP.cAMP, 45-55
  and glucose effect, 46
CRP.cAMP and activation of transcription, 51
  and inhibition of transcription, 52
  and lac transcription, 51
  and repression, 52
  binding to DNA, 51
CYA gene, 46, 52
CYA gene mutations, 46
Cyclic AMP receptor protein, 45-55
Cytokeratin proteins, 178
Cytokeratins, 149
Cytoskeletal proteins, 149
Cytoskeleton, 147

Dense amino acid labeling, 203
Deo gene, 46
Dexamethasone response mutants, 251-252
DFMO (see also α-difluoromethyornithine), 193
DNA regulatory element, 273-274
DNA sequences in steroid hormone regulation, 259-267
DNA methyltransferase, 156
DNA cleavage, intercalator induced, 185-187
DNA cleaving activity, in cell nucleus, 187
DNA intercalating agents, 184-199
DNA gyrase, 69, 189, 191
DNA, repetitive sequences, 33
DNA-acceptor protein complexes, 260
DNA-binding proteins, steroid hormone receptor
  proteins, 270-279
DNA-dependent RNA polymerase, 67-74
DNase I, 176
DNase I footprinting, 159
DNase I hypersensitive site, 71-72
DNase protection, of pre DNA, 32
Dyad symmetry, 72

E. coli lac repressor, 90
E. coli NusA protein, 78
E. coli ribosomal RNA operons, 81
Ecdysone responsive gene, 272
EGF (see also epidermal growth factor), 248
Ehrlich, M., 155-161
Ellipticine, 184-185
Enhancers, cis-acting constitutive, 239-240
Epidermal growth factor, 248
Epipodophyllotoxins, 192
17-β-estradiol, 193
Estrogen receptor, 260

Ethidium bromide, 188
Evolution, 263
Evolutionary conservation of nuclear acceptor sites
  for PRov, 263
Expression of LTR H-21^d gene, 249

Footprinting analysis, 70
Friedman, D.I., 77-85

GALIO gene, 269
GALI gene, 269
GAL4 gene, 269
GA1P2 gene, 52
Garges, S., 45-55
GC cells transfected with pHG-xgpt, 209
Geiduschek, E.P., 57-65
Gene regulation by 5'flanking DNA sequences, 207
Glucksmann, A., 67-76
Glucocorticoid, 245-258, 260
  dependent enhancer, 237
  dependent MMTV LTR, 245-258
  enhancer binding proteins, 251-252
  hormone responsive element, 237
  induction of LTR H-2L(d), 249
  receptor, 245-258, 276
  receptor binding proteins, 251-252
  receptor binding sites in MMTV-LTR, 271-279
  receptor binding sites in MMTV-LTR, 219-231
  receptor mutants, 249
  receptors, 219-231, 233, 237
  regulatory element of MMTV-LTR, 276
  regulatory elements, 271-279
  responsive enhancer, 219-231
Glucocorticoids, regulation by, 233-243
Goldberger, A., 259-267
GRE (see also glucocorticoid responsive enhancer), 219-231
Greene, J.R., 57-65
Groner, B., 219-231, 245-258
Growth hormone, 201
  gene, 272-273
  gene expression, regulation by thyroid hormone, 205
  gene expression, 205-208
  gene, 5'flanking DNA sequences, 207
  gene transcription, 205
  mRNA, half life, 206
  synthesis, 205
Guy, H.R., 129-146

H-2L^d, 248
H-2L^d gene, 248
H-ras gene, 246
Hager, G.L., 233-243

HaMuSV (see also Harvey murine sarcoma virus), 234
HaMuSV enhancer, 234
Harvey murine sarcoma virus, 246
Harvey murine sarcoma virus LTR, 234
HAT medium, 250-251
HDP (see helix destabilizing protein)
Heat-shock transcription factor, 167
Heat shock gene, 167
Heavy metals, interaction with DNA-protein complex, 175-181
HeLa S(3) cell cycle, 177
HeLa soluble transcription extracts, 168
Helix-turn-helix DNA-binding unit, 23
Helix-turn-helix motif, 7, 36, 50
Helix-turn-helix unit, 8, 27
Helix destabilizing protein, 129-146
  cDNA, 130
  mRNA, 131, 133-135
Hellwig, R.J., 163-173
Heterogeneous nuclear RNA, 117-127
Hnilica, L.S., 175-181
hnRNA, 97-106, 141
hnRNA (see also heterogeneous nuclear RNA), 117-127
30s hnRNP, 129-146
hnRNP, 30s particles, 98
hnRNP, A proteins, 98
hnRNP, B proteins, 98
hnRNP, core polypeptides, 98
hnRNP, core proteins, 100
hnRNP, C proteins, 100
hnRNP, 97-106, 129-146
hnRNP complexes, 97-106
hnRNP core proteins, 118-119, 122-123
hnRNP fibril, 141
hnRNP particle core protein, 129-146
hnRNP particle core protein A1, 141
hnRNP particles, 118-127
Holoubek, V., 117-127
Hora, J., 259-267
Hormonal regulation of oncogene expression, 246-248
Hormonal response element, 245-258
Hormonal response mutants, 252-252
Hormone receptor complex, 239-240
Hormone regulatory element, 239-240
Horowitz, Z.D., 210-217
Horton, M., 259-267
HRE (see also hormonal response element), 245-258
Human growth hormone gene, 272
Hydroxyurea, 193

Intermediate filament proteins, 149
Intrastrand crosslinking, 176
Inverted repeats, 71-72
IPTG, 13

Janocko, L., 210-217
Jordan, S.R., 3-12

Keene, J.D., 107-116
Kuijpers, H., 147-154

L-T3 (see also L-triiodothyronine), 201-217
L-T4 (see also L-thyroxine), 201-217
L-thyroxine, 201-217
L-triiodothyronine, 201-217
La antisera, 107-108
La gene, 11
La protein, 107-116
  binding to oligonucleotides, 110
  binding to RNA, 109
  cDNA, 111
  RNA 3' termination, 109
La RNP, 108
Lac repressor, 13-19, 23, 90, 92-93
Lac repressor protein, domain structure, 13-19
Lac operon, 269
Lactose repressor protein, 13-19
Lambda
  N gene, 81
  N gene product, 77
  nutR region, 79
  pN, 78
  punA1, 133 phage, 80
  punA1, 79-80
  repressor, 3-12, 23, 27
Lamins, 147-154
Leser, G., 97-106
Lichtler, A., 233-243
Littlefield, B., 259-267
LTR chromatin, 239
LTR chromatin, and hormone receptor interaction, 239
LTR promoters, 237
LTR-Hras proto-oncogene, 246
Lysosome gene, 272, 274, 276

m-AMSA (see 4'-(9-acridinylamino)-methanesulfon-m-anisidine), 183-199
m-AMSA, effect on chromatin conformation, 188
m-AMSA-stimulated DNA-protein crosslinking, 190-192
Major histocompatibility antigen H-2L^{d}, 248
Malic enzyme, 201
Mammary tumor virus, 271-279
Mariman, E., 147-154
Martin, B., 107-116
Martin, T., 97-106
Matthews, B.W., 21-28
Matthews, K.S., 13-19

MDBP (see methylated DNA-binding protein), 158
Metallothionein IIA promoter, 272
4'-(9-acridinylamino)-methanesulfon-m-anisidide, 184
(see also m-AMSA)
Methylated DNA-binding protein, sequence-specific
binding, 158
Methylated DNA-binding protein, 155-161
Methylation, DNA, 32, 155-161
Methylation of hnRNP particle proteins, 119-122
Methylmethanethiosulfonate, modification of repressor, 14
Methylmethanethiosulfonate, 14
MHC class I molecules, 248
Micrococcus luteus DNA, 156
Mitosis, 152
MMTV, 219, 271-279
enhancer, 219
hormone regulatory element, 273
MMTV LTR, 219-231, 233-243, 245-258, 271-279
activation of transcription, 238
and glucocorticoid receptor binding sites, 271-279
chromatin structure of, 238-239
hormone response element, 233-243
promoter, 233-234
repression of transcription, 238
tissue specificity of, 236-238
-c-myc, 247
-H-2L^d, 248
-Ha enhancer, 234
-mos, 247
-polyoma middle T, 247
-ras, 246-247
-src, 247
-tk, 251
Moloney Sarcoma Virus enhancer region, 272
Mouse mammary tumor virus (see also MMTV), 219, 233-243
Mouse mammary tumor virus LTR, 219-231, 245-258
Mutation, effect on cII protein activity, 30
Mutation, effect on pRE function, 30

N-mediated antitermination, 78
N4 DNA replication, 68
N4 single-stranded DNA binding protein, 70
N4 virion RNA polymerase, 44, 69, 72
N4 virion RNA polymerase promoters, 74
N4 virions, 67
Naldixic acid, 69, 191
Negative transcriptional element, 234
NIH3T3 cells, 237
Niyogi, S.K., 163-173
NMR, 87-94
NOESY, 89
Non-histone protein-DNA interactions, 175-181

Non-histone proteins, 175-181
Novobiocin, 189
Nuclear acceptor sites, 259-267
  chromosomal proteins, 261-263
  for PRov, 260-267
  for steroid receptors, 259-267
Nuclear magnetic resonance, 87-94
Nuclear matrix, 147-154, 263
Nuclear Overhauser emhancement, 88
Nuclear protein matrix, 147-154
Nuclear ribonucleoprotein particles, 117-127
Nuclease hypersensitivity sites, 194
Nucleic acid locus, nut, 77-78
Nucleoids (see also protein-DNA complexes), 188-189
Nucleolar proteins, 149
Nucleoprotein complexes, 168
Nucleoprotein complexes in transcription, 169
Nucleoprotein organization of LTR sequences, 238-239
Nucleoprotein structural organization, 238-239
Nucleosome structures, 238-239
Nus mutations, 78
$NusA^s$, 81
$NusA^s$ host
NusA, 81
NusA gene, 78-79
NusA protein, 77-85
NusA recognition site, 79
Nut regions, 78
Nut sites, 78
NutR region, 80

Oligonucleotide-directed mutagenesis, 235
Oligo scanning mutagenesis, 236
Olinski, R., 175-181
Olson, E.R., 77-85
OmpA gene, 46, 52
Oncogene, LTR-Hras proto-oncogene, 246
Oncogenes, 246-248
Or3, cro binding site, 23
Ornithine decarboxylase mRNA, 247
Ovalbumin gene, 263
Ovalbumin gene CAAT box, 165
Ovalbumin gene DNA-protein complex, 263
Ovalbumin gene promoter, 272
Oviduct acceptor protein, 263

P21 transforming protein, 246
P22 pRE promoter, 29
P37 protein, 247
Pabo, C.O., 3-12
Pascual, A., 201-217
PDGF (see also platelet derived growth factor), 247

Pfahl, M., 245-258
Phage 21, 78
Phage P22, 80
Photo-CIDNP, 89
Photochemically induced nuclear polarization, 89
Plasmid pGH-xgpt, 207
Plasmid pSV2-gpt, 207
Platelet derived growth factor, 247
PN action, 78
PN-nut interaction, 78
Poly(ADP-ribose)polymerase, 165
Polyoma middle T, 247
Ponta, H., 245-258
PP36 protein, 247
PP60 c-src protein, 247
PP60 phosphoprotein, 247
PRE, 30
PRE promoter, 29
PRE, RNA polymerase recognition, 30
PRE-mRNA, 117-127
PRE-mRNA processing, 149
PRE-mutation, 31
Premessenger RNA, 117-127
Progesterone receptor, 260, 276
Prokaryotic chromatin, 57-65
Promoter, 37
  adenovirus, 168
  alcohol dehydrogenase, 168
  pL, 77
  pR, 77
  recognition by RNA polymerase II, 167
  recognition factors, 164
  prokaryotic, 68
  68-69, 73
  specificity, 165
Protein-associated DNA cleavage, 184, 186
Protein-DNA-protein interaction, 38-39
Protein-DNA complexes, 188
Protein-DNA interactions, 74, 87
Protein-protein interactions, 70, 74
Protein A1, 129-146
PRov (see also progesterone receptor), 260, 276
PRov nuclear acceptor site proteins, 261-263
PRov nuclear acceptor sites, 260-267
PsiE gene, 52
PsiO gene, 52
PunA133, 79-80
PunA1, 80

Raaka, B.M., 201-217
Ramaekers, F., 147-154
Rat pituitary cells (GH1, GH3, GC), 201-203

Rat pituitary cells, 201-203
Rat growth hormone gene plasmid, pGH-xgpt, 207-208
RecA mediated cleavage, 5
Receptor
  dimer, 273
  mutant selection, 250-251
  specificity, 260
Receptor, steroid hormone (see also steroid hormone)
  complex, 270
  DNA binding proteins, 270-279
  receptors, 259-267, 269-279
  regulation, 259-267
  regulation of transcription, 219-231
Receptor, thyroid hormone (see also thyroid hormone)
  nuclear, dense amino acid labeling, 203
  nuclear, half life, 203-205
  nuclear receptors, 201-217
  photoaffinity labeling, 203
  receptors, 201-217
  regulatory element, 208
Repressor, 3-12, 52
  cro, 27
  cro, interaction with DNA, 21-28
  cro, structure, 21-28
  protein structure, model, 16-17
Ribonucleoprotein particles, 149
Ribosomal promoter, 272
Richard-Foy, H., 233-243
Rifampicin, 67
RNA chain elongation, 168
  polymerase-initiation complex, 70
  polymerase-promoter interaction, 67-76
  polymerase, 77-78
  polymerase II eukaryotic, 163-173
  eukaryotic nucleoprotein complexes, 162-173
  eukaryotic protein factors, 163-173
  eukaryotic transcription by, 163-173
  polymerase II, 139, 163-173
  promoter recognition by, 167
  specific transcription by, 163-164
  polymerase III, 107-108
RNA processing, 117-127, 147-154
RNA-protein interactions, 117-127
  in hnRNP, 123-125
RNase, 165
Ro protein, 110
Rothman-Denes, L.B., 67-76

Sahnoun, H., 201-217
Salmonella NusA, 80
Salmonella NusA gene, 79
Samuels, H>H., 201-217
Schauer, A.T., 77-85
Schmidt, W.N., 175-181

Seelke, R., 259-267
Sequence-specific resonance, 89
Single-stranded DNA binding protein, 69, 71, 74
Site-specific mutagenesis, 72
Sjogren's syndrome, 107, 110
SnRNA, 110
SnRNP, 97, 152
SnRNP antibodies, 104
SP01, 58
2D NOE spectroscopy, 89
Spelsberg, T., 259-267
Stanley, F., 201-217
Steroid hormone, 270-279
Steroid hormone receptor (see also receptor, steroid)
  complex, 270
  DNA binding proteins, 270-279
  receptors, 259-267, 269-279
  regulation, 259-267
  regulation of transcription, 219-231
Steroid hormones, mechanism of action, 259-267
Steroid receptor complex, 259-267
Steroid receptors, 219-231
Streptolidigin, 67
Super induction, 239-240
SV40 polyadenylation signal, 207
Systemic lupus erythematosus, 107

T-41 protein, 13
Termination sites tLi, 77
TF1 binding site, in hmUra-DNA, 60
TF1 binding site, in T-containing DNA, 60
TF1 binding to SP01 DNA, 59
TF1, type II DNA-binding protein, 58-61
Thymidine kinase gene, 250-251
Thyroid hormone, regulation of growth hormone gene
  expression, 205
Thyroid hormone nuclear receptors (see also receptors,
  hormone)
  dense amino acid labeling, 203
  half life, 203-205
  receptors, 201-217
Thyroid hormone receptors (see also receptors, hormone)
  photoaffinity labeling, 203
  receptors, 201-217
  regulatory element, 208
Thyroid hormones, 201-217
Tissue specificity of MMTV LTR, 236-238
Tk gene, 250-251
Tna gene, 46
Topoisomerase II-DNA complexes, 183-199
Topoisomerase II-mediated DNA cleavability, in
  proliferation 194
Toyoda, H., 259-267
Trans-acting factors, 207
Trans control elements, 118

Transcription, 68, 74
  cell-free extracts, 163
  elongation complex, 170
  initiation complex, 170
  in vitro, 69
  regulation by steroid hormone, 219-231
  activation in cis, 237
  antitermination, 77
  antitermination protein (pN), 77
  complexes, 168
  elongation, 169
  factors, 164-166
  initiation, 233, 239-240
  initiation, 68, 72, 77, 163-164, 169
  initiation site, 167
  termination, 79
  termination signal, 73
  terminators tR3, 77
  terminator tR2, 77
Transformants, 208
Transient expression assay, 234
Trp repressor, 23
Tryptophan oxygenase gene, 272
Type II DNA-binding protein, TF1, 58-61
Type II DNA-binding protein, 57-65

U1-RNA, 150-151
U1-RNP, 151-152
U1-snRNA, 150-151
U4 RNA, 109
U6 RNA, 109-110
Unwinding protein 1, 130
UP1 (see unwinding protein 1), 130
Uteroglobin gene, 272-273

V-src, 247
VA RNAs, 108, 110
Van Venrooij, W.J., 147-154
Verheijen, R., 147-154
Vimentin, 149
Virion RNA polymerase, 67-76
Vooijs, P., 147-154

Wang, RY.-H, 155-161
Wilson, A.H., 129-146
Wulff, D.L., 29-43
Wuthrich, K., 87-94

XGPT mRNA, 210
XP 12 DNA, 156

Yaffee, B.M., 201-217

Zwelling, L.A., 183-199